Water Treatment

Grade 2

American Water Works
Association

ABC
Association of
Boards of Certification

Disclaimer
Many of the photographs and illustrative drawings that appear in this book
have been furnished through the courtesy of various product distributors and
manufacturers. Any mention of trade names, commercial products, or services does
not constitute endorsement or recommendation for use by the American Water
Works Association or the US Environmental Protection Agency. In no event will
AWWA be liable for direct, indirect, special, incidental, or consequential damages
arising out of the use of information presented in this book. In particular, AWWA
will not be responsible for any costs, including, but not limited to, those incurred as
a result of lost revenue. In no event shall AWWA's liability exceed the amount paid
for the purchase of this book.

Library of Congress Cataloging-in-Publication Data
CIP data has been applied for.

ISBN: 9781625761248

000200010272023365

6666 West Quincy Avenue
Denver, CO 80235-3098
303.794.7711

Contents

Foreword

This book is part of the *Water System Operations* (WSO) series. This water operator education series was designed by the American Water Works Association (AWWA) to address core test content on certification exams by operator certification type (treatment or distribution) and certification grade level.

The current books in the series are:

WSO Water Treatment, Grade 1

WSO Water Treatment, Grade 2

WSO Water Treatment, Grades 3 & 4

WSO Water Distribution, Grades 1 & 2

WSO Water Distribution, Grades 3 & 4

Acknowledgments

The WSO series was developed by AWWA with the help of a volunteer steering committee of subject matter experts.

We would like to extend our thanks to the following individuals for their invaluable help and expertise:

- William Lauer, Project Manager, QualQuest, LLC (Lakewood, CO)
- Zac Bertz, Joint Water Commission (Hillsboro, OR)
- Mary Howell, Backflow Management, Inc. (Portland, OR)
- Bob Hoyt, City of Worcester Department of Public Works and Parks (Worcester, MA)
- Ted Kenney, New England Water Works Association (Holliston, MA)
- Darin LaFalam, City of Worcester Department of Public Works and Parks (Worcester, MA)
- Ray Olson, Distribution System Resources (Littleton, CO)
- Paul Riendeau, New England Water Works Association (Holliston, MA)

Additionally, we would like to thank Barbara Martin, AWWA/Partnership for Safe Water (Denver, CO), for her technical assistance and review.

How to Use This Book

Thank you for purchasing this volume in the Water System Operations series. AWWA's WSO series conforms to the latest Association of Boards of Certification (ABC) Need-to-Know criteria. ABC administers water operator certification testing for most of North America. Some states and provinces utilize different certification testing authorities, but this book covers all of the fundamentals every water operator needs to effectively pass their certification test and do their job effectively.

To help you advance in your career as a water operator, the WSO series is divided by subject areas (water treatment and water distribution) and certification grades (1, 2, 3, and 4, following the most common practice among states and provinces). Reference material, including basic science and mathematics concepts and information on water sources and quality, is included throughout all volumes in the series.

New features have been added to help you get the most from this book:

Key words   Important terms are highlighted in blue when they are first introduced, with a corresponding definition box in the margin of the page. You can also look up key words in the glossary at the end of the book.

The first pretreatment provided in most surface water treatment systems is screening. Coarse screens located on an intake structure are usually called *trash racks* or *debris racks*. Their function is to prevent clogging of the intake by removing sticks, logs, and other large debris in a river, lake, or reservoir. Finer screens may then be used at the point where the water enters the treatment system to remove smaller debris that has passed the trash racks.

The two basic types of screens used by water systems are bar screens and wire-mesh screens. Both types are available in models that are manually cleaned or automatically cleaned by mechanical equipment.

Bar Screens

Bar screens are made of straight steel bars, welded at both ends to two horizontal steel members. The screens are usually ranked by the open distance between bars as follows:

- Fine: spacing of ¹⁄₁₆ to ½ in. (1.5–13 mm)
- Medium: spacing of ½ to 1 in. (13–25 mm)
- Coarse: spacing of 1¼ to 4 in. (32–100 mm)

> **screening**
> A pretreatment method that uses coarse screens to remove large debris from the water to prevent clogging of pipes or channels to the treatment plant.

> **bar screens**
> A series of straight steel bars, welded at their ends to horizontal steel beams, forming a grid. Bar screens are placed on intakes or in waterways to remove large debris.

Video clips   A number of subjects in this book are linked to helpful video clips available on the AWWA WSO website (www.awwa.org/wsovideoclips). The videos present visual, hands-on information to supplement the descriptions in the text.

 WATCH THE VIDEO
Sedimentation and Clarifiers (www.awwa.org/wsovideoclips)

End-of-chapter questions   To test your understanding of the core concepts and applications in the book, questions are provided at the end of each chapter. The answers are provided at the end of the book, with the steps worked-out for math problems.

End-of-book questions   In the higher-level WSO books, questions are included at the end of the book to test your understanding of topics covered in previous grades. Keep in mind that certification tests will cover material from both lower-level and higher-level books.

Visit our website at www.awwa.org/wso to check out the additional books in the series and to access the free online resources associated with this book.

Chapter 1

Basic Microbiology and Chemistry

Chemical Formulas and Equations

A group of chemically bonded atoms forms a particle called a molecule. The simplest molecules contain only one type of atom, such as when two atoms of oxygen combine (O_2) or when two atoms of chlorine combine (Cl_2). Molecules of compounds are made up of the atoms of at least two different elements; for example, one oxygen atom and two hydrogen atoms form a molecule of the compound water (H_2O). "H_2O" is called the chemical formula of water. The formula is a shorthand way of writing *what* elements are present in a molecule of a compound, and *how many* atoms of each element are present in each molecule.

Reading Chemical Formulas

The following are examples of chemical formulas and what they indicate.

Example 1

The chemical formula for calcium carbonate is

$$CaCO_3$$

According to the formula, what is the chemical makeup of the compound?

First, the letter symbols given in the formula indicate the three elements that make up the calcium carbonate compound:

$$Ca = calcium$$
$$C = carbon$$
$$O = oxygen$$

Second, the subscripts (the small numbers at the lower right corners of the letter symbols) in the formula indicate how many atoms of each element are present in a single molecule of the compound. There is no number just to the right of the Ca or C symbols; this indicates that only one atom of each element is present in the molecule. The subscript 3 to the right of the O symbolizing oxygen indicates that there are three oxygen atoms in each molecule.

$$CaCO_3$$

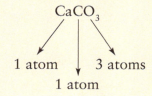

1 atom   3 atoms

1 atom

chemical formula
Using the chemical symbols for each element, a shorthand way of writing what elements are present in a molecule and how many atoms of each element are present in each of the molecules.

Determining Percent by Weight of Elements in a Compound

If 100 lb of sodium chloride (NaCl) were separated into the elements that make up the compound, there would be 39.3 lb of pure sodium (Na) and 60.7 lb of pure chlorine (Cl). We say that sodium chloride is 39.3 *percent* sodium *by weight* and that it is 60.7 *percent* chlorine *by weight*. The percent by weight of each element in a compound can be calculated using the compound's chemical formula and atomic weights from the periodic table.

The first step in calculating percent by weight of an element in a compound is to determine the molecular weight (sometimes called *formula weight*) of the compound. The molecular weight of a compound is defined as the sum of the atomic weights of all the atoms in the compound.

For example, to determine the molecular weight of sodium chloride, first count how many atoms of each element a single molecule contains:

Na	Cl
1 atom	1 atom

Next, find the atomic weight of each atom, using the periodic table:

atomic weight of Na = 22.99

atomic weight of Cl = 35.45

Finally, multiply each atomic weight by the number of atoms of that element in the molecule, and total the weights:

	Number of Atoms		Atomic Weight		Total Weight
sodium (Na)	1	×	22.99	=	22.99
chlorine (Cl)	1	×	35.45	=	35.45
			molecular weight of NaCl	=	58.44

Once the molecular weight of a compound is determined, the percent by weight of each element in the compound can be found with the following formula:

$$\text{percent element by weight} = \frac{\text{weight of element in compound}}{\text{molecular weight of compound}} \times 100$$

Using the formula, first calculate the percent by weight of sodium in the compound:

$$\text{percent Na by weight} = \frac{\text{weight of Na in compound}}{\text{molecular weight of compound}} \times 100$$

$$= \frac{22.99}{58.44} \times 100$$

$$= 0.393 \times 100$$

$$= 39.3\% \text{ sodium by weight}$$

percent by weight

The proportion, calculated as a percentage, of each element in a compound.

molecular weight

The sum of the atomic weights of all the atoms in the compound. Also called formula weight.

Then, calculate percent by weight of chlorine in the compound:

$$\text{percent Cl by weight} = \frac{\text{weight of Cl in compound}}{\text{molecular weight of compound}} \times 100$$

$$= \frac{35.45}{58.44} \times 100$$

$$= 0.607 \times 100$$

$$= 60.7\% \text{ chlorine by weight}$$

To check the calculations, add the percentages. The total should be 100:

$$
\begin{array}{rl}
39.3\% & \text{Na} \\
+\ \ 60.7\% & \text{Cl} \\
\hline
100.0\% & \text{NaCl}
\end{array}
$$

Chemical Equations

A chemical equation is a shorthand way, through the use of chemical formulas, to write the reaction that takes place when certain chemicals are brought together. As shown in the following example, the left side of the equation indicates the *reactants*, or chemicals that will be brought together; the arrow indicates which direction the reaction occurs; and the right side of the equation indicates the *products*, or results, of the chemical reaction.

calcium bicarbonate	plus	calcium hydroxide	react to form	calcium carbonate	plus	water
$Ca(HCO_3)_2$	+	$Ca(OH)_2$	$\longrightarrow$	$2CaCO_3$	+	$2H_2O$
	Reactants				Products	

The *2* in front of $CaCO_3$ is called a coefficient. A coefficient indicates the relative number of molecules of the compound that are involved in the chemical reaction. If no coefficient is shown, then only one molecule of the compound is involved. For example, in the preceding equation, one molecule of calcium bicarbonate reacts with one molecule of calcium hydroxide to form two molecules of calcium carbonate and two molecules of water. Without the coefficients, the equation could be written

$$Ca(HCO_3)_2 + Ca(OH)_2 \longrightarrow CaCO_3 + CaCO_3 + H_2O + H_2O$$

If you count the atoms of calcium (Ca) on the left side of the equation and then count the ones on the right side, you will find that the numbers are the same. In fact, for each element in the equation, as many atoms are shown on the left side as on the right. An equation for which this is true is said to be *balanced*. A balanced equation accurately represents what really happens in a chemical reaction: because matter is neither created nor destroyed, the number of atoms of each element going into the reaction must be the same as the number coming out. Coefficients allow balanced equations to be written compactly.

chemical equation
A shorthand way, using chemical formulas, of writing the reaction that takes place when chemicals are brought together. The left side of the equation indicates the chemicals brought together (the reactants); the arrow indicates in which direction the reaction occurs; and the right side of the equation indicates the results (the products) of the chemical reaction.

coefficient
An indication of the relative number of molecules of the compound that are involved in the chemical reaction.

Coefficients and subscripts can be used to calculate the molecular weight of each term in an equation, as illustrated in the following example.

Example 2

Calculate the molecular weights for each of the four terms in the following equation:

$$Ca(HCO_3)_2 + Ca(OH)_2 \longrightarrow 2CaCO_3 + 2H_2O$$

First, calculate the molecular weight of $Ca(HCO_3)_2$:

	Number of Atoms		Atomic Weight		Total Weight
calcium (Ca)	1	×	40.08	=	40.08
hydrogen (H)	2	×	1.01	=	2.02
carbon (C)	2	×	12.01	=	24.02
oxygen (O)	6	×	16.00	=	96.00
	molecular weight of $Ca(HCO_3)_2$			=	162.12

The molecular weight for $Ca(OH)_2$ is determined as follows:

	Number of Atoms		Atomic Weight		Total Weight
calcium (Ca)	1	×	40.08	=	40.08
oxygen (O)	2	×	16.00	=	32.00
hydrogen (H)	2	×	1.01	=	2.02
	molecular weight of $Ca(OH)_2$			=	74.10

The coefficient 2 in front of the next term of the equation ($2CaCO_3$) indicates that two molecules of $CaCO_3$ are involved in the reaction. First find the weight of *one molecule*, then double that weight to determine the weight of *two molecules*:

	Number of Atoms		Atomic Weight		Total Weight
calcium (Ca)	1	×	40.08	=	40.08
carbon (C)	1	×	12.01	=	12.01
oxygen (O)	3	×	16.00	=	48.00
	weight of one molecule of $CaCO_3$			=	100.09
	weight of two molecules of $CaCO_3$			=	(2)(100.09)
				=	200.18

The coefficient in front of the fourth term in the equation ($2H_2O$) also indicates that two molecules are involved in the reaction. As in the last calculation, first determine the weight of one molecule of H_2O, then the weight of two molecules:

	Number of Atoms		Atomic Weight		Total Weight
hydrogen (H)	2	×	1.01	=	2.02
oxygen (O)	1	×	16.00	=	16.00
weight of one molecule of H_2O				=	18.02
weight of two molecules of H_2O				=	(2)(18.02)
				=	36.04

In summary, the weights that correspond to each term of the equation are

$$Ca(HCO_3)_2 + Ca(OH)_2 \longrightarrow 2CaCO_3 + 2H_2O$$

$$162.12 \qquad 74.10 \qquad 200.18 \quad 36.04$$

Notice that the total weight on the left side of the equation (236.22) is equal to the total weight on the right side of the equation (236.22), meaning the equation is balanced.

The practical importance of the weight of each term of the equation is that the chemicals shown in the equation will always react in the proportions indicated by their weights. For example, from the calculation above, you know that $Ca(HCO_3)_2$ reacts with $Ca(OH)_2$ in the ratio 162.12:74.10. This means that, given 162.12 lb of $Ca(OH)_2$, you must add 74.10 lb of $Ca(HCO_3)_2$ for a complete reaction. Given twice the amount of $Ca(HCO_3)_2$ (that is, 324.24 lb), you must add twice the amount of $Ca(OH)_2$ (equal to 148.20 lb) to achieve complete reaction. The next two examples illustrate more complicated calculations using the same principle.

Example 3

If 25 g of $Ca(OH)_2$ were added to some $Ca(HCO_3)_2$, how many grams of $Ca(HCO_3)_2$ would react with the $Ca(OH)_2$?

Remember, the molecular weights indicate the weight ratio in which the two compounds will react. The molecular weight of $Ca(HCO_3)_2$ is 162.12, and the molecular weight of $Ca(OH)_4)_2$ is 74.10. Use this information to set up a proportion in order to determine how many grams of $Ca(HCO_3)_2$ will react with the $Ca(OH)_2$:

known ratio   *desired ratio*

$$\frac{74.10 \text{ g } Ca(OH)_2}{162.12 \text{ g } Ca(HCO_3)_2} = \frac{25 \text{ g } Ca(OH)_2}{x \text{ g } Ca(HCO_3)_2}$$

Next, solve for the unknown value:

$$\frac{74.10}{162.12} = \frac{25}{x}$$

$$\frac{(x)(74.10)}{162.12} = 25$$

$$x = \frac{(25)(162.12)}{74.10}$$

$$x = 54.7 \text{ g Ca(HCO}_3)_2$$

Given the molecular weights and the chemical equation indicating the ratio by which the two chemicals would combine, we were able to calculate that 54.7 g of $Ca(HCO_3)_2$ would react with 25 g of $Ca(OH)_2$.

Moles and Molarity

You may sometimes find chemical reactions described in terms of **moles** of a substance. The measurement *mole* (an abbreviation for gram-mole) is closely related to molecular weight. The molecular weight of water, for example, is 18.02—and 1 mol of water is defined to be 18.02 g of water. (The abbreviation for mole is *mol*.) The general definition of a mole is as follows:

A mole of a substance is a number of grams of that substance, where the number equals the substance's molecular weight.

In example 2, you saw that the following equation and molecular weights were correct:

$$Ca(HCO_3)_2 + Ca(OH)_2 \longrightarrow 2CaCO_3 + 2H_2O$$

$$162.12 \qquad 74.10 \qquad 2(100.09) \; 2(18.02)$$

If 162.12 g of $Ca(HCO_3)_2$ were used in the reaction, then the ratio equations given in example 3 would show that the weights of each of the substances in the reaction were as follows:

$$Ca(HCO_3)_2 + Ca(OH)_2 \longrightarrow 2CaCO_3 + 2H_2O$$

$$162.12 \text{ g} \qquad 74.10 \text{ g} \; (2)(100.09) \text{ g} \; (2)(18.02) \text{ g}$$

Because of the way a mole is defined, this could also be written in a more compact form:

$$Ca(HCO_3)_2 + Ca(OH)_2 \longrightarrow 2CaCO_3 + 2H_2O$$

$$1 \text{ mole} \qquad 1 \text{ mole} \qquad 2 \text{ moles} \; 2 \text{ moles}$$

Reading this information, a chemist could state, "One mole of $Ca(HCO_3)_2$ is needed to react with one mole of $Ca(OH)_2$, and the reaction yields two moles of $CaCO_3$ and two moles of water."

mole
The quantity of a compound or element that has a weight in grams equal to the substance's molecular or atomic weight. Used in this text generally as an abbreviation for gram-mole.

When measuring chemicals in moles, always remember that *the weight of a mole of a substance depends on what the substance is*. One mole of water weighs 18.02 g—but one mole of calcium carbonate weighs 100.09 g.

Example 4

A lab procedure calls for 3.0 mol of sodium bicarbonate ($NaHCO_3$) and 0.10 mol of potassium chromate (K_2CrO_4). How many grams of each compound are required?

To find the grams required of $NaHCO_3$, first determine the weight of 1 mol of the compound:

	Number of Atoms		Atomic Weight		Total Weight
sodium (Na)	1	×	22.99	=	22.99
hydrogen (H)	1	×	1.01	=	1.01
carbon (C)	1	×	12.01	=	12.01
oxygen (O)	3	×	16.00	=	48.00
molecular weight of $NaHCO_3$				=	84.01

Therefore, 1 mol of $NaHCO_3$ weighs 84.01 g.

Next, multiply the weight of 1 mol by the number of moles required. The required amount of $NaHCO_3$ is 3 mol. The weight of 3 mol $NaHCO_3$ is (3)(84.01 g) = 252.03 g.

To find grams required of K_2CrO_4, first determine the weight of 1 mol of the compound:

	Number of Atoms		Atomic Weight		Total Weight
potassium (K)	2	×	39.10	=	78.20
chromium (Cr)	1	×	52.00	=	52.00
oxygen (O)	4	×	16.00	=	64.00
molecular weight of K_2CrO_4				=	194.20

Therefore, 1 mol of K_2CrO_4 weighs 194.20 g.

Next, multiply the weight of 1 mol by the number of moles required. The required amount of K_2CrO_4 is 0.10 mol. This amount weighs (0.10) (194.20 g) = 19.42 g.

Equivalent Weights and Normality

Another method of expressing the concentration of a solution is normality. Normality depends in part on the valence of an element or compound. An element or compound may have more than one valence, and it is not always clear which valence (and therefore what concentration) a given normality represents. Because of this problem, normality is being replaced by molarity as the expression of concentration used for chemicals in the lab.

In the lab, you will often have detailed, step-by-step instructions for preparing a solution of a needed normality. Nonetheless, it is useful to have a basic idea of what the measurement means. To understand normality, you must first understand equivalent weights.

Equivalent Weights

The equivalent weight of an element or compound is the weight of that element or compound that, in a given chemical reaction, has the same combining capacity as 8 g of oxygen or as 1 g of hydrogen. The equivalent weight may vary with the reaction being considered. However, one equivalent weight of a reactant will always react with one equivalent weight of the other reactant in a given reaction.

Although you are not expected to know how to determine the equivalent weights of various reactants at this level, it will help if you remember one characteristic: the equivalent weight of a reactant either will be *equal* to the reactant's molecular weight or will be a *simple fraction* of the molecular weight. For example, if the molecular weight of a compound is 60.00 g, then the equivalent weight of the compound in a reaction will be 60.00 g or a simple fraction (usually ½, ⅓, ¼, ⅕, or ⅙) of 60.00 g.

Normality

Normality is defined as the number of equivalent weights of solute per liter of solution. Therefore, to determine the normality of a solution, you must first determine how many equivalent weights of solute are contained in the total weight of dissolved solute. Use the following equation:

$$\text{number of equivalent weights} = \frac{\text{total weight}}{\text{equivalent weight}}$$

equivalent weight
The weight of a compound that contains one equivalent of a proton (for acids) or one equivalent of a hydroxide (for bases). The equivalent weight can be calculated by dividing the molecular weight of a compound by the number of H⁺ or OH⁻ present in the compound.

normality
The number of equivalent weights of solute per liter of solution.

Dilution Calculations

Sometimes a particular strength of solution will be created by diluting a strong solution with a weak solution of the same chemical. The new solution will have a concentration somewhere between the weak and the strong solutions.

Although there are several methods available for determining what amounts of the weak and strong solutions are needed, perhaps the easiest is the *rectangle method* (sometimes called the *dilution rule*), shown in Figure 1-1.

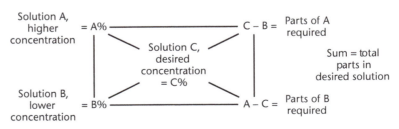

Figure 1-1 Schematic for rectangle method

The following example illustrates how the rectangle method is used.

Example 5

What volumes of a 3 percent solution and an 8 percent solution must be mixed to make 400 gal of a 5 percent solution?

Use the rectangle method to solve the problem. First write the given concentrations into the proper places in the rectangle:

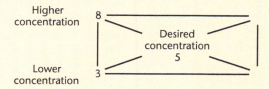

Now complete the rectangle by subtraction:

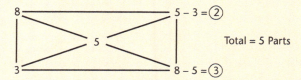

The circled numbers on the right side of the rectangle indicate the volume *ratios* of the solutions to be mixed. The sum of the circled numbers is a total of five parts to be added (2 parts + 3 parts = 5 parts total).

Two parts out of the five parts (2/5) of the new solution should be made up of the 8 percent solution. And three parts out of the five parts (3/5) of the new solution should be made up of the 3 percent solution.

Using the ratios, determine the *number of gallons* of each solution to be mixed:

$$\left(\frac{2}{5}\right)(400 \text{ gal}) = 160 \text{ gal of 8\% solution}$$

$$\left(\frac{3}{5}\right)(400 \text{ gal}) = 240 \text{ gal of 3\% solution}$$

Mixing these amounts will result in 400 gal of a 5 percent solution.

Solutions

A solution consists of two parts: a *solvent* and a *solute*. These parts are completely and evenly mixed, forming what is referred to as a *homogeneous* mixture. The solute part of the solution is dissolved in the solvent (Figure 1-2).

In a true solution, the solute will remain dissolved and will not settle out. Salt water is a true solution; salt is the solute and water is the solvent. In contrast, sand mixed into water does not form a solution—the sand will settle out when the water is left undisturbed.

In water treatment, the most common solvent is water. Before it is dissolved, the solute may be solid (such as dry alum), liquid (such as sulfuric acid), or gaseous (such as chlorine).

The concentration of a solution is a measure of the amount of solute dissolved in a given amount of solvent. A concentrated (strong) solution is a solution

solution

A liquid containing a dissolved substance. The liquid alone is called the solvent, the dissolved substance is called the solute. Together they are called a solution.

concentration

In chemistry, a measurement of how much solute is contained in a given amount of solution. Concentrations are commonly measured in milligrams per liter (mg/L).

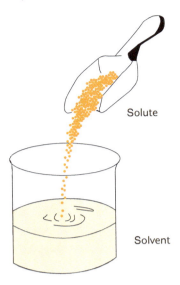

Figure 1-2 Solution, composed of a solute and a solvent

in which a relatively great amount of solute is dissolved in the solvent. A dilute (or weak) solution is one in which a relatively small amount of solute is dissolved in the solvent.

There are many ways of expressing the concentration of a solution, including the following:

- Milligrams per liter
- Grains per gallon
- Percent strength
- Molarity
- Normality

Molarity and normality were discussed previously in the chapter. The other expressions of concentration are briefly described next.

Milligrams per Liter and Grains per Gallon

The measurements milligrams per liter (mg/L) and grains per gallon (gpg) each express the *weight* of solute dissolved in a given *volume* of solution. The mathematics involved in using milligrams per liter are covered in Chapter 2.

Percent Strength

The percent strength of a solution can be expressed as percent by weight or percent by volume. The percent-by-weight calculation is used more often in water treatment. Conversions between milligrams per liter and percentages are discussed in Chapter 2.

Percent Strength by Weight

The equation used to calculate percent by weight is as follows:

$$\text{percent strength (by weight)} = \frac{\text{weight of solute}}{\text{weight of solution}} \times 100$$

where

weight of solutions = weight of solute + weight of solvents

Use of the equation is illustrated in the following examples.

Example 6

If 25 lb of chemical is added to 400 lb of water, what is the percent strength of the solution by weight?

Recall the formula:

$$\text{percent strength (by weight)} = \frac{\text{weight of solute}}{\text{weight of solution}} \times 100$$

The weight of the solute is given as 25 lb of chemical, but the weight of the solution is not given. Instead, the weight of the solvent (400 lb of water) is given. To determine the weight of the solution, combine the weights of the solute and the solvent:

$$\text{weight of solution} = \text{weight of solute} + \text{weight of solvent}$$

$$= 25 \text{ lb} + 400 \text{ lb}$$

$$= 425 \text{ lb}$$

Using this information, calculate the percent concentration:

$$\text{percent strength (by weight)} = \frac{\text{weight of solute}}{\text{weight of solution}} \times 100$$

$$= \frac{25 \text{ lb chemical}}{425 \text{ lb solution}} \times 100$$

$$= 0.059 \times 100$$

$$= 5.9\% \text{ strength}$$

Example 7

If 40 lb of chemical is added to 120 gal of water, what is the percent strength of the solution by weight?

First, calculate the weight of the solution. The weight of the solution is equal to the weight of the solute plus the weight of solvent. To calculate this, first convert gallons of water to pounds of water using the following formula:

$$\text{volume of water (in gallons)} \times 8.34 = \text{weight of water (in pounds)}$$

Therefore:

$$(120 \text{ gal})(8.34 \text{ lb/gal}) = 1,001 \text{ lb water}$$

Then calculate the weight of solution:

$$\text{weight of solution} = \text{weight of solute} + \text{weight of solvent}$$

$$= 40 \text{ lb} + 1,001 \text{ lb}$$

$$= 1,041 \text{ lb}$$

Now calculate the percent strength of the solution:

$$\text{percent strength (by weight)} = \frac{\text{weight of solute}}{\text{weight of solution}} \times 100$$

$$= \frac{40 \text{ lb chemical}}{1{,}041 \text{ lb solution}} \times 100$$

$$= 0.038 \times 100$$

$$= 3.8\% \text{ strength}$$

Standard Solutions

A standard solution is any solution that has an accurately known concentration. Although there are many uses of standard solutions, they are often used to determine the concentration of substances in other solutions. Standard solutions are generally made up based on one of three characteristics:

- Weight per unit volume
- Dilution
- Reaction

Weight per Unit Volume

When a standard solution is made up by weight per unit volume, a pure chemical is accurately weighed and then dissolved in some solvent. By the addition of more solvent, the amount of solution is increased to a given volume. The concentration of the standard is then determined in terms of molarity or normality, as discussed previously.

Dilution

When a given volume of an existing standard solution is diluted with a measured amount of solvent, the concentration of the resulting (more dilute) solution can be determined from the following equation:

$$\left(\begin{array}{c}\text{normality of}\\\text{solution 1}\end{array}\right)\left(\begin{array}{c}\text{volume of}\\\text{solution 1}\end{array}\right) = \left(\begin{array}{c}\text{normality of}\\\text{solution 2}\end{array}\right)\left(\begin{array}{c}\text{volume of}\\\text{solution 2}\end{array}\right)$$

This equation can be abbreviated as follows:

$$(N_1)(V_1) = (N_2)(V_2)$$

When this equation is being used, it is important to remember that both volumes—V_1 and V_2—must be expressed in the same units. That is, both must be in liters (L) or both must be in milliliters (mL).

Example 8

You have a standard 1.4N solution of H_2SO_4. How much water must be added to 100 mL of the standard solution to produce a 1.2N solution of H_2SO_4?

First determine the total volume of the new solution by using the relationship between solution concentration and volume:

$$(N_1)(V_1) = (N_2)(V_2)$$

$$(1.4N)(100 \text{ mL}) = (1.2N)(x \text{ mL})$$

Solve for the unknown value:

$$\frac{(1.4)(100)}{1.2} = x \text{ mL}$$

$$116.67 = x \text{ mL}$$

Therefore, the total volume of the new solution will be 116.67 mL. Since the volume of the original solution is 100 mL, 16.67 mL of water (that is, 116.67 mL – 100 mL) needs to be added to obtain the 1.2N solution:

100 mL of 1.4N solution + 16.67 mL of water = 116.67 mL of 1.2N solution

Reaction

A similar equation can be used for calculations involving reactions between samples of two solutions, as illustrated in the following example.

Example 9

You know that 32 mL of a 0.1N solution of HCl is required to react with (neutralize) 30 mL of a certain base solution. What is the normality of the base solution?

$$(N_1)(V_1) = (N_2)(V_2)$$

$$(0.1N)(32 \text{ mL}) = (xN)(30 \text{ mL})$$

$$\frac{(0.1)(32)}{30} = x \text{ normality}$$

$$0.1 = x \text{ normality}$$

WATCH THE VIDEOS

Operator Chemistry: Formulas and Chemical Solutions
(www.awwa.org/wsovideoclips)

Study Questions

1. Molarity is the number of
   a. gram equivalent weights of solute per liter of solution.
   b. gram atomic weight per oxidation number.
   c. moles of solute per kilogram of solvent.
   d. moles of solute per liter of solution.

2. "CaCO$_3$" is the _____ of calcium carbonate.
   a. chemical symbol
   b. atomic formula
   c. chemical formula
   d. chemical equation

3. The first step in calculating percent by weight of an element in a compound is to determine the
   a. percent by weight.
   b. molecular weight.
   c. chemical equation.
   d. coefficient.

4. What term refers to the number of equivalent weights of solute per liter of solution?
   a. Normality
   b. Equivalent weight
   c. Molecular weight
   d. Coefficient

5. A solution consists of two parts: a solvent and a
   a. diluent.
   b. solute.
   c. concentrate.
   d. suspension.

6. The chemical formula for calcium hydroxide (lime) is Ca(OH)$_2$. Determine the number of atoms of each element in a molecule of the compound.

7. The equation of the reaction between calcium carbonate (CaCO$_3$) and carbonic acid (H$_2$CO$_3$) is shown. If 10 lb of H$_2$CO$_3$ is used in the reaction, how many pounds of CaCO$_3$ will react with the H$_2$CO$_3$?

8. You need to prepare 25 gal of a solution with 2.5 percent strength. Assume the solution will have the same density as water: 8.34 lb/gal. How many pounds of chemical will you need to dissolve in the water?

9. How much water should be added to 80 mL of a 2.5N solution of H$_2$CO$_3$ to obtain a 1.8N solution of H$_2$CO$_3$?

10. How many atoms of oxygen are present in Ca(HCO$_3$)$_2$?

Chapter 2
Operator Math

Volume Measurements

A volume measurement defines the amount of space that an object occupies. The basis of this measurement is the *cube*, a square-sided box with all edges of equal length, as shown in the diagram below. The customary units commonly used in volume measurements are cubic inches, cubic feet, cubic yards, gallons, and acre-feet. The metric units commonly used to express volume are cubic centimeters, cubic meters, and liters.

The calculations of surface area and volume are closely related. For example, to calculate the surface area of one of the below cubes, you would multiply two of the dimensions (length and width) together. To calculate the volume of that cube, however, a *third dimension* (depth) is used in the multiplication.

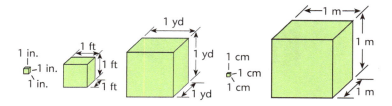

The concept of volume can be simplified as follows:

$$\text{volume} = (\text{area of surface})(\text{third dimension})$$

The *area of surface* to be used in the volume calculation is the *representative surface* area, the side that gives the object its basic shape. For example, suppose you begin with a rectangular area. Notice the shape that would be created by stacking a number of those same rectangles one on top of the other:

Because the rectangle gives the object its basic shape in this example, it is considered the representative area. Note that the same volume could have been created by stacking a number of smaller rectangles one behind the other:

Although an object may have more than one representative surface area, as illustrated in the two preceding diagrams, sometimes only one surface is the representative area. Consider, for instance, the following shape:

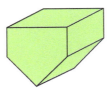

Let's compare two different sides of this shape—the top and the front—to determine if they are representative areas. In the first case, a number of the top shapes (rectangles) stacked together does not result in the same shape volume. Therefore, this rectangular area is not a representative area. In the second case, however, a number of front shapes (pentagons) stacked one behind the other results in the same shape volume as the original object. Therefore, the pentagonal area may be considered a representative surface area.

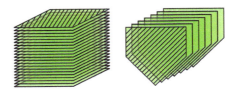

Rectangles, Triangles, and Circles

For treatment plant calculations, representative surface areas are most often rectangles, triangles, circles, or a combination of these. The following diagrams illustrate the three basic shapes for which volume calculations are made.

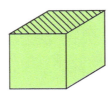

In the first diagram, the rectangle defines the shape of the object; in the second diagram, the triangle, rather than the rectangle, gives the trough its basic shape; in the third, the surface that defines the shape of the cylinder is the circle.

The formulas for calculating the volume of each of these three shapes are given below. Note that they are closely associated with the area formulas given previously.

Volume Formulas

$$\text{Rectangle tank volume} = \left(\begin{array}{c}\text{area of}\\\text{rectangle}\end{array}\right)\left(\begin{array}{c}\text{third}\\\text{dimension}\end{array}\right)$$

$$= hw\left(\begin{array}{c}\text{third}\\\text{dimension}\end{array}\right)$$

$$\text{Trough volume} = \left(\begin{array}{c}\text{area of}\\\text{triangle}\end{array}\right)\left(\begin{array}{c}\text{third}\\\text{dimension}\end{array}\right)$$

$$= \left(\frac{bh}{2}\right)\left(\begin{array}{c}\text{third}\\\text{dimension}\end{array}\right)$$

$$\text{Cylinder volume} = \left(\begin{array}{c}\text{area of}\\\text{circle}\end{array}\right)\left(\begin{array}{c}\text{third}\\\text{dimension}\end{array}\right)$$

$$= (0.785\ D^2)\left(\begin{array}{c}\text{third}\\\text{dimension}\end{array}\right)$$

$$\text{Cone volume} = 1/3(\text{volume of a cylinder})$$

$$\text{Sphere volume} = \left(\frac{\pi}{6}\right)(\text{diameter})^3$$

Use of the volume formulas is demonstrated in the examples that follow.

Example 1

Calculate the volume of water contained in the tank illustrated below if the depth of water (called *side water depth*, or SWD) in the tank is 10 ft.

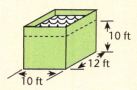

First, recall the volume formula:

$$\text{volume} = (\text{area of surface})(\text{third dimension})$$

In this example, the *rectangle* is the representative surface, and the area of a rectangle can be calculated as length times width. The dimension not used in the area calculation is the *depth*. Thus, the volume calculation is as follows:

$$\text{volume} = (\text{length})(\text{width})(\text{depth})$$

$$= (12\ \text{ft})(10\ \text{ft})(10\ \text{ft})$$

$$= 1{,}200\ \text{ft}^3$$

Example 2

What is the volume (in cubic inches) of water in the trough shown below if the depth of water is 8 in.?

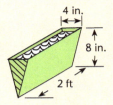

First, all dimensions must be expressed in the same terms. Since the answer is desired in cubic inches, the 2-ft dimension should be converted to inches:

$$(2\ \text{ft})(12\ \text{in./ft}) = 24\ \text{in.}$$

Then, we apply the volume formula:

$$\text{volume} = (\text{area of surface})(\text{third dimension})$$

The triangle is the representative surface. The area of a triangle is calculated as follows:

$$\frac{(\text{base})(\text{height})}{2}$$

The third dimension may be considered length or width—a difference in terminology. Thus, the volume calculation is as follows:

$$\text{volume} = \left(\frac{bh}{2}\right)(\text{length})$$

$$= \frac{(4\ \text{in.})(8\ \text{in.})}{2}(24\ \text{in.})$$

$$= \frac{(4\ \text{in.})(8\ \text{in.})(24\ \text{in.})}{2}$$

$$= 384\ \text{in.}^3$$

Example 3

What is the volume of water contained in the tank shown below if the depth of water (SDW) is 28 ft?

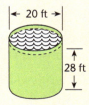

Recall the volume formula:

$$\text{volume} = (\text{area of surface})(\text{third dimension})$$

The circle is the representative surface, and the area of a circle is calculated as follows:

$$(0.785)(D^2)$$

The third dimension is depth. Thus, the volume calculation is as follows:

$$\text{volume} = (0.785)(D^2)(\text{depth})$$

$$= (0.785)(20 \text{ ft})^2(28 \text{ ft})$$

$$= (0.785)(20 \text{ ft})(20 \text{ ft})(28 \text{ ft})$$

$$= 8{,}792 \text{ ft}^3$$

Example 4

A tank with a cylindrical bottom has dimensions as shown below. What is the capacity of the tank? (Assume that the cross section of the bottom of the tank is a half circle.)

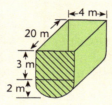

In problems involving a representative surface area that is a combination of shapes, it is often easier to calculate the representative surface area first, then calculate the volume:

$$\text{representative surface area} = \text{area of rectangle} + \text{area of half-circle}$$

$$= (4 \text{ m})(3 \text{ m}) + \frac{(0.785)(4 \text{ m})(4 \text{ m})}{2}$$

$$= 12 \text{ m}^2 + 6.28 \text{ m}^2$$

$$= 18.28 \text{ m}^2$$

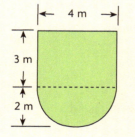 =  +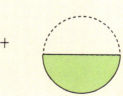

And now calculate the volume of the tank:

$$\text{volume} = (\text{area of surface})(\text{third dimension})$$

$$= (18.28 \text{ m}^2)(20 \text{ m})$$

$$= 365.6 \text{ m}^3$$

Cones and Spheres

There are many shapes (though very few in water treatment calculations) for which the concept of a "representative surface" does not apply. The cone and sphere are notable examples of this. That is, we cannot "stack" areas of the same size on top of one another to obtain a cone or sphere.

Calculating the volume of a cylinder was discussed earlier. The volume of a cone represents 1/3 of that volume:

$$\text{volume of a cone} = \tfrac{1}{3}(\text{volume of a cylinder})$$

or

$$= \frac{(0.785)(D^2)(\text{depth})}{3}$$

The volume of a sphere is more difficult to relate to the other calculations or even to a diagram. In this case, the formula should be memorized. To express the volume of a sphere mathematically, use the following formula:

$$\text{volume of a sphere} = \left(\frac{\pi}{6}\right)(\text{diameter})^3$$

The symbol π, or pi, represents the relationship between the circumference and diameter of a circle. The number 3.14 is used for pi. Thus, the equation may be re-expressed as follows:

$$\text{volume of a sphere} = \left(\frac{3.14}{6}\right)(\text{diameter})^3$$

Example 5

Calculate the volume of a cone that is 3 m tall and has a base diameter of 2 m.

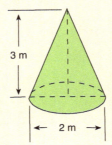

3 m

2 m

To begin, recall the formula used for calculating the volume of a cone:

volume of a cone = 1/3 (volume of a cylinder)

Thus, the volume is calculated as follows:

$$\text{volume} = \frac{(0.785)(D^2)(\text{third dimension})}{3}$$

$$= \frac{(0.785)(2\text{ m})(2\text{ m})(3\text{ m})}{3}$$

$$= 3.14\text{ m}^3$$

Example 6

If a spherical tank is 30 ft in diameter, what is its capacity?

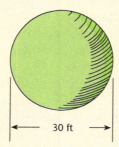

30 ft

Recall the formula used for calculating the volume of a sphere:

$$\text{volume of a sphere} = \left(\frac{\pi}{6}\right)(\text{diameter})^3$$

The volume is calculated as follows:

$$\text{volume} = \left(\frac{3.14}{6}\right)(30\text{ ft})(30\text{ ft})(30\text{ ft})$$

$$= 14{,}130\text{ ft}^3$$

Occasionally it is necessary to calculate the volume of a tank that consists of two distinct shapes. In other words, there is no representative surface area for the entire shape. In this case, the volumes should be calculated separately, then the two volumes added. The diagrams below illustrate this method:

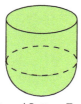

Round Bottom Tank   =   Cylinder   +   Half Sphere

Conversions

In making the conversion from one unit to another, you must know

- the number that relates the two units and
- whether to multiply or divide by that number.

For example, in converting from feet to inches, you must know that in 1 ft there are 12 in., and you must know whether to multiply or divide the number of feet by 12.

Although the number that relates the two units of a conversion is usually known or can be looked up, there is often confusion about whether to multiply or divide. *Dimensional analysis*, discussed previously, is one method to help decide whether to multiply or divide for a particular conversion.

Conversion Tables

Usually the fastest method of converting units is to use a *conversion table*, and to follow the instructions indicated by the table headings. For example, if you want to convert from feet to inches, look in the *Conversion* column of the table for *From* "feet" *To* "inches." Read across this line and perform the operation indicated by the headings of the other columns; that is, multiply the number of feet by 12 to get the number of inches.

Suppose, however, that you want to convert inches to feet. Look in the *Conversion* column for *From* "inches" *To* "feet," and read across this line. The headings tell you to multiply the number of inches by 0.08333 (which is the decimal equivalent of 1/12) to get the number of feet. Multiplying by either 1/12 or 0.08333 is the same as dividing by 12.

The instruction to *multiply* by certain numbers (called conversion factors) is used throughout the conversion table. There is no column headed *Divide by* because the fractions representing division (such as 1/12) were converted to decimal numbers (such as 0.08333) when the table was prepared.

To use the conversion table, remember the following three steps:

1. In the *Conversion* column, find the units you want to change *From* and *To*. (Go *From* what you have *To* what you want.)
2. Multiply the *From* number you have by the conversion factor given.
3. Read the answer in *To* units.

Example 7

Convert 288 in. to feet.

In the *Conversion* column of the table, find *From* "inches" *To* "feet." Reading across the line, perform the multiplication indicated; that is, multiply the number of inches (288) by 0.08333 to get the number of feet:

$$(288 \text{ in.})(0.08333) = 24 \text{ ft}$$

conversion

The process of changing from one unit of measure to another (e.g., from gallons to liters).

Example 8

A tank holds 50 gal of water. How many cubic feet of water is this, and what does it weigh?

First, convert gallons to cubic feet. Using the table, you find that to convert *From* "gallons" *To* "cubic feet," you must multiply by 0.1337 to get the number of cubic feet:

$$(50 \text{ gal})(0.1337) = 6.69 \text{ ft}^3$$

Note that this number of cubic feet is actually a rounded value (6.685 is the actual calculated number). Rounding helps simplify calculations.

Next, convert gallons to pounds of water. Using the table, you find that to convert *From* "gallons" *To* "pounds of water," you must multiply by 8.34 to get the number of pounds of water:

$$(50 \text{ gal})(8.34 \text{ lb/gal}) = 417 \text{ lb of water}$$

Notice that you could have arrived at approximately the same weight by converting 6.69 ft³ to pounds of water. Using the table, we get

$$(6.69 \text{ ft}^3)(62.4) = 417.46 \text{ lb of water}$$

This slight difference in the two answers is due to rounding numbers both when the conversion table was prepared and when the numbers are used in solving the problem. You may notice the same sort of slight difference in answers if you have to convert from one kind of units to two or three other units, depending on whether you round intermediate steps in the conversions.

Box Method

Another method that may be used to determine whether multiplication or division is required for a particular conversion is called the *box method*. This method is based on the relative sizes of different squares ("boxes"). The box method can be used when a conversion table is not available (such as during a certification exam). This method of conversion is often slower than using a conversion table, but many people find it simpler.

Because multiplication is usually associated with an *increase* in size, moving from a smaller box to a larger box corresponds to using multiplication in the conversion:

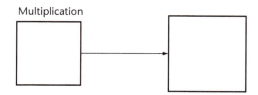

Multiplication

Division, on the other hand, is usually associated with a *decrease* in size. Therefore, moving from a larger box to a smaller box corresponds to using division in the conversion:

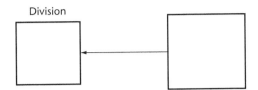

To use the box method to determine whether to multiply or divide in making a conversion, set up and label the boxes according to the following procedure:

1. Write the equation that relates the two types of units involved in the conversion. One of the two numbers in the equation must be a 1 (for example, 1 ft = 12 in., or 1 ft = 0.305 m).
2. Draw a small box on the left, a large one on the right, and connect them with a line.
3. In the *smaller* box, write the name of the units associated with the 1 (for example, 1 ft = 12 in.—*ft* should be written in the smaller box). Note that the name of the units next to the 1 must be written in the smaller box; otherwise, the box method will give incorrect results.
4. In the larger box, write the name of the remaining units. Those units will also have a number next to them, a number that is not 1. Write that number over the line between the boxes.

Suppose, for example, that you want to make a box diagram for feet-to-inches conversions. First, write the equation that relates feet to inches:

$$1 \text{ ft} = 12 \text{ in.}$$

Next, draw the conversion boxes (smaller box on the left) and the connecting line:

Now label the diagram. Because the number 1 is next to the units of feet (1 ft), write *ft* in the smaller box. Write the name of the other units, inches (*in.*), in the larger box. And write the number that is next to inches, 12, over the line between the boxes.

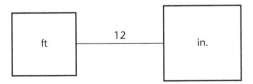

To convert from feet to inches, then *multiply by 12* because you are moving from a smaller box to a larger box. And to convert from inches to feet, *divide by 12* because you are moving from a larger box to a smaller box.

Let's look at another example of making and using the box diagram. Suppose you want to convert cubic feet to gallons. First write down the equation that relates these two units:

$$1 \text{ ft}^3 = 7.48 \text{ gal}$$

Then draw the smaller and larger boxes and the connecting line; label the boxes, and write in the conversion number:

The smaller box corresponds to cubic feet and the larger box to gallons. To convert from cubic feet to gallons according to this box diagram, *multiply* by 7.48 because you are moving from a smaller to a larger box. And to convert from gallons to cubic feet, *divide* by 7.48 because you are moving from a larger to a smaller box.

Conversions of US Customary Units

This section discusses important conversions between terms expressed in US customary units (based on the box method).

Conversions From Cubic Feet to Gallons to Pounds

In making the conversion from cubic feet to gallons to pounds of water, you must know the following relationships:

$$1 \text{ ft}^3 = 7.48 \text{ gal}$$
$$1 \text{ gal} = 8.34 \text{ lb}$$

You must also know whether to multiply or divide, and which of the above numbers are used in the conversion. The following box diagram should assist in making these decisions:

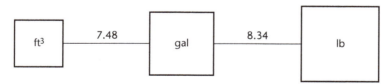

Example 9

Convert 1 ft³ to pounds.

First write down the diagram to aid in the conversion:

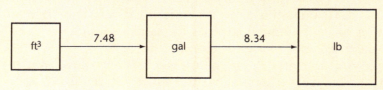

When you are converting from cubic feet to pounds, you are moving from smaller to larger boxes. Therefore, *multiplication* is indicated in both conversions:

$$(1 \text{ ft}^3)(7.48 \text{ gal/ft}^3)(8.34 \text{ lb/gal}) = 62.38 \text{ lb}$$

This total can be rounded to 62.4 lb, the number commonly used for water treatment calculations.

Flow Conversions

The relationships among the various US customary flow units are shown by the following diagram. Note the abbreviations commonly used in discussing flow:

- gps = gallons per second
- gpm = gallons per minute
- gpd = gallons per day

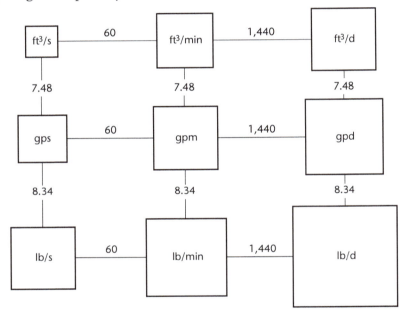

Since flows of cubic feet per hour, gallons per hour, and pounds per hour are less frequently used, they have not been included in the diagram. However, some chemical feed rate calculations require converting from or to these units.

The lines that connect the boxes and the numbers associated with them can be thought of as *bridges* that relate two units directly and all other units in the diagram indirectly. The relative sizes of the boxes are an aid in deciding whether multiplication or division is appropriate for the desired conversion.

The relationship among the boxes should be understood, not merely memorized. The principle is basically the same as that described in the preceding section. For example, notice how every box in a single vertical column has the same *time* units; a conversion in this direction corresponds to a change in volume units. Every box in a single horizontal row has the same *volume* units; a conversion in this direction corresponds to a change in time units.

Although you need not draw the nine boxes each time you make a flow conversion, it is useful to have a mental image of these boxes to make the calculations. For the example that follows, however, the boxes are used in analyzing the conversions.

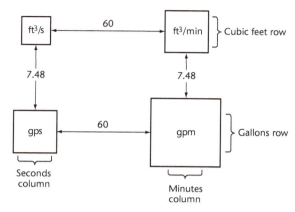

Example 10

The flow rate in a water line is 2.3 ft³/s. What is this rate expressed as gallons per minute? (Assume the flow is steady and continuous.)

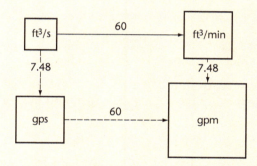

There are two possible paths from cubic feet per second to gallons per minute. Either will give the correct answer. Notice that each path has factors of 60 and 7.48, with only a difference in order. In each case, you are moving from a smaller to a larger box, and thus *multiplication* by both 60 and 7.48 is indicated:

$$(2.3 \text{ ft}^3/\text{s})(60 \text{ s/min})(7.48 \text{ gal/ft}^3) = 1{,}032 \text{ gpm}$$

Notice that you can write both multiplication factors into the same equation; you do not need to write one equation for converting cubic feet per second to cubic feet per minute and another for converting cubic feet per minute to gallons per minute.

Linear Measurement Conversions

Linear measurement defines the distance along a line; it is the measurement between two points. The US customary units of linear measurement include the inch, foot, yard, and mile. In most treatment plant calculations, however, the mile is not used. Therefore, this section discusses conversions of inches, feet, and yards only. The box diagram associated with these conversions is

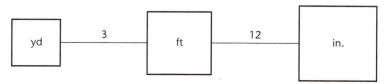

Example 11

The maximum depth of sludge drying beds is 14 in. How many feet is this?

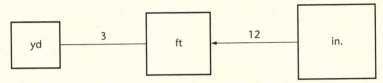

In converting from inches to feet, you are moving from a larger to a smaller box. Therefore, *division* by 12 is indicated.

$$\frac{14\ \text{in.}}{12\ \text{in./ft}} = 1.17\ \text{ft}$$

Area Measurement Conversions

To make area conversions in US customary units, you work with units such as square yards, square feet, or square inches. These units are derived from the following multiplications:

$$(\text{yards})(\text{yards}) = \text{square yards, or yd}^2$$

$$(\text{feet})(\text{feet}) = \text{square feet, or ft}^2$$

$$(\text{inches})(\text{inches}) = \text{square inches, or in.}^2$$

By examining the relationship of yards, feet, and inches in linear terms, you can recognize the relationship between yards, feet, and inches in square terms. For example,

$$
\begin{aligned}
1\ \text{yd} &= 3\ \text{ft} \\
(1\ \text{yd})(1\ \text{yd}) &= (3\ \text{ft})(3\ \text{ft}) \\
1\ \text{yd}^2 &= 9\ \text{ft}^2
\end{aligned}
$$

This method of comparison may be used whenever you wish to compare linear terms with square terms. Compare the diagram used for linear conversions with that used for square measurement conversions:

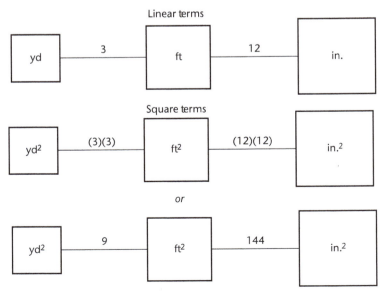

Example 12

The surface area of a sedimentation basin is 170 yd². How many square feet is this?

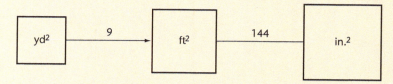

When converting from square yards to square feet, you are moving from a smaller to a larger box. Therefore, *multiplication* by 9 is indicated:

$$(170 \text{ yd}^2)(9 \text{ ft}^2/\text{yd}^2) = 1{,}530 \text{ ft}^2$$

Volume Measurement Conversions

To make volume conversions in US customary unit terms, you work with such units as cubic yards, cubic feet, and cubic inches. These units are derived from the following multiplications:

$$(\text{yards})(\text{yards})(\text{yards}) = \text{cubic yards, or yd}^3$$

$$(\text{feet})(\text{feet})(\text{feet}) = \text{cubic feet, or ft}^3$$

$$(\text{inches})(\text{inches})(\text{inches}) = \text{cubic inches, or in.}^3$$

By examining the relationship of yards, feet, and inches in linear terms, you can recognize the relationship between yards, feet, and inches in cubic terms. For example,

$$\begin{aligned} 1 \text{ yd} &= 3 \text{ ft} \\ (1 \text{ yd})(1 \text{ yd})(1 \text{ yd}) &= (3 \text{ ft})(3 \text{ ft})(3 \text{ ft}) \\ 1 \text{ yd}^3 &= 27 \text{ ft}^3 \end{aligned}$$

The box diagram associated with these cubic conversions is

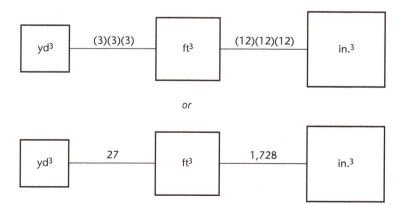

Example 13

Convert 15 yd³ to cubic inches.

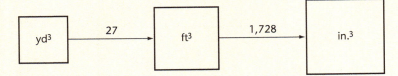

In converting from cubic yards to cubic feet and from cubic feet to cubic inches, you are moving from smaller to larger boxes. Thus, *multiplication* by 27 and 1,728 is indicated:

$$(15 \text{ yd}^3)(27 \text{ ft}^3/\text{yd}^3)(1,728 \text{ in.}^3/\text{ft}^3) = 699,840 \text{ in.}^3$$

Average Daily Flow

The amount of water a community uses every day can be expressed in terms of an **average daily flow (ADF)**. The ADF is the average of the actual daily flows that occur within a period of time, such as a week, a month, or a year. It is expressed mathematically as

$$\text{ADF} = \frac{\text{sum of all daily flows}}{\text{total number of daily flows used}}$$

The ADF can reflect a week's data (weekly ADF), a month's (monthly ADF), or a year's data (annual ADF, or AADF—the most commonly calculated average).

ADF is important because it is used in several treatment plant calculations. The following examples illustrate the calculation of ADF.

Example 14

A water treatment plant reported that the total volume of water treated for the calendar year 1995 was 152,655,000 gal. What was the annual ADF for 1995?

In this problem, the sum of all daily flows has already been determined—a total of 152,655,000 gal was treated during the year. Knowing that there are 365 days in a year, calculate the ADF:

$$\text{ADF} = \frac{\text{sum of all daily flows}}{\text{total number of daily flows used}}$$

$$= \frac{152,655,000 \text{ gal}}{365 \text{ days}}$$

$$= 418,233 \text{ gpd}$$

average daily flow (ADF)

A measurement of the amount of water treated by a plant each day. It is the average of the actual daily flows that occur within a period of time, such as a week, a month, or a year. Mathematically, it is the sum of all daily flows divided by the total number of daily flows used.

Example 15

In January 1993, a total of 68,920,000 L of water was treated at a plant. What was the ADF at the treatment plant for this period?

As in the previous example, the sum of all daily flows has already been determined: a *total* of 68,920,000 L was treated for the month. January has 31 days; fill the information into the ADF formula:

$$ADF = \frac{\text{sum of all daily flows}}{\text{total number of daily flows used}}$$

$$= \frac{68,920,000 \text{ L}}{31 \text{ days}}$$

$$= 2,223,226 \text{ L/d}$$

Example 16

The following daily flows (in million gallons per day [mgd]) were treated during June 2000 at a water treatment plant. What was the ADF for this period?

June	1	6.21	June	11	7.59	June	21	6.43
	2	6.68		12	7.01		22	6.26
	3	7.31		13	6.85		23	6.87
	4	7.80		14	6.43		24	7.27
	5	6.77		15	6.52		25	7.95
	6	6.32		16	6.79		26	7.33
	7	5.96		17	6.91		27	6.72
	8	5.83		18	7.37		28	6.51
	9	6.09		19	7.02		29	5.92
	10	7.22		20	6.88		30	5.90

To calculate the ADF for this period, first add all daily flows. The total of these flows is 202.72 mil gal. With this information, calculate the ADF for the period:

$$ADF = \frac{\text{sum of all daily flows}}{\text{total number of daily flows used}}$$

$$= \frac{202.72 \text{ mil gal}}{30 \text{ days}}$$

$$= 6.76 \text{ mgd}$$

In some problems, you will not know actual daily flows for each day during a particular period, but you will know the ADF for that period. ADF information can be used in calculating other ADFs. The following example contrasts the two methods of calculating ADFs.

Example 17

The volume of water (in megaliters [ML]) treated for each day during a 2-week period is listed below. What is the ADF for this 2-week period?

Week 1		Week 2	
Sunday	2.41	Sunday	2.52
Monday	3.37	Monday	3.39
Tuesday	3.44	Tuesday	3.48
Wednesday	3.61	Wednesday	3.88
Thursday	3.23	Thursday	3.19
Friday	2.86	Friday	2.82
Saturday	2.75	Saturday	2.70

Since you are given the actual flows for each day in the 2-week period, you could calculate the ADF in a manner similar to that used in the previous examples. That is,

$$\text{ADF} = \frac{\text{sum of all daily flows}}{\text{total number of daily flows used}}$$

$$= \frac{43.65 \text{ ML}}{14 \text{ days}}$$

$$= 3.12 \text{ ML/d}$$

Suppose, however, that in this problem you do not know the actual flow for each day in the week, but you know the ADF for week 1 (3.10 ML/d) and the ADF for week 2 (3.14 ML/d). The ADF for the 2-week period can still be calculated, but *not* using the ADF equation given above (because in this case you do not know the *sum* of all daily flows). Instead, a similar formula is used:

$$\text{ADF} = \frac{\text{sum of all weekly ADFs}}{\text{total number of weekly ADFs used}}$$

In this case,

$$\text{ADF} = \frac{3.10 \text{ ML/d} + 3.14 \text{ ML/d}}{2}$$

$$= \frac{6.24 \text{ ML/d}}{2}$$

$$= 3.12 \text{ ML/d}$$

Notice that, although only ADF information was used, the answer was the same as when all 14 actual daily flows were used in the calculation.

In a similar manner, ADFs for each of 12 months can be used to determine the ADF for a year. The equation that is used is

$$ADF = \frac{\text{sum of all monthly ADFs}}{\text{total number of monthly ADFs used}}$$

Example 18

The ADF (in million gallons per day) at a treatment plant for each month in the year is given below. Using this information, calculate the annual ADF.

January	10.71	July	11.96
February	9.89	August	12.24
March	10.32	September	11.88
April	10.87	October	11.53
May	11.24	November	11.36
June	11.58	December	10.98

If you knew all 365 flows for the year, the annual ADF would be calculated using the formula

$$ADF = \frac{\text{sum of all daily flows}}{\text{total number of daily flows used}}$$

In this problem, however, the ADF for each *month* of the year is given. Therefore, the ADF is calculated using the formula

$$ADF = \frac{\text{sum of all monthly ADFs}}{\text{total number of monthly ADFs used}}$$

Filling in the information given in this problem,

$$ADF = \frac{13.56 \text{ mgd}}{12}$$

$$= 11.21 \text{ mgd}$$

Surface Overflow Rate

The faster the water leaves a sedimentation tank, or clarifier, the more turbulence is created and the more suspended solids are carried out in the effluent. Overflow rate—the speed with which water leaves the sedimentation tank—is controlled by an increase or decrease in the flow rate into the tank.

The surface overflow rate measures the amount of water leaving a sedimentation tank per square foot of tank surface area. The treatment plant operator must be able to determine the surface overflow rate that produces the best-quality effluent leaving the tank.

surface overflow rate
A measurement of the amount of water leaving a sedimentation tank per unit of tank surface area. Mathematically, it is the flow rate from the tank divided by the tank surface area.

Since surface overflow is the gallons per day flow *up and over* each square foot of tank surface, the corresponding mathematical equation is

$$\text{surface overflow rate} = \frac{\text{flow (gpd)}}{\text{tank surface (ft}^2)}$$

Notice that the depth of the sedimentation tank is not a consideration in the calculation of surface overflow rate. Figures 2-1 and 2-2 depict surface overflow for rectangular and circular tanks, respectively.

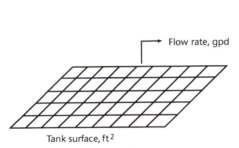

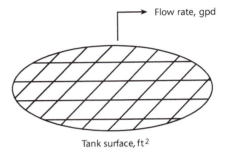

Figure 2-1 Surface overflow for rectangular tank

Figure 2-2 Surface overflow for circular tank

Example 19

The flow to a treatment plant is 4.8 mgd. If the sedimentation tank is 120 ft long, 20 ft wide, and 8 ft deep, what is the surface overflow rate?

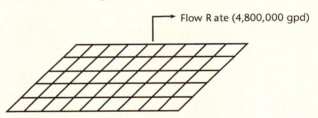

Solve by first determining the surface area, then apply the formula for calculating the surface overflow rate:

$$\text{surface area} = (\text{length})(\text{width})$$
$$= (120 \text{ ft})(20 \text{ ft})$$
$$= 2,400 \text{ ft}^2$$

$$\text{surface overflow rate} = \frac{\text{flow (gpd)}}{\text{tank surface (ft}^2)}$$
$$= \frac{4,800,000 \text{ gpd}}{2,400 \text{ ft}^2}$$
$$= 2,000 \text{ gpd/ft}^2$$

Example 20

A circular sedimentation basin with a 55-ft diameter receives a flow of 2,075,000 gpd. What is the surface overflow rate?

$$\text{surface area} = 0.785\,D^2$$

$$= (0.785)(55\text{ ft})(55\text{ ft})$$

$$= 2{,}375\text{ ft}^2$$

$$\text{surface overflow rate} = \frac{\text{flow (gpd)}}{\text{tank surface (ft}^2)}$$

$$= \frac{2{,}075{,}000\text{ gpd}}{2{,}375\text{ ft}^2}$$

In the previous two examples, the surface overflow rate was the unknown factor. However, *any one* of three factors (surface overflow rate, water flow rate, or surface area) can be unknown. If the other two factors are known, the same mathematical setup can be used to solve for the unknown value.

Example 21

A 20-m diameter tank has a surface overflow area of 0.35 L/m²/s. What is the daily flow to the tank?

$$\text{surface area} = 0.785\,D^2$$

$$= (0.785)(20\text{ m})(20\text{ m})$$

$$= 314\text{ m}^2$$

$$\text{surface overflow rate} = \frac{\text{flow (L/s)}}{\text{tank surface (m}^2)}$$

$$0.35\text{ L/m}^2\text{/s} = \frac{x\text{ L/s}}{314\text{ m}^2}$$

$$(0.35)(314) = x$$

$$109.9\text{ L/s} = x$$

Example 22

A sedimentation tank receives a flow of 7.6 mgd. If the surface overflow rate is 845 gpd/ft², what is the surface area of the sedimentation tank?

$$\text{surface overflow rate} = \frac{\text{flow (gpd)}}{\text{tank surface (ft}^2)}$$

$$845 \text{ gpd/ft}^2 = \frac{7{,}600{,}000 \text{ gpd}}{x \text{ ft}^2}$$

$$(845)(x) = 7{,}600{,}000$$

$$x = \frac{7{,}600{,}000}{845}$$

$$x = 8{,}994 \text{ ft}^2$$

Weir Overflow Rate

The calculation of **weir overflow rate** (Figures 2-3 and 2-4) is important in detecting high velocities near the weir, which adversely affect the efficiency of the sedimentation process. With excessively high velocities, the settling solids are pulled over the weirs and into the effluent troughs.

Since weir overflow rate is gallons-per-day flow over each foot of weir length, the corresponding mathematical equation for the weir overflow rate is

$$\text{weir overflow rate} = \frac{\text{flow (gpd)}}{\text{weir length (ft)}}$$

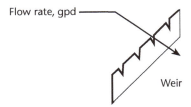
Flow rate, gpd

Weir

Figure 2-3 Weir overflow for rectangular clarifier

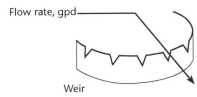
Flow rate, gpd

Weir

Figure 2-4 Weir overflow for circular clarifier

weir overflow rate
A measurement of the flow rate of water over each foot of weir in a sedimentation tank or circular clarifier. Mathematically, it is the flow rate over the weir divided by the total length of the weir.

Example 23

A circular clarifier, which has a continuous weir around its circumference, receives a flow of 3.6 mgd. If the diameter is 80 ft, what is the weir overflow rate?

Before you can calculate the weir overflow rate, you must know the total length of the weir. The relationship of the diameter and circumference of a circle is the key to determining this measurement:

$$\text{circumference} = (3.14)(\text{diameter})$$

In this problem, the diameter is 80 ft. Therefore, the length of weir (circumference) is calculated as follows:

$$\text{circumference} = (3.14)(80\ \text{ft})$$

$$= 251.2\ \text{ft}$$

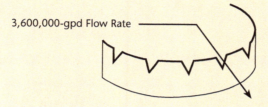

3,600,000-gpd Flow Rate

251.2-ft Weir

Now solve for the weir overflow rate:

$$\text{weir overflow rate} = \frac{\text{flow (gpd)}}{\text{weir length (ft)}}$$

$$= \frac{3,600,000\ \text{gpd}}{251.2\ \text{ft}}$$

$$= 14,331\ \text{gpd/ft}$$

Example 24

If a sedimentation tank has a total of 25 m of weir over which the water flows, what is the weir overflow rate when the flow is 40.5 L/s?

40.5-L/s Flow Rate

25-m Weir

$$\text{weir overflow rate} = \frac{\text{flow (L/s)}}{\text{weir length (m)}}$$

$$= \frac{40.5\ \text{L/s}}{25\ \text{m}}$$

$$= 1.62\ \text{(L/m)/s}$$

Example 25

A clarifier has a continuous weir around the perimeter and a diameter of 60 ft. It receives a flow of 1.87 mgd. What is the weir overflow rate?

If the diameter is 60 ft, then the weir length (circumference) is calculated as follows:

$$\text{circumference} = (3.14)(60 \text{ ft})$$

$$= 188.4 \text{ ft}$$

Now solve for the weir overflow rate:

$$\text{weir overflow rate} = \frac{\text{flow (gpd)}}{\text{weir length (ft)}}$$

$$= \frac{1,870,000 \text{ gpd}}{188.4 \text{ ft}}$$

$$= 9,926 \text{ gpd/ft}$$

Filter Loading Rate

The **filter loading rate** is measured in gallons of water applied to each square foot of surface area. Or, stated more precisely, filter loading rate measures the amount of water flowing down through each square foot of filter area. Whereas the surface overflow rate is measured in gallons per *day* per square foot, the filter loading rate is measured in gallons per *minute* per square foot. Figure 2-5 depicts filter loading.

The mathematical equation associated with filter loading rate is

$$\text{filter loading rate} = \frac{\text{flow}}{\text{filter area}}$$

Sometimes the filter loading rate is measured as the fall of water (in inches) through a filter per minute. This is expressed mathematically as

$$\text{filter loading rate} = \frac{\text{inches of water fall}}{\text{minutes}}$$

filter loading rate
A measurement of the volume of water applied to each unit of filter surface area. Mathematically, it is the flow rate into the filter divided by the total filter area.

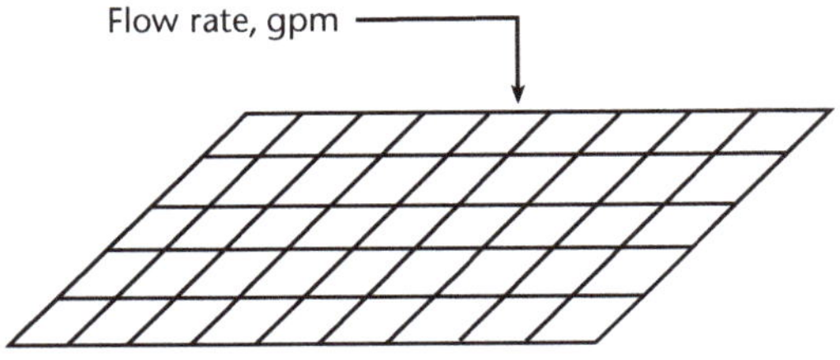

Flow rate, gpm

Filter, ft²

Figure 2-5 Filter loading

Some typical loading rates for various types of filters are shown in the following table:

Type of Filter	Common Loading Rate
Slow sand	0.016–0.16 gpm/ft^2
Rapid sand	2 gpm/ft^2
Dual media (coal, sand)	2–5 gpm/ft^2
Multimedia (coal, sand, and garnet; or coal, sand, and ilmenite)	5–10 gpm/ft^2

Example 26

A slow sand filter is 15 ft wide and 25 ft long. If the filter receives a flow of 125,600 gpd, what is the filter loading rate in gallons per minute per square foot?

Convert the flow from gallons per day to gallons per minute:

$$\frac{125,600 \text{ gpd}}{1,440 \text{ min/d}} = 87.2 \text{ gpm}$$

Fill this information into the following equation:

$$\text{filter overflow rate} = \frac{\text{flow (gpm)}}{\text{filter area (ft}^2)}$$

$$= \frac{87.2 \text{ gpm}}{(15 \text{ ft})(25 \text{ ft})}$$

$$= 0.23 \text{ gpm/ft}^2$$

Filter loading rates, as previously noted, are sometimes expressed as the vertical movement of water in the filter in one minute—that is, as inches-per-minute (in./min) fall in the water level.

The gallons-per-minute-per-square-foot and inches-per-minute measurements are directly related, as noted by the following equation:

$$1 \text{ gpm/ft}^2 = 1.6 \text{ in./min}$$

You can use the box method of conversion when converting from one term to another.

The following two examples illustrate conversion between these two terms.

Example 27

A drop of how many inches per minute corresponds to a filter loading rate of 2.6 gpm/ft²?

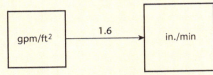

In converting from gallons per minute per square foot to inches per minute filter loading rate, you are moving from a smaller box to a larger box. Therefore, multiplication by 1.6 is indicated:

$$(2.6 \text{ gpm/ft}^2)(1.6) = 4.16 \text{ in./min}$$

Example 28

What is the filter loading rate in gallons per minute per square foot of a sand filter in which the water level dropped 15 in. in 3 min after the influent valve was closed?

First, determine the inches per minute drop in water level:

$$\frac{15 \text{ in.}}{3 \text{ min}} = 5 \text{ in./min}$$

Next, convert to gallons per minute per square foot.

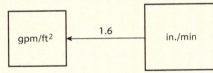

Converting from inches per minute to gallons per day per square foot, you move from a larger box to a smaller box and divide by 1.6:

$$\frac{5 \text{ in./min}}{1.6} = 3.13 \text{ gpm/ft}^2$$

Example 29

A rapid sand filter is 20 ft wide and 35 ft long. If the flow through the filter is 2.3 mgd, what is the filter loading rate in gallons per minute per square foot?

First, convert the gallons per day to gallons per minute:

$$\frac{2,300,000 \text{ gpd}}{1,440 \text{ min/day}} = 1,597.2 \text{ gpm}$$

Then express the filter loading rate mathematically as follows:

$$\text{filter loading rate} = \frac{\text{flow (gpm)}}{\text{filter area (ft}^2)}$$

$$= \frac{1,597.2 \text{ gpm}}{(20 \text{ ft})(35 \text{ ft})}$$

$$= 2.28 \text{ gpm/ft}^2$$

Filter Backwash Rate

The filter backwash rate is measured in gallons of water flowing upward (backward) each minute through a square foot of filter surface area. The units of measure for backwash rate are the same as those for filter loading rate—that is, gallons per *minute* per square foot of filter area. Figure 2-6 depicts filter backwash.

The mathematical equation associated with the filter backwash rate is

$$\text{filter backwash rate} = \frac{\text{flow (gpm)}}{\text{filter area (ft}^2)}$$

Sometimes the backwash rate is measured as the rise of water in inches that occurs each minute. This can be expressed mathematically as

$$\text{filter backwash rate} = \frac{\text{inches of water rise}}{\text{minutes}}$$

Typically, backwash rates will vary from 15 gpm/ft² to 22.5 gpm/ft². These rates are equivalent to a rise in the water level from 24 to 36 in./min.

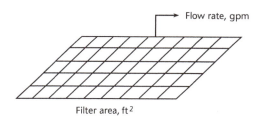

Filter area, ft²

Figure 2-6 Filter backwash

Example 30

A mixed-media filter is 25 ft wide and 32 ft long. If the filter receives a backwash flow of 17,300,000 gpd, what is the filter backwash flow rate in gallons per minute per square foot?

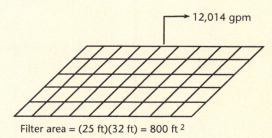

Filter area = (25 ft)(32 ft) = 800 ft²

First convert the backwash flow from gallons per day to gallons per minute:

$$\frac{17,300,000 \text{ gpd}}{1,440 \text{ min/d}} = 12,014 \text{ gpm}$$

With this information, calculate filter backwash rate:

$$\text{filter backwash rate} = \frac{\text{flow (gpm)}}{\text{filter area (ft}^2)}$$

$$= \frac{12{,}014 \text{ gpm}}{(25 \text{ ft})(32 \text{ ft})}$$

$$= \frac{12{,}014 \text{ gpm}}{800 \text{ ft}^2}$$

$$= 15.02 \text{ gpm/ft}^2$$

Filter backwash rates, as noted earlier, are sometimes expressed in terms of vertical rise of water in a time interval measured in minutes—for example, inches per minute (in./min). The gallons per minute per square foot and inches per minute units of measure are directly related to each other, as shown in the following example.

Example 31

How many inches rise per minute corresponds to a filter backwash rate of 18 gpm/ft^2?

Use the box method to convert from gallons per minute per square feet to inches per minute. Moving from a smaller to a larger box indicates multiplication:

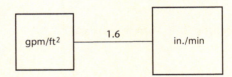

Calculate the inches rise per minute as follows:

$$(18 \text{ gpm/ft}^2)(1.6) = 28.8 \text{ in./min}$$

Example 32

A rapid sand filter is 3 m wide and 5 m long. If backwash water is flowing upward at a rate of 18 ML/d, what is the backwash rate in liters per square meter per second?

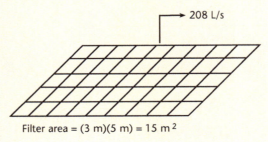

Filter area = (3 m)(5 m) = 15 m^2

First convert flow from liters per day to liters per second:

$$\frac{18,000,000 \text{ L/d}}{86,400 \text{ s/d}} = 208 \text{ L/s}$$

The filter backwash rate can be written mathematically as follows:

$$\text{filter backwash rate} = \frac{\text{flow (L/s)}}{\text{filter area (m}^2)}$$

$$= \frac{208 \text{ L/s}}{15 \text{ m}^2}$$

$$= 13.9 \text{ (L/m}^2)/\text{s}$$

Mudball Calculation

Mudballs, a conglomeration of floc particles and sand, are a common problem in sand filters. The presence of mudballs is the basic cause of many filter malfunctions. Therefore, measuring and controlling the amount of accumulated mudballs are of prime importance in sand filter operation.

To measure the amount of mudballs in a filter, a tube about 3 in. in diameter and 6 in. long is pushed down into the sand of the filter and then lifted, full of sand (Figure 2-7). A sieve is used to separate the sand from any mudballs present.

The volume of mudballs can then be determined by placing the balls in a graduated cylinder containing a known amount of water. As mudballs are put in, the water level rises. The volume of mudballs, therefore, is the *increase in volume* in the graduated cylinder. The percent volume of mudballs can then be determined using the following formula:

$$\text{percent volume of mudballs} = \frac{\text{volume of mudballs}}{\text{total volume of sample collected}} \times 100$$

The following example problems illustrate the calculation of percent mudballs.

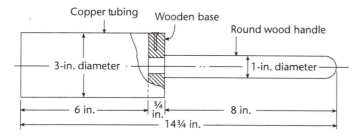

Figure 2-7 Mudball sampling tube

> **mudball**
> An accumulation of media grains and suspended material that creates clogging problems in filters.

Example 33

A sample of sand taken from the filter for mudball determination has a volume of 1,500 mL. If the volume of mudballs in the sample is found to be 37 mL, what is the percent volume of mudballs in the sample?

$$\text{percent volume of mudballs} = \frac{\text{volume of mudballs}}{\text{total volume of sample collected}} \times 100$$

$$= \frac{37 \text{ mL}}{1,500 \text{ mL}} \times 100$$

$$= 0.0246 \times 100$$

$$= 2.46\% \text{ mudballs}$$

Example 34

A sample of sand taken from the filter for mudball determination has a volume of 5,600 mL. If the volume of mudballs in the sample is found to be 30 mL, what is the percent volume of mudballs in the sample?

$$\text{percent volume of mudballs} = \frac{\text{volume of mudballs}}{\text{total volume of sample collected}} \times 100$$

$$= \frac{30 \text{ mL}}{5,600 \text{ mL}} \times 100$$

$$= 0.0054 \times 100$$

$$= 0.54\% \text{ mudballs}$$

Detention Time

detention time
The average length of time a drop of water or a suspended particle remains in a tank or chamber. Mathematically, it is the volume of water in the tank divided by the flow rate through the tank. The units of flow rate used in the calculation are dependent on whether the detention time is to be calculated in minutes, hours, or days.

The concept of **detention time** is used in conjunction with many treatment plant processes, including flash mixing, coagulation–flocculation, and sedimentation. Detention time refers to the length of time a drop of water or a suspended particle remains in a tank or chamber. For example, if water entering a 50-ft long tank has a flow velocity of 1 ft/s, an average of 50 s would elapse from the time a drop entered the tank until it left the tank.

Detention time may also be thought of as the number of minutes or hours required for each tank to empty. You may find it helpful to form a mental image of the flow from the time water enters the tank until it leaves the tank completely ("plug flow"), as shown in Figure 2-8.

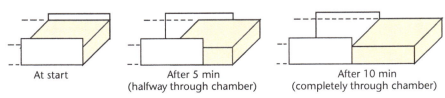

At start   After 5 min   After 10 min
            (halfway through chamber)   (completely through chamber)

Figure 2-8 Simplified schematic for a detention time of 10 min

Typical ranges of detention times for various treatment processes are as follows:

Tank	Detention Time Range
Flash-mixing basin	30–60 s
Flocculation basin	20–60 min
Sedimentation basin	1–12 h

The equation used to calculate detention time is

$$\text{detention time} = \frac{\text{volume of tank}}{\text{flow rate}}$$

If the flow rate used in the equation is gallons *per day*, then the detention time calculated will be expressed in *days*. If the flow rate used in the equation is gallons *per minute*, the detention time calculated will be expressed in *minutes*. The calculation method is illustrated in the following examples.

Example 35

A sedimentation tank has a capacity of 140,000 gal. If the hourly flow to the clarifier is 31,500 gph, what is the detention time?

Since the flow rate is expressed in hours, the detention time calculated is also in hours:

$$\text{detention time} = \frac{\text{volume of tank}}{\text{flow rate}}$$

$$= \frac{140{,}000 \text{ gal}}{31{,}500 \text{ gph}}$$

$$= 4.44 \text{ h}$$

Example 36

A flocculation basin is 48 ft long, is 23 ft wide, and has a side water depth of 17 ft. If the flow to the basin is 3,500 gpm, what is the detention time?

Since the flow rate is expressed in minutes, the detention time calculated is also in minutes:

$$\text{detention time} = \frac{\text{volume of tank}}{\text{flow rate}}$$

$$= \frac{(48 \text{ ft})(23 \text{ ft})(17 \text{ ft})(7.48 \text{ gal/ft}^3)}{520 \text{ gpm}}$$

$$= \frac{146{,}384.6 \text{ gal}}{3{,}500 \text{ gpm}}$$

$$= 40 \text{ min}$$

Example 37

A flash mixing basin has a capacity of 6,800 L. If the flow to the mixing basin is 140 L/s, what is the detention time?

Since the flow rate is expressed in seconds, the detention time is expressed in seconds:

$$\text{detention time} = \frac{\text{volume of tank}}{\text{flow rate}}$$

$$= \frac{6,800 \text{ L}}{140 \text{ L/s}}$$

$$= 48.57 \text{ s}$$

Example 38

The flow rate through a circular clarifier is 4,752,000 gpd. If the clarifier is 65 ft in diameter and 12 ft deep, what is the clarifier detention time in hours?

To calculate a detention time in hours, first express the gallons-per-day flow rate as gallons per hour:

$$\frac{4,752,000 \text{ gpd}}{24 \text{ h/d}} = 198,000 \text{ gph}$$

Then calculate the detention time (being certain to express tank volume in gallons):

$$\text{detention time} = \frac{\text{volume of tank}}{\text{flow rate}}$$

$$= \frac{(0.785)(65 \text{ ft})(65 \text{ ft})(12 \text{ ft})(7.48 \text{ gal/ft}^3)}{198,000 \text{ gph}}$$

$$= \frac{297,700 \text{ gal}}{198,000 \text{ gph}}$$

$$= 1.5 \text{ h}$$

It should be noted that the calculations for detention time illustrated in this chapter are only theoretical. For many situations, these calculations are sufficient for use in normal water system operations. The flow of water through a pipeline, for instance, is for the most part plug flow; that is to say, essentially every drop of water enters a section of pipe and leaves the other end in the same amount of time.

Tanks and basins, in contrast, will not actually be plug flow. Some of the reasons why some particles of water are detained, while others speed from the inlet to the outlet (called short-circuiting) are poor distribution of water at the inlet, dead spaces in the corners of the basin, stratification due to temperature differentials, inadequate baffling, and surface effects from wind.

The contact time of disinfectants and water as they pass through treatment before the water is delivered to the first customer must now be considered under the federal Surface Water Treatment Rule (SWTR). Contact time is the average time that a disinfectant is in contact with the water as it passes through a basin. For a well-designed basin, contact time may be relatively close to the theoretical detention time. For a poorly designed or unbaffled basin, it may be only a fraction of the theoretical time.

Additional information on methods of determining contact time through basins may be found in AWWA publications describing details of the SWTR.

Pressure

As shown in Figure 2-9, pressure may be expressed in different ways depending on the unit area selected. Normally, however, the unit area of a square inch and the expression of pressures in pounds per square inch (psi) are preferred. In metric units, pressure is generally expressed in kilopascals (kPa).

In the operation of a water treatment system, you will be primarily concerned with the pressures exerted by water. Water pressures are directly related to the height (depth) of water. Suppose, for example, you have a container 1 ft by 1 ft by 1 ft (a cubic-foot container) that is filled with water. What is the pressure on the square-foot bottom of the container?

Pressure in this case is expressed in pounds per unit area. In this case, since the density of water is 62.4 lb/ft^2, the force of the water pushing down on the square foot surface area is 62.4 lb (Figure 2-10).

From this information, the pressure in pounds per square inch can also be determined. Convert pounds per square foot to pounds per square inch (psi):

$$\frac{62.4 \text{ lb}}{\text{ft}^2} = \frac{62.4 \text{ lb}}{(1 \text{ ft})(1 \text{ ft})}$$

$$= \frac{62.4 \text{ lb}}{(12 \text{ in.})(12 \text{ in.})}$$

$$= \frac{62.4 \text{ lb}}{144 \text{ in.}^2}$$

$$= 0.433 \frac{\text{lb}}{\text{in.}^2}$$

$$= 0.433 \text{ psi}$$

> **pressure**
> The force pushing on a unit area. Normally pressure can be measured in pascals (Pa), pounds per square inch (psi), or feet of head.
>
> **pounds per square inch (psi)**
> A measure of pressure.

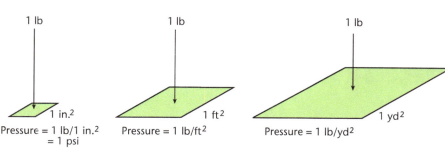

Figure 2-9 Forces acting on areas

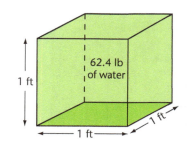

Figure 2-10 Cubic foot of water

This means that a foot-high column of water over a square-inch surface area weighs 0.433 lb, resulting in a pressure of 0.433 psi (Figure 2-11). The factor 0.433 allows you to convert from pressure measured in feet of water to pressure measured in pounds per square inch, as shown later in this section. A conversion factor can also be developed for converting from pounds per square inch to feet.

Since 1 ft is equivalent to 0.433 psi, set up a ratio to determine how many feet of water are equivalent to 1 psi (that is, how many feet high a water column must be to create a pressure of 1 psi):

$$\frac{1 \text{ ft}}{0.433 \text{ psi}} = \frac{x \text{ ft}}{1 \text{ psi}}$$

Then solve for the unknown value:

$$\frac{(1)(1)}{0.433} = x$$

$$2.31 \text{ ft} = x$$

Therefore, 1 psi is equivalent to the pressure created by a column of water 2.31 ft high (Figure 2-12).

Since the density of water is assumed to be a constant 62.4 lb/ft³ in most hydraulic calculations, the height of the water is the most important factor in determining pressures. It is the height of the water that determines the pressure over the square-inch area. Pressure measured in terms of the height of water (in meters or feet) is referred to as **head**. As long as the height of the water stays the same, changing the shape of the container does not change the pressure at the bottom or any other level in the container. For example, see Figure 2-13. Each of the containers is filled with water to the same height. Therefore, the pressures against the bottoms of the containers are the same. And the pressure at any depth in one container is the same as the pressure at the same depth in either of the other containers.

In water system operation, the shape of the container can help to maintain a usable volume of water at higher pressures. For example, suppose you have an elevated storage tank and a standpipe that contain equal amounts of water. When the water levels are the same, the pressures at the bottom of the tanks are the

head (pressure)

(1) A measure of the energy possessed by water at a given location in the water system, expressed in feet (or meters). (2) A measure of the pressure (force) exerted by water, expressed in feet (or meters).

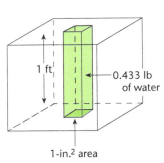

Figure 2-11 A 1-foot column of water over a 1-in.² area

1 ft

0.433 lb of water

1-in.² area

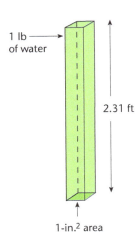

Figure 2-12 A 2.31-ft water column (which creates a pressure of 1 psi at the bottom surface)

1 lb of water

2.31 ft

1-in.² area

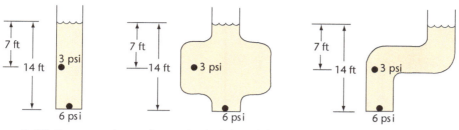

Figure 2-13 Pressure depends on the height of the water, not the shape of the container

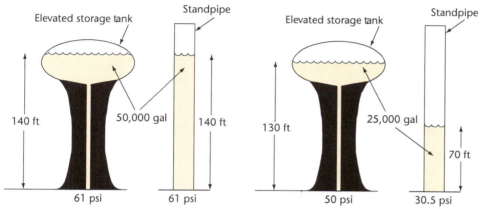

A. Same pressures at bottoms   **B. Different pressures at bottoms**

Figure 2-14 Comparison of maximum pressure values in containers with different shapes but the same volume of water

same (Figure 2-14A). However, if half of the water is withdrawn from each tank, the pressure at the bottom of the elevated tank will be greater than the pressure at the bottom of the standpipe (Figure 2-14B).

Because of the direct relationship between the pressure in *pounds per square inch* at any point in water and the *height in feet* of water above that point, pressure can be measured either in pounds per square inch or in feet (of water), called head.

When the water pressure in a main or in a container is measured by a gauge, the pressure is referred to as gauge pressure. If measured in pounds per square inch, the gauge pressure is expressed as pounds per square inch gauge (psig). The gauge pressure is not the total pressure within the main. Gauge pressure does not show the pressure of the atmosphere, which is equal to approximately 14.7 psi at sea level. Because atmospheric pressure is exerted everywhere (against the outside of the main as well as the inside, for instance), it can generally be neglected in water system calculations. However, for certain calculations, the total (or absolute) pressure must be known. The absolute pressure (expressed as pounds per square inch absolute [psia]) is obtained by adding the gauge and atmospheric pressures. For example, in a main under 50-psi gauge pressure, the absolute pressure would be 50 psig + 14.7 psi = 64.7 psia. A line under a partial vacuum, with a gauge pressure of –2, would have an absolute pressure of (14.7 psi) + (–2 psig) = 12.7 psia (Figure 2-15).

Pressure gauges can also be calibrated in feet of head. A pressure gauge reading of 14 ft of head, for example, means that the pressure is equivalent to the pressure exerted by a column of water 14 ft high. The equations that relate gauge

gauge pressure
The water pressure as measured by a gauge. Gauge pressure is not the total pressure. Total water pressure (absolute pressure) also includes the atmospheric pressure (about 14.7 psi at sea level) exerted on the water. However, because atmospheric pressure is exerted everywhere (against the outside of the main as well as the inside, for example), it is generally not written into water system calculations. Gauge pressure in pounds per square inch is expressed as "psig."

pounds per square inch gauge (psig)
Pressure measured by a gauge and expressed in terms of pounds per square inch.

absolute pressure
The total pressure in a system, including both the pressure of water and the pressure of the atmosphere (about 14.7 psi, at sea level). Compare with gauge pressure.

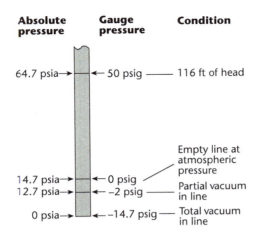

Figure 2-15 Gauge versus absolute pressure

pressure in pounds per square inch to pressure in feet of head are given below. In this book, conversions from one term to another will use the first equation only.

$$1 \text{ psig} = 2.31 \text{ ft head}$$
$$1 \text{ ft head} = 0.433 \text{ psig}$$

The following examples illustrate conversions from feet of head to pounds per square inch gauge and from pounds per square inch gauge to feet of head. The method used is similar to the conversion approach (box method).

Example 39

Convert a gauge pressure of 14 ft to pounds per square inch gauge.

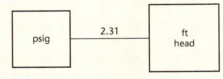

1 psig = 2.31 ft head

Using the diagram, in moving from feet of head to pounds per square inch gauge, you are moving from a larger box to a smaller box. Therefore, you should divide by 2.31:

$$\frac{14 \text{ ft}}{2.31 \text{ ft/psig}} = 6.06 \text{ psig}$$

Example 40

A head of 250 ft of water is equivalent to what pressure in pounds per square inch gauge?

1 psig = 2.31 ft head

When you move from a larger box to a smaller box, division by 2.31 is indicated:

$$\frac{250 \text{ ft}}{2.31 \text{ ft/psig}} = 108.23 \text{ psig}$$

Example 41

A pressure of 210 kPa (gauge) is equivalent to how many meters of head?

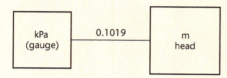

When you move from a smaller box to a larger box, multiplication by 0.1019 is indicated:

$$[210 \text{ kPa(gauge)}](0.1019 \text{ m/kPa}) = 21.4 \text{ m}$$

Example 42

What would the pounds-per-square-inch gauge pressure readings be at points A and B in the diagram in Figure 2-16?

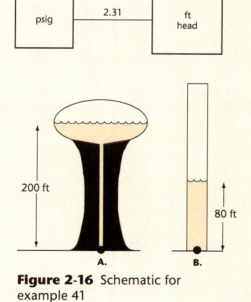

Figure 2-16 Schematic for example 41

In each case, the conversion is from feet of head to pressure in pounds per square inch gauge. Therefore, when you move from a larger box to a smaller box, division by 2.31 is indicated:

$$\text{pressure at A} = \frac{200 \text{ ft}}{2.31 \text{ ft/psig}}$$

$$= 86.58 \text{ psig}$$

$$\text{pressure at B} = \frac{80 \text{ ft}}{2.31 \text{ ft/psig}}$$

$$= 34.63 \text{ psig}$$

Flow Rate Problems

The measurement of water flow rates is an important part of efficient water supply operations. Flow rate information is used in determining chemical dosages, assessing the cost of treatment, evaluating the plant efficiency, determining the amounts of water used by various commercial and industrial sources, and planning for future expansion of the treatment facility and its distribution system.

Two types of flow rates are important in water supply operations: (1) **instantaneous flow rates** (flow rates at a particular moment) and (2) **average flow rates** (the average of the instantaneous flow rates over a given period of time, such as a day). Instantaneous flow rates are calculated based on the cross-sectional area and velocity of water in a channel or pipe. Average flow rates are calculated from records of instantaneous flow rates or from records of time and total volume.

Flow rates in water supply operations are measured by various metering devices, such as weirs or Parshall flumes for open-channel flow and venturi or orifice meters for closed conduit flow.

Instantaneous Flow Rate Calculations

The flow rate of water through a channel or pipe at a particular moment depends on the cross-sectional area and the velocity of the water moving through it. This relationship is stated mathematically as follows:

$$Q = AV$$

where

$$Q = \text{flow rate}$$

$$A = \text{area}$$

$$V = \text{velocity}$$

instantaneous flow rate

A flow rate of water measured at one particular instant, such as by a metering device, involving the cross-sectional area of the channel or pipe and the velocity of the water at that instant.

average flow rate

The average of the instantaneous flow rates over a given period of time, such as a day.

Figure 2-17A illustrates the $Q = AV$ equation as it pertains to flow in an open channel. Since flow rate and velocity must be expressed for the same unit of time, the flow rate in the open channel is expressed as follows:

$$Q \qquad\qquad A \qquad\qquad V$$

$$\text{(flow rate)} = \underset{\text{ft}^3/\text{time}}{\text{(width)}} \quad \underset{\text{ft}}{\text{(length)}} \quad \underset{\text{ft}}{\text{(velocity)}} \atop \text{ft/time}$$

In using the $Q = AV$ equation, the time units given for the velocity must match the time units for cubic-feet flow rate. For example, if the velocity is expressed as feet per second (ft/s), then the resulting flow rate must be expressed as cubic feet per second (ft³/s). If the velocity is expressed as feet per minute (ft/min), then the resulting flow rate must be expressed as cubic feet per minute (ft³/min). And if the velocity is expressed as feet per day (ft/d), then the resulting flow rate must be expressed as cubic feet per day (ft³/d). Figures 2-17B through 2-17D illustrate this concept.

Using the $Q = AV$ equation, if you know the cross-sectional area of the water in the channel (the width and depth of the rectangle), and you know the velocity, then you will be able to calculate the flow rate. This approach to flow rate calculations can be used if the problem involves an open-channel flow with a rectangular cross section, as shown previously, or if the problem involves a pipe flow with a circular cross section, as shown in Figure 2-17E. Only the calculation of the cross-sectional area will differ.

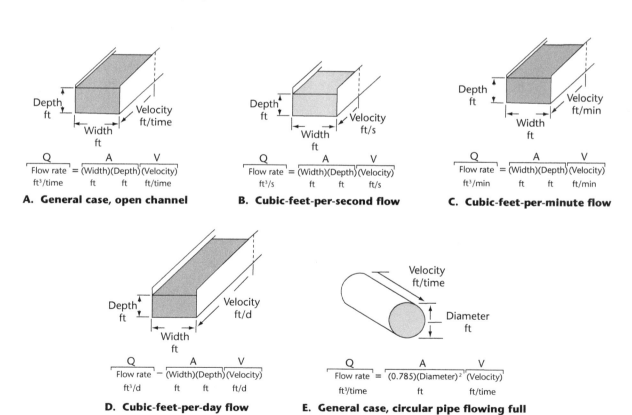

Figure 2-17 The $Q = AV$ equation as it pertains to flow in an open channel

If a circular pipe is flowing full (as it will be in most situations in water supply and treatment), the resulting flow rate is expressed as follows:

Q		A	V
(flow rate) ft³/time	=	(0.785)(diameter)² ft²	(velocity) ft/time

As in the case of the rectangular channel, if the velocity is expressed as feet per second, then the resulting flow rate must be expressed as cubic feet per second. If the velocity is expressed as feet per minute, then the resulting flow rate must be expressed as cubic feet per minute. If the velocity is expressed as feet per day, then the resulting flow rate must be expressed as cubic feet per day.

Example 43

A 15-in.-diameter pipe is flowing full. What is the gallons-per-minute flow rate in the pipe if the velocity is 110 ft/min?

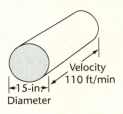

To use the $Q = AV$ equation, the diameter and velocity terms should be expressed using feet. Therefore, convert the 15-in. diameter to feet:

$$\frac{15 \text{ in.}}{12 \text{ in./ft}} = 1.25 \text{ ft}$$

Now use the $Q = AV$ equation to calculate the flow rate. Since the velocity is expressed in feet per minute, first calculate the cubic-feet-per-minute flow rate, then convert to gallons per minute:

$$
\begin{array}{ccc}
Q & A & V
\end{array}
$$

$$
\begin{aligned}
\text{(flow rate)} &= (0.785)(\text{diameter})^2 \ (\text{velocity}) \\
&= (0.785)(1.25 \text{ ft})(1.25 \text{ ft})(110 \text{ ft/min}) \\
&= 134.92 \text{ ft}^3/\text{min}
\end{aligned}
$$

Convert cubic feet per minute to gallons per minute:

$$(134.92 \text{ ft}^3/\text{min})(7.48 \text{ gal/ft}^3) = 1{,}009 \text{ gpm}$$

Example 44

A 305-mm-diameter pipe flowing full is carrying 35 L/s. What is the velocity of the water (in meters per second) through the pipe?

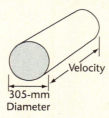

You are asked to determine the velocity V, given the flow rate Q and diameter. Note first that the diameter of 305 mm is equivalent to 0.305 m. To use the $Q = AV$ formula to calculate velocity in a pipe, you should use the following mathematical setup:

$$Q \qquad\qquad A \qquad\qquad V$$

$$(\text{flow rate}) \;=\; (0.785)(\text{diameter})^2 \;\; (\text{velocity})$$

Since you want to know velocity in meters per second, the flow rate must also be expressed in terms of meters and seconds (cubic meters per second). The information given in the problem expresses the flow rate as liters per second. Therefore, the flow rate must be converted to cubic meters per second before you begin the $Q = AV$ calculation:

$$35 \text{ L/s} \times 0.001 \text{ m}^3/\text{L} = 0.035 \text{ m}^3/\text{s}$$

Now determine the velocity using the $Q = AV$ equation:

$$Q \qquad\qquad A \qquad\qquad V$$

$$0.035 \text{ m}^3/\text{s} \;=\; (0.785)(0.305 \text{ m})^2 \quad (x)$$

And then solve for the unknown value:

$$\frac{0.035 \text{ m}^3/\text{s}}{(0.785)(0.305 \text{ m})^2} = x$$

$$0.48 \text{ m/s} = x$$

Chemical Dosage Problems

One of the more common uses of mathematics in water treatment practices is for performing chemical dosage calculations. As a basic- or intermediate-level operator, there are several common types of dosage calculations that you may be required to perform:

- Milligrams-per-liter to pounds-per-day conversions
- Milligrams-per-liter to percentage conversions
- Feed rate conversions

- Calculations for chlorine dosage, demand, and residual
- Percent strength calculations
- Solution dilution calculations

In addition to the general types of calculations, the operator may be required to perform calculations dealing with specific treatment processes. Some of these operations, such as ion exchange softening and fluoridation, are discussed in later chapters of this book.

Milligrams-per-Liter to Pounds-per-Day Conversions

The formula for converting milligrams per liter to pounds per day is derived from the formula for converting parts per million (ppm) to pounds per day, which is as follows:

$$\begin{array}{l} \text{feed} \\ \text{rate} \\ \text{(lb/d)} \end{array} = \begin{array}{l} \text{dosage} \\ \text{(ppm)} \end{array} \times \begin{array}{l} \text{flow rate} \\ \text{(mgd)} \end{array} \times \begin{array}{l} \text{conversion} \\ \text{factor} \\ \text{(8.34 lb/gal)} \end{array}$$

In the range of 0 to 2,000 mg/L, milligrams per liter are approximately equal to parts per million (ppm). For example, 150 mg/L of calcium is approximately equal to 150 ppm of calcium. Therefore, "mg/L" can be substituted for "ppm" in the equation just given. The substitution yields the following equation, which is used to convert between milligrams per liter and pounds per day:

$$\begin{array}{l} \text{feed} \\ \text{rate} \\ \text{(lb/d)} \end{array} = \begin{array}{l} \text{dosage} \\ \text{(mg/L)} \end{array} \times \begin{array}{l} \text{flow rate} \\ \text{(mgd)} \end{array} \times \begin{array}{l} \text{conversion} \\ \text{factor} \\ \text{(8.34 lb/gal)} \end{array}$$

Converting feed rate in milligrams per liter to pounds per day is a common water treatment calculation. Therefore, you should *memorize the conversion formula given above*. The following examples illustrate how the formula is used showing the units to clarify how the formula was derived.

Example 45

The dry alum dosage rate is 12 mg/L at a water treatment plant. The flow rate at the plant is 3 mgd. How many pounds per day of alum are required?

This is a milligrams-per-liter to pounds-per-day conversion problem. You will need to apply the conversion equation:

feed rate (lb/d) = (dosage mg/L)(flow rate mgd)(conversion factor 8.34 lb/gal)

In this scenario, to arrive at pounds per day, we must employ a few other conversion factors:

$$1 \text{ mL} = 1 \text{ gram}$$

$$1 \text{ L} = 1000 \text{ gram}$$

$$1 \text{ gram} = 1000 \text{ mg}$$

$$\text{Feed rate (lb/d)} = (12 \text{ mg/L})(1 \text{ gr}/1000 \text{ mg})(1\text{L}/1000 \text{ gr})$$
$$(3{,}000{,}000 \text{ gal/day})(8.34 \text{ lb/gal})$$

You can see that in converting mg/L, the units cancel out, resulting in a unitless number divided by 1 million—that is, parts per million. We are left with the following:

$$1 \text{ mg/L} = 1 \text{ ppm, or } 1/1,000,000$$

All that remains in our formula is gal/day × lb/gal. The gallons cancel and we have lb/day. You can now use the conversion equation and understand that the units do indeed produce the desired result.

Example 46

A fluoride compound is added at a concentration of 1.5 mg/L. The flow rate at the treatment plant is 2 mgd. How many pounds per day of the compound are added?

First write the conversion equation:

$$\text{feed rate} = (\text{dosage})(\text{flow rate})(\text{conversion factor})$$

Fill in the information given in the problem, and solve for the unknown value:

$$x = (1.5)(2)(8.34)$$
$$= 25.02 \text{ lb/d}$$

Example 47

The chlorine dosage rate at a water treatment plant is 2 mg/L. The flow rate at the plant is 700,000 gpd. How many pounds per day of chlorine are required?

First write the conversion formula:

$$\text{feed rate} = (\text{dosage})(\text{flow rate})(\text{conversion factor})$$

Before the information given in the problem can be filled in, 700,000 gpd must be converted to million gallons per day. To do this, locate the position of the "millions comma" and replace it with a decimal point. For example, in the number 2,400,000, the millions comma is between the 2 and 4. To express 2,400,000 gpd as million gallons per day, replace the millions comma with a decimal point:

$$2,400,000 \text{ gpd} = 2.4 \text{ mgd}$$

In this example, converting 700,000 gpd to million gallons per day appears as follows:

$$,700,000 \text{ gpd}$$
$$\uparrow$$
millions comma

$$700,000 \text{ gpd} = 0.7 \text{ mgd}$$

Now write the information into the equation and solve for the unknown value:

$$x = (2)(0.7)(8.34)$$
$$= 11.68 \text{ lb/d}$$

Example 48

A pump discharges 0.025 m³/s. What chlorine feed rate (in kilograms per day) is required to provide a dosage of 2.5 mg/L?

This problem involves a flow rate in cubic meters per second, a dosage in milligrams per liter, and a feed rate in kilograms per day. First, write the conversion formula:

feed rate = (dosage)(flow rate)(conversion factor)

Fill in the equation and solve for the unknown value:

$$x = (2.5 \text{ mg/L})(0.025 \text{ m}^3/\text{s})(86,400 \text{ s/d})(1,000 \text{ L})$$

$$= 5.4 \text{ kg/d}$$

In the four preceding examples, the unknown value was always pounds or kilograms per day. In Equation 2-1, however, any one of the three variables may be the unknown value:

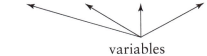

feed rate = (dosage)(flow rate)(conversion factor)

variables

Occasionally, for example, you may know how many pounds per day of chemicals are added and what the plant flow rate is, and need to know what chemical concentration in milligrams per liter this represents.

The next two examples illustrate the procedure used.

Example 49

In a treatment plant, 250 lb/d of dry alum is added to a flow of 1,550,000 gpd. What is this dosage in milligrams per liter?

From the statement of the problem, you can see that this is a milligrams-per-liter and pounds-per-day problem. First write the conversion equation that relates the two terms:

feed rate = (dosage)(flow rate)(conversion factor)

Next, the flow rate in gallons per day must be converted to million gallons per day:

1,550,000 gpd

↑

millions comma

1,550,000 gpd = 1.55 mgd

Now complete the problem by filling in the information given and solving for the unknown value:

$$250 = (x)(1.55)(8.34)$$

$$\frac{250}{(1.55)(8.34)} = x$$

$$19.34 \text{ mg/L} = x$$

Example 50

On one day at a treatment plant, the coagulation process was operated for 15 hours. The average flow during the period was 0.013 m³/s; 15 kg of alum was fed into the water. What was the alum dosage in milligrams per liter during that period?

The problem involves flow rate, concentration, and feed rate. Write the equation that relates the terms:

feed rate = (dosage)(flow rate)(conversion factor)

Now write the information into the equation and solve for the unknown:

$$15 = (x)(0.013)(86.4)$$

$$\frac{15}{(0.013)(86.4)} = x$$

$$13.35 \text{ mg/L} = x$$

In some problems the flow variable might be the unknown value. The following example illustrates the procedure used to solve such a problem.

Example 51

If 100 lb/d of dry alum were fed into the flow at a treatment plant to achieve a chemical dosage of 20 mg/L, what was the flow rate at the plant in million gallons per day?

First write the equation that relates the terms:

feed rate = (dosage)(flow rate)(conversion factor)

Then fill in the information given and solve for the unknown value.

$$100 = (20)(x)(8.34)$$

$$\frac{100}{(20)(8.34)} = x$$

$$0.60 \text{ mgd} = x$$

One variation of the milligrams-per-liter to pounds-per-day problem involves the calculation of hypochlorite dosage. Calcium hypochlorite (usually 65 or 70 percent available chlorine) or sodium hypochlorite (usually 5 to 15 percent available chlorine) is sometimes used for chlorination instead of chlorine gas (100 percent available chlorine).

To provide the same disinfecting power as chlorine gas, a greater weight of calcium hypochlorite is required, and an even greater weight of sodium hypochlorite is needed. For example, assume that water flowing through a treatment plant requires a chlorine dosage of 300 lb/d. If chlorine gas were used for disinfection, the proper dosage rate would be 300 lb/d. If hypochlorites were used, however, about 430 lb/d of calcium hypochlorite or 3,000 lb/d of sodium hypochlorite would be required to provide the same disinfecting power.

In calculating the amount of hypochlorite required, first calculate the amount per day of chlorine (100 percent available) required, and then determine the amount of hypochlorite required. The following examples illustrate the calculations.

Example 52

Disinfection at a treatment plant requires 280 lb/d of chlorine. If calcium hypochlorite (65 percent available chlorine) is used, how many pounds per day will be required?

Since there is only 65 percent available chlorine in the calcium hypochlorite compound, *more* than 280 lb/d of calcium hypochlorite will have to be added to the water to obtain the same disinfecting power as 280 lb/d of 100 percent available chlorine. To be specific, in this case, *65 percent of some number greater than 280* should equal 280:

$$(65\%)(\text{greater number of lb/d}) = 280 \text{ lb/d}$$

This equation can be restated as follows:

$$(0.65)(x \text{ lb/d}) = 280 \text{ lb/d}$$

Then solve for the unknown value:

$$x = \frac{280 \text{ lb/d}}{0.65}$$

$$= 430.77 \text{ lb/d calcium hypochlorite required}$$

Example 53

A water supply requires 30 lb/d of chlorine for disinfection. If sodium hypochlorite with 10 percent available chlorine is used for the disinfection, how many pounds per day of sodium hypochlorite are required?

Since there is only 10 percent available chlorine in the sodium hypochlorite, considerably more than 30 lb/d will be required to accomplish the disinfection. Specifically, 10 percent of some greater number should equal 30 lb/d:

$$(10\%)(\text{greater number of lb/d}) = 30 \text{ lb/d}$$

This equation can be restated as

$$(0.10)(x \text{ lb/d}) = 30 \text{ lb/d}$$

Then solve for the unknown value:

$$x = \frac{30 \text{ lb/d}}{0.10}$$

$$= 300 \text{ lb/d sodium hypochlorite required}$$

Normally, hypochlorite problems such as those in the two preceding examples are part of a milligrams-per-liter to pounds-per-day calculation. The following example illustrates the combined calculations.

Example 54

How many pounds per day of hypochlorite (70 percent available chlorine) are required for disinfection in a plant where the flow rate is 1.4 mgd and the chlorine dosage is 2.5 mg/L?

To solve this problem, you will need to calculate how many pounds per day of 100 percent chlorine are required; then calculate the hypochlorite requirement.

First write the equation to convert milligrams-per-liter concentration to pounds per day:

$$\text{feed rate} = (\text{dosage})(\text{flow rate})(\text{conversion factor})$$

Fill in the equation with the information given in the problem:

$$x = (2.5)(1.4)(8.34)$$

$$= 29.19 \text{ lb/d chlorine required}$$

Since 70 percent hypochlorite is to be used, *more* than 29.19 lb/d of hypochlorite will be required. Specifically, 70 percent of some greater number should equal 29.19 lb/d:

$$(70\%)(\text{greater number of lb/d}) = 29.19 \text{ lb/d}$$

This equation can be restated as follows:

$$(0.7)(x \text{ lb/d}) = 29.19 \text{ lb/d}$$

Then solve for the unknown value:

$$x = \frac{29.19 \text{ lb/d}}{0.7}$$

$$= 41.7 \text{ lb/d hypochlorite required}$$

Sometimes chlorine is used to disinfect tanks or pipelines. In such cases, a certain concentration must be achieved by the one-time addition of chlorine to a volume of water. The equation used to convert from milligrams per liter to pounds of chlorine required is similar to the equation used to convert to pounds-per-day feed rate; however, the volume of water in the container to be disinfected replaces the flow rate through the plant. Thus, the equation to calculate pounds of chlorine needed, given the disinfection dosage to be achieved, is as follows:

$$\begin{array}{c}\text{chlorine} \\ \text{weight} \\ \text{(lb)}\end{array} = \begin{array}{c}\text{dosage} \\ \text{(mg/L)}\end{array} \times \begin{array}{c}\text{volume of} \\ \text{container} \\ \text{(mil gal)}\end{array} \times \begin{array}{c}\text{conversion} \\ \text{factor} \\ \text{(8.34 lb/gal)}\end{array}$$

The following two examples illustrate this idea.

Example 55

How many pounds of hypochlorite (65 percent available chlorine) are required to disinfect 4,000 ft of 24-in. water line if an initial dose of 40 mg/L is required?

In this problem, the volume of the water line must be calculated:

$$(0.785)(2 \text{ ft})(2 \text{ ft})(4,000 \text{ ft})(7.48 \text{ gal/ft}^3) = 93,949 \text{ gal}$$

The volume must be expressed in terms of million gallons:

$$93,949 \text{ gal} = 0.094 \text{ mil gal}$$

Next, use the equation that relates milligrams per liter to weight of chlorine required:

$$\text{number of pounds} = (\text{dosage})(\text{tank volume})(\text{conversion factor})$$

Plugging in the numbers, we can solve as follows:

$$x = (40)(0.094)(8.34)$$

$$= 31.36 \text{ lb/d chlorine required}$$

Since 65 percent hypochlorite is to be used, more than 31.36 lb/d of hypochlorite will be required. In this case, 65 percent of some greater number should equal 31.36 lb/d:

$$(65\%)(\text{greater number of lb}) = 31.36 \text{ lb}$$

This equation can be restated as

$$(0.65)(x \text{ lb}) = 31.36 \text{ lb}$$

Then solve for the unknown value:

$$x = \frac{31.36 \text{ lb}}{0.65}$$

$$= 48.25 \text{ lb hypochlorite required}$$

Example 56

How many kilograms of chlorine (100 percent available) are required to disinfect a 500,000-L tank if the tank is to be disinfected with 50 mg/L of chlorine?

First write the conversion equation:

$$\text{number of kilograms} = \frac{(\text{dosage})(\text{tank volume})}{1,000,000}$$

Then fill in the equation and solve:

$$x = \frac{(50 \text{ mg/L})(500,000 \text{ L})}{1,000,000}$$

$$= 25 \text{ kg chlorine required}$$

Milligrams-per-Liter to Percentage Conversions

A concentration or dosage expressed as milligrams per liter can also be expressed as a percentage. Milligrams per liter are approximately equal to parts per million, and *percent* means "parts per hundred."

$$\text{milligrams per liter} \approx \text{parts per million} = \frac{\text{parts}}{1,000,000}$$

$$\text{percent} = \text{parts per hundred} = \frac{\text{parts}}{100}$$

Because 1,000,000 ÷ 100 = 10,000, converting from parts per million (or milligrams per liter) to a percentage is accomplished by dividing by 10,000. The following box diagram can be used:

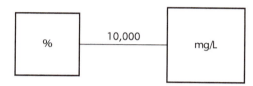

It will help to remember that you can *divide* by 10,000 by moving the decimal point four places to the left (four places because 10,000 has four zeros), yielding a much smaller number. For example,

$$31,400 \div 10,000 = 3.1400, \text{ or } 3.14$$

You can *multiply* by 10,000 by moving the decimal point four places to the right, yielding a much larger number. For example,

$$2.31 \times 10,000 = 23,100$$

The following examples illustrate the use of the box diagram.

Example 57

A chemical is to be dosed at 25 mg/L. Express the dosage as a percentage.

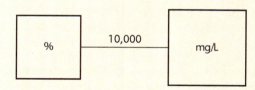

You are moving from a larger to a smaller box, so division by 10,000 is indicated:

$$\frac{25}{10,000} = 0.0025\%$$

Example 58

Express 120 ppm as a percentage.

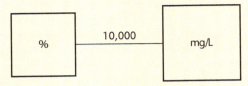

Referring to the box diagram, you are moving from a large box to a small box, so division is indicated:

$$\frac{120}{10,000} = 0.0120\%$$

Example 59

The sludge in a clarifier has a total solids concentration of 2 percent. Express the concentration as milligrams per liter of total solids.

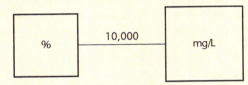

Referring to the box diagram, you are moving from a smaller box to a larger box, so multiply by 10,000:

$$(2)(10,000 \text{ mg/L}) = 20,000 \text{ mg/L}$$

Table 2-1 shows the relationship between percent concentration and milligrams-per-liter concentration.

Table 2-1 Percent versus milligrams-per-liter concentration

Percent	mg/L or ppm	Percent	mg/L or ppm
100.000	1,000,000	0.100	1,000
50.000	500,000	0.050	500
10.000	100,000	0.010	100
5.000	50,000	0.005	50
1.000	10,000	0.001	10
0.500	5,000	0.0001	1

Feed Rate Conversions

Some chemical dosage problems involve a conversion of feed rates from one term to another, such as from pounds per day to pounds per hour, or from gallons per day to gallons per hour. As with many of the conversions discussed in the math section, the box method can be used in making the conversions. The following box diagram will be used, where gph stands for gallons per hour, gpd for gallons per day, lb/h for pounds per hour, and lb/d for pounds per day:

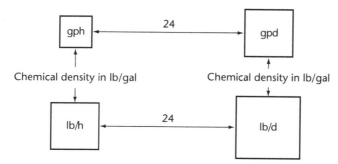

Examples of feed rate conversions follow.

Example 60

The feed rate for a chemical is 230 lb/d. What is the feed rate expressed in pounds per hour?

In moving from pounds per day to pounds per hour, you are moving from a larger box to a smaller box. Therefore, division by 24 is indicated:

$$\frac{230 \text{ lb/d}}{24} = 9.58 \text{ lb/h feed rate}$$

Example 61

A chemical has a density of 11.58 lb/gal. The desired feed rate for the chemical is 0.6 gal/h. How many pounds per day is this?

Converting from gallons per hour to pounds per day, you are moving from a smaller box to a larger box (gph → gpd) and then to a still larger box (gpd → lb/d). Therefore, multiplication by the density of the chemical and by 24 is indicated:

$$x = (0.6)\left(\begin{array}{c}\text{density of} \\ \text{chemical}\end{array}\right)(24)$$

$$= (0.6)(11.58)(24)$$

$$= 166.75 \text{ lb/d feed rate}$$

Calculations for Chlorine Dosage, Demand, Residual, and Contact Time

The chlorine requirement, or chlorine dosage, is the sum of the chlorine demand and the desired chlorine residual. This can be expressed mathematically as follows:

chlorine dosage (mg/L)	=	chlorine demand (mg/L)	+	chlorine residual (mg/L)

Example 62

A water is tested and found to have a chlorine demand of 6 mg/L. The desired chlorine residual is 0.2 mg/L. How many pounds of chlorine will be required daily to chlorinate a flow of 8 mgd?

Apply the chorine demand formula:

chlorine dosage (mg/L) = chlorine demand (mg/L) + chlorine residual (mg/L)

Fill in the information given in the problem:

$$dosage = 6 \text{ mg/L} + 0.2 \text{ mg/L}$$

$$= 6.2 \text{ mg/L}$$

Now use Equation 2-1 to convert the dosage in milligrams per liter to a feed rate in pounds per day:

$$feed \ rate = (dosage)(flow \ rate)(conversion \ factor)$$

$$= (6.2)(8)(8.34)$$

$$= 413.66 \text{ lb/d}$$

Example 63

The chlorine demand of a water is 5.5 mg/L. A chlorine residual of 0.3 mg/L is desired. How many pounds of chlorine will be required daily for a flow of 28 mgd?

chlorine dosage (mg/L) = chlorine demand (mg/L) + chlorine residual (mg/L)

$$= 5.5 \text{ mg/L} + 0.3 \text{ mg/L}$$

$$= 5.8 \text{ mg/L}$$

Convert dosage in milligrams per liter to feed rate in pounds per day:

$$(5.8)(28)(8.34) = 1,354 \text{ lb/d chlorine feed rate}$$

Note that if prechlorination is performed, it is not usually necessary to achieve a residual. Setting the residual equal to zero produces the following equivalencies:

chlorine dosage (mg/L) = chlorine demand (mg/L) + chlorine residual (mg/L)

chlorine dosage (mg/L) = chlorine demand (mg/L) + 0

chlorine dosage (mg/L) = chlorine demand (mg/L)

Therefore, when no residual is required, as in most prechlorination, the dosage is equal to the chlorine demand of the water.

The equation used to calculate chlorine dosage can also be used to calculate chlorine demand when dosage and residual are given, or to calculate chlorine residual when dosage and demand are given. The following examples illustrate the calculations.

Example 64

The chlorine dosage for a water is 5 mg/L. The chlorine residual after 30 min contact time is 0.6 mg/L. What is the chlorine demand in milligrams per liter?

chlorine dosage (mg/L) = chlorine demand (mg/L) + chlorine residual (mg/L)

$$5 \text{ mg/L} = x \text{ mg/L} + 0.6 \text{ mg/L}$$

Now solve for the unknown value:

$$5 \text{ mg/L} - 0.6 \text{ mg/L} = x \text{ mg/L}$$

$$4.4 \text{ mg/L} = x$$

Example 65

The chlorine dosage of a water is 7.5 mg/L, and the chlorine demand is 7.1 mg/L. What is the chlorine residual?

chlorine dosage (mg/L) = chlorine demand (mg/L) + chlorine residual (mg/L)

Fill in the given information and solve for the unknown value:

$$7.5 \text{ mg/L} = 7.1 \text{ mg/L} + x \text{ mg/L}$$

$$7.5 \text{ mg/L} - 7.1 \text{ mg/L} = x \text{ mg/L}$$

$$0.4 \text{ mg/L} = x$$

Sometimes the dosage is not given in milligrams per liter, but the feed rate setting of the chlorinator and the daily flow rate through the plant are known. In such cases, feed rate and flow rate should be used to calculate the dosage in milligrams per liter (following the procedure shown in Examples 49 and 50). The dosage in milligrams per liter can then be used in Equation 2-1, as in Examples 64 and 65.

Example 66

Adequate chlorine (or other disinfectant) contact time is a requirement for surface water treatment plants and is called $C \times T$ value. It determines disinfection sufficiency and is the product of the disinfectant residual times and the effective contact time for the chamber or clearwell in which disinfection takes place. In order to calculate $C \times T$ values, operators need to know the chlorine residual in the chamber, the flow rate of water through the chamber, the baffling factor (T10) of the chamber, and the volume of the chamber. The results are shown in mg-min/L.

A disinfection chamber with a chlorine residual of 1.6 mg/L measures 44 ft by 30 ft and has a baffling factor of 0.34. What is the $C \times T$ value at 300 gpm if the chamber holds 8.3 ft of water?

Effective volume of chamber = 44 ft × 30 ft × 8.3 ft × 7.48 gal/ft^3 × 0.34

= 27,863.3 gallons

Effective contact time (ECT) = 27,863.3 gal/300 gpm = 92.9 min

$C \times T$ value = ECT × residual = 92.9 min × 1.6 mg/L

= 148.6 mg-min/L

The baffling factor of a chamber is always less than 1 and is an indicator of the short-circuiting characteristics encountered in the chamber. Chambers with larger baffling factors are preferred because they yield higher $C \times T$ values.

WATCH THE VIDEOS
Operator Math: Volume Measurement and Surface Overflow Rate
(www.awwa.org/wsovideoclips)

Study Questions

1. Find the detention time in hours for a clarifier that has an inner diameter of 112.2 ft and a water depth of 10.33 ft if the flow rate is 7.26 mgd.
   a. 2.10 hr
   b. 2.14 hr
   c. 2.52 hr
   d. 2.96 hr

2. A chemical metering pump is pumping 26.3 gal/day. How many mL/min is this?
   a. 47.6 mL/min
   b. 60.8 mL/min
   c. 69.1 mL/min
   d. 75.4 mL/min

3. A water treatment plant has an emergency shutdown. How many water supply hours are left in a 119.8-ft diameter tank given the following data?
   ▪ Tank's water level = 27.6 ft
   ▪ Water cannot go below 16.0 ft at any time to comply with fire control
   ▪ Water usage averages 483 gpm
   a. 33.7 hr
   b. 35.0 hr
   c. 35.4 hr
   d. 36.2 hr

4. How many lb/day of sodium fluorosilicate (Na_2SiF_6) is required given the following parameters?

- Flow rate is 1,750 gpm.
- Fluoride desired is 1.20 mg/L.
- Fluoride in raw water is 0.15 mg/L.
- Sodium fluorosilicate is 98.1% pure.
- Fluoride (F) ion percent is 60.6%.
    a. 34 lb/day, F
    b. 37 lb/day, F
    c. 42 lb/day, F
    d. 48 lb/day, F

5. A filter is 24 ft by 28 ft. Calculate the filtration rate in gpm, if it receives a flow of 3,250 gpm.
    a. 4.4 gpm/ft²
    b. 4.8 gpm/ft²
    c. 5.0 gpm/ft²
    d. 5.1 gpm/ft²

6. Pressure measured in terms of the height of water (in meters or feet) is referred to as
    a. gauge pressure.
    b. head.
    c. barometric pressure.
    d. absolute pressure.

7. In the equation to convert milligrams-per-liter concentration to pounds per day, (dosage)(flow rate)(conversion factor) =
    a. feed rate.
    b. feed demand.
    c. feed capacity.
    d. total feed.

8. In the volume formula, how is volume calculated?

9. What is the formula for calculating the area of a circle?

10. If a spherical tank is 20 ft in diameter, what is its capacity?

Chapter 3
USEPA Water Regulations

Regulations that govern US water supply and treatment are developed by the **US Environmental Protection Agency (USEPA)** under the Safe Drinking Water Act (SDWA). Most states administer USEPA regulations after adopting regulations that are no less stringent than federal rules; and in some cases, states have adopted stricter regulations or have developed regulations for additional contaminants not regulated by USEPA.

This chapter discusses current and anticipated USEPA regulations and the challenges that operators face in their efforts to comply with the regulations. Water system operators should consult their local and state regulatory agencies to verify applicable regulations that may be different than the federal regulations listed in this chapter. The chapter concludes with a discussion of selected contaminants that are commonly found in water, their significance, and the methods for their removal.

Types of Water Systems

The SDWA defines a public water system (PWS) as a supply of piped water for human consumption that has at least 15 service connections, or serves 25 or more persons 60 or more days each year. By that definition, private homes, groups of homes with a single water source but having fewer than 25 residents, and summer camps with their own water source that operate less than 60 days per year are not PWSs. They may, however, be subject to state or local regulations. Such systems may also be subject to state and local well construction and water quality requirements.

PWSs are classified into three categories based on the type of customers served:

- *Community PWS:* a system whose customers are full-time residents
- *Nontransient noncommunity PWS:* an entity having its own water supply, serving an average of at least 25 persons who do not live at the location but who use the water for more than 6 months per year
- *Transient noncommunity PWS:* an establishment having its own water system, where an average of at least 25 people per day visit and use the water occasionally or for only short periods of time

These classifications are based on the differences in exposure to contaminants experienced by persons using the water. Most chemical contaminants are believed

US Environmental Protection Agency (USEPA)

A US government agency responsible for implementing federal laws designed to protect the environment. Congress has delegated implementation of the Safe Drinking Water Act to the USEPA.

to potentially cause adverse health effects from long-term exposure. Short-term exposure to low-level chemical contamination may not carry the same risk as long-term exposure.

Therefore, the monitoring requirements for both community and noncommunity water systems apply to all contaminants that are considered a health threat. The transient and nontransient noncommunity systems must monitor only for nitrite and nitrate, as well as biological contamination (those sources that pose an immediate threat from brief exposure). The remaining community systems, about 52,000 in the United States, have more stringent and frequent monitoring requirements.

Before examining the specific regulations that govern contaminants, the operator needs to know the difference between the two concepts used in the contaminant monitoring process: the maximum contaminant level goal (MCLG) and the maximum contaminant level (MCL):

- The MCLG is set for most substances at a level where there are no known, or anticipated, health effects. For those substances that are suspected carcinogens, the MCLG is set at zero.
- The MCL is set as close as feasible to the MCLG for substances regulated under the SDWA. The MCL is a level that is reasonably and economically achievable. This is the enforceable regulated level. Water systems that exceed an MCL must take steps to install treatment to reduce the contaminant concentration to below the MCL. Where USEPA has found it impractical to set an MCL, a treatment technique has been established instead of an MCL.

With these concepts in mind, the various regulations can be examined. This discussion is not meant to be all-inclusive. Because the regulatory process is an ever-evolving one, the reader is cautioned that some of the stated facts presented in this discussion may have changed since the writing of this chapter. For up-to-date information, it is best to contact the local office of the regulatory authority in the district or state where the utility operates.

Table 3-1 contains some of the more common regulated contaminants and their respective MCL or treatment technique descriptions. These standards are provided for illustration only and are not intended to be used for regulatory purposes (see the official USEPA regulatory information on the agency website).

Operations personnel are expected to know the regulatory limits for compounds encountered in their water supply. However, the number and variety of regulated substances make it unlikely that operators would know all of the regulatory limits. Operators must rely on current references for the most accurate information. These are available from the regulatory agency responsible for the location of the treatment plant.

Disinfection By-product and Microbial Regulations

Drinking water treatment, including use of chemical disinfectants such as chlorine, ozone, and chlorine dioxide, has been an important step in protecting drinking water consumers from exposure to harmful microbial contaminants. However, these chemical disinfectants can also react with organic and inorganic substances in the water to produce by-products that may be harmful to

maximum contaminant level goal (MCLG)

Nonenforceable health-based goals published along with the promulgation of an MCL. Originally called recommended maximum contaminant levels (RMCLs).

maximum contaminant level (MCL)

The maximum permissible level of a contaminant in water as specified in the regulations of the Safe Drinking Water Act.

Table 3-1 Selected USEPA drinking water standards

Contaminant	MCL or Treatment Technique (mg/L)*
Total coliform	5% (monthly positives)
Turbidity	0.3 ntu monthly or 1 ntu[†]
Chlorite	1.0
Haloacetic acids (HAA5)	0.060
Total trihalomethanes (TTHM)	0.080
Chloramines (as Cl_2)	4.0
Chlorine (as Cl_2)	4.0
Chlorine dioxide (as ClO_2)	0.8
Arsenic	0.010
Copper	Treatment technique, action level = 1.3
Cyanide (as free cyanide)	0.2
Fluoride	4.0
Lead	Action level = 0.015
Mercury (inorganic)	0.002
Nitrate (measured as nitrogen)	10
Nitrite (measured as nitrogen)	1
Radium 226 and 228 (combined)	5 pCi/L
Uranium	10 µg/L

*The listed standards are numerical representations of the current USEPA drinking water standard and do not include the sample frequency or location and other important compliance information. For a complete definition of the standards, consult USEPA Drinking Water Standards.

[†]Turbidity less than or equal to 0.3 nephelometric turbidity units (ntu) for the combined filter effluent for 95% of the monthly samples. At no time can turbidity be above 1 ntu.

drinking water consumers, particularly some susceptible segments of the population. Therefore, drinking water treatment using chemical disinfectants involves a delicate balancing act—i.e., adding enough disinfectant to control harmful microorganisms but not enough to produce unacceptably high levels of regulated **disinfection by-products (DBPs)**.

USEPA has enacted several regulations impacting microbial control and production of DBPs in groundwater and surface water supplies for small and large public drinking water systems. These rules are referred to collectively as the Microbial/Disinfection By-Products (M/DBP) Rules. Microbial protection for consumers of drinking water from public supplies is provided by provisions of current or pending rules listed below and discussed in more detail later in this chapter:

- Filter Backwash Recycling Rule (FBRR)
- Ground Water Rule (GWR)
- Interim Enhanced Surface Water Treatment Rule (IESWTR)
- Long-Term 1 Enhanced Surface Water Treatment Rule (LT1ESWTR)
- Long-Term 2 Enhanced Surface Water Treatment Rule (LT2ESWTR)
- Stage 1 Disinfectants and Disinfection By-products Rule (Stage 1 DBPR)
- Stage 2 Disinfectants and Disinfection By-products Rule (Stage 2 DBPR)
- Surface Water Treatment Rule (SWTR)
- Total Coliform Rule (TCR)

disinfection by-product (DBP) A new chemical compound formed by the reaction of disinfectants with organic compounds in water. At high concentrations, many DBPs are considered a danger to human health.

Provisions of the Disinfectants and Disinfection By-products Rule (DBPR) are intended to protect drinking water consumers against the unintended public health consequences associated with consumption of treated drinking water containing residual disinfectants and DBPs produced from degradation of these residual disinfectants or reaction of disinfectants with organic and inorganic DBP precursors.

More details regarding the DBPR, including the current Stage 1 DBPR and Stage 2 DBPR, are described in this chapter. Also included in the DBPR description is a brief discussion of some currently unregulated DBPs that are being heavily researched and may be the subject of future regulation. In the following discussion, the DBPR will be discussed first, followed by the microbial protection rules (SWTR, GWR, and TCR).

Disinfection By-product Rule (DBPR)

The Stage 1 DBPR and Stage 2 DBPR requirements discussed in the following sections focus first on two specific contaminants (TTHM and HAA5) and then on other aspects of these regulations dealing with control or removal of DBP precursors ("enhanced coagulation"), bromate, chlorite, and residual disinfectants.

Stage 1 DBPR—HAA5 and TTHM Provisions

The Stage 1 DBPR was published in 1998 and established an MCL of 0.080 mg/L for TTHM (the sum of four trihalomethanes, which are chloroform, bromodichloromethane, dibromochloromethane, and bromoform) and 0.060 mg/L for HAA5 (the sum of five specific haloacetic acids, which are mono-, di-, and trichloroacetic acids plus mono- and dibromoacetic acids). Although the MCLs for TTHM and HAA5 were officially written as 0.080 mg/L and 0.060 mg/L, respectively, the limits are commonly referred to as "80/60," or 80 µg/L and 60 µg/L. While knowing the numerical value of each MCL is important in understanding compliance with the DBPR, it is equally important to understand the methodology, in all its subtleties, used to calculate the compliance value that will be compared to this MCL.

For TTHM and HAA5, the compliance value is determined by monitoring the distribution system. Compliance monitoring locations need to be representative of the distribution system. Systems serving more than 10,000 persons who use surface water sources are required to monitor at least four locations per plant, meaning that distribution systems fed by more than one treatment plant must have at least four monitoring locations designated for each plant entry point.

The compliance monitoring location for systems with only one monitoring point must be representative of maximum residence time in the distribution system. A minimum of one out of every four compliance monitoring locations for systems with more locations must also be representative of maximum residence time. The other locations must be far enough away from the plant entry points to be representative of average residence time in the distribution system.

Unlike acute toxicity risks, for which the exposure could be a single glass of water, cancer risks like those believed to be linked to TTHM and HAA5 involve longer periods of exposure (daily glasses of water spanning decades). For chronic exposures such as these, exposure to an excessively high concentration of a given cancer-causing agent will not necessarily result in the consumer getting cancer from this source. Conversely, a consumer exposed to a lower concentration every day for a lifetime could be more likely to develop cancer. Therefore, regulation of DBPs to reduce cancer risks is *not* based on limiting exposure to a single incident (i.e., not a "single hit"), but rather is aimed at reducing the repeated exposure over time. In other words, DBP exposure needs to be evaluated on an average basis over time.

Under the Stage 1 DBPR, the compliance value for TTHM and HAA5 is determined by calculating a running annual average (RAA) during the previous 12 months for each DBP for all monitoring locations at each plant. Most systems are required to monitor quarterly (i.e., 4 times per year), although small groundwater systems (<10,000 persons) may be allowed to sample once a year. Typically, the RAA is based on 4 monitoring locations sampled quarterly, meaning RAA will be the average of 16 monitoring results each for HAA5 and TTHM.

Table 3-2 illustrates one facility's calculations of RAA for HAA5 that were used for Stage 1 compliance (this table also shows calculation of values for Stage 2 DBPR, which will be discussed later). It is important to reemphasize that compliance is based solely on the RAA, not on a single quarterly result at any one monitoring location. Consequently, it is *not* correct to refer to a single quarterly monitoring result above 60 µg/L for HAA5 or above 80 µg/L for TTHM as being above the MCL. Therefore, even though several individual monitoring values in Table 3-2 are greater than 60 µg/L, the facility is in compliance with the HAA5 MCL because the RAA is 45 µg/L for HAA5.

Utility personnel should be consistent and rigorous in their use of terminology when dealing with the general public or with state and local health officials, and should ensure that all people participating in these discussions are consistent in applying the MCL only to RAA values and do not make the common mistake of referring to a single quarterly monitoring value as being "above the MCL."

Stage 2 DBPR—HAA5 and TTHM Provisions

The Stage 2 DBPR, published in 2006, is now in effect. This rule tightened requirements for DBPs, but compliance is not achieved by modifying the numerical value of the MCLs or by requiring monitoring of new constituents. Instead, the rule makes compliance more challenging than under the Stage 1 DBPR by (1) changing the way the compliance value is calculated and (2) changing the compliance monitoring locations to sites representative of the greatest potential for THM and HAA formation. These changes were made to ensure uniform compliance with the DBP standards across all areas of the distribution system; that is, compliance is required at each sampling location.

The compliance value in the Stage 2 DBPR is called the *locational running annual average* (LRAA), and it is calculated by separately averaging the four quarterly samples at each monitoring location. Compliance is based on the maximum LRAA value (see Table 3-2). Furthermore, the Stage 2 DBPR included several interim steps that led to the replacement of many existing Stage 1 DBPR monitoring locations with new locations representative of the greatest potential for consumer exposure to high levels of TTHM and HAA5.

Table 3-2 Example RAA and LRAA calculations for Stage 1 DBPR and Stage 2 DBPR

| Year | Quarter | Sampling Location, µg/L | | | |
		A	B	C	D
1	3rd	52	68	63	66
1	4th	35	42	38	41
2	1st	47	49	42	43
2	2nd	18	42	45	37
LRAA		38	50	47	47
Maximum LRAA			50		
RAA			45		

running annual average (RAA)

The average of four quarterly samples at each monitoring location to ensure compliance with the Stage 1 DBPR.

The Stage 2 DBPR required that facilities maintain compliance with the Stage 1 DBPR using the existing monitoring locations during the first three years after the final version of the Stage 2 DBPR was published. In the time period between the third and sixth year after the Stage 2 DBPR was published, compliance continues to be based on maintaining 80/60 (TTHM and HAA5) or lower for RAA; it also includes a requirement for maximum LRAA at existing Stage 1 monitoring locations. The long-term goal of the Stage 2 DBPR was to identify locations within the distribution system with the greatest potential for either TTHM or HAA5 formations and then base compliance on the LRAA at or below 80/60 for each of these locations. Many of these locations were identified during the initial distribution system evaluation (IDSE).

The IDSE included monitoring, modeling, and/or other evaluations of drinking water distribution systems to identify locations representative of the greatest potential for consumer exposure to high levels of TTHM and HAA5. The goal of the IDSE was to evaluate a number of potential monitoring locations to justify selection of monitoring locations for long-term compliance (i.e., Stage 2B) with the Stage 2 DBPR.

One item to note regarding the Stage 2 DBPR as it applies to TTHM and HAA5 is that the goal was to find the locations in the distribution system where average annual levels of these DBPs are highest. TTHM formation increases as contact time with free or combined chlorine increases, although formation in the presence of combined chlorine is limited. Therefore, establishing points in the distribution system with the highest potential for TTHM formation is related to knowing the points with maximum water age. Utilities that have not performed a tracer study in the distribution system to determine water age should consider doing so.

By contrast, peak locations for HAA5 are more complicated because microorganisms in biofilm attached to distribution system pipe surfaces can biodegrade HAA5. Consequently, increasing formation of HAA5 over time is offset by biodegradation, eventually reaching a point where HAA5 levels decrease over time, even to the point where they drop to zero. Figure 3-1 shows a gradual reduction in HAA5 formation over time in a distribution system, followed by an eventual decrease of HAA5 as water age increases (water age measured in tracer test). In chlorination systems, HAA5 formation is limited. In fact, ammonium chloride is added as a quenching agent in HAA5 compliance samples in order to halt HAA5 formation prior to analysis (*Standard Methods for the Examination of Water and Wastewater*, latest edition). Therefore, little additional HAA5 formation occurs after chloramination to offset HAA5 biodegradation occurring in the distribution system.

WATCH THE VIDEO
Disinfection Byproducts: TTHMs & HAA5 (www.awwa.org/wsovideoclips)

Enhanced Coagulation Requirement of the Stage 1 DBPR and Stage 2 DBPR

The enhanced coagulation requirement has been developed to promote optimization of coagulation processes in conventional surface water treatment systems as required to improve removal of organic DBP precursors. The focus of the SWTR is separate from that of the enhanced coagulation requirement, with the former directed toward optimizing particle removal and the latter toward optimizing removal of natural organic matter (DBP precursors). Both promote efforts by water utilities to properly control and optimize coagulation processes and reduce DBP formation.

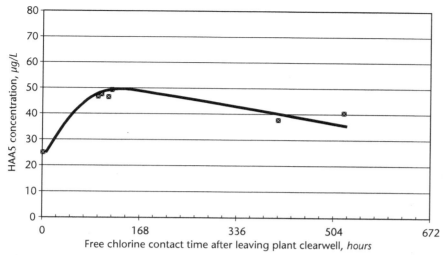

Figure 3-1 Formation and decay of HAA5 in a distribution system (time estimated by fluoride tracer test—T_{100})

Under the enhanced coagulation requirements, treatment plants must remove specific percentages of total organic carbon (TOC) based on their source water TOC and alkalinity levels. Facilities must meet the enhanced coagulation requirements unless they meet any of the following exemptions (USEPA Stage 1 DBPR Guidance):

1. The PWS's source water TOC level is <2.0 mg/L, calculated quarterly as an RAA.
2. The PWS's treated water TOC level is <2.0 mg/L, calculated quarterly as an RAA.
3. The PWS's source water TOC level is <4.0 mg/L, calculated quarterly as an RAA; the source-water alkalinity is >60 mg/L (as calcium carbonate [$CaCO_3$]), calculated quarterly as a running annual average; and either the TTHM and HAA5 running annual averages are no greater than 0.040 mg/L and 0.030 mg/L, respectively, or the PWS has made a clear and irrevocable financial commitment to use technologies that will limit the levels of TTHMs and HAA5 to no more than 0.040 mg/L and 0.030 mg/L, respectively.
4. The PWS's TTHM and HAA5 RAAs are no greater than 0.040 mg/L and 0.030 mg/L, respectively, and the PWS uses only chlorine for primary disinfection and maintenance of a residual in the distribution system.
5. The PWS's source water specific ultraviolet absorption at 254 nm (SUVA), prior to any treatment and measured monthly, is ≤2.0 L/mg-m, calculated quarterly as an RAA.
6. The PWS's finished-water SUVA, measured monthly, is ≤2.0 L/mg-m, calculated quarterly as an RAA.

Additionally, alternative compliance criteria for softening systems include the following:

7. Softening that results in lowering the treated water alkalinity to <60 mg/L (as $CaCO_3$), measured monthly and calculated quarterly as an RAA.
8. Softening that results in removing at least 10 mg/L of magnesium hardness (as $CaCO_3$), measured monthly and calculated quarterly as an RAA.

Utilities that cannot meet these avoidance criteria should know their enhanced coagulation endpoint, identified as the coagulant dosage and/or pH value that, when achieved, no longer produces significant TOC reduction. Specifically, when the source water TOC is not reduced by at least 0.3 mg/L with an incremental dosage increase of 10 mg/L alum (or equivalent ferric salt) and the pH value of the source reaches a value listed in Table 3-3, the enhanced coagulation endpoint has been reached.

If a utility is not exempt, a number of steps have to be evaluated relating to TOC removal, alkalinity of source water, range of source water TOC, required TOC removal for given source water characteristics, and several other factors.

Bromate

The bromate MCL from the Stage 1 DBPR remained at 0.010 mg/L for the Stage 2 DBPR. Bromate can be present in systems using ozone that have bromide present at the ozone application point. Bromate is also potentially formed during manufacture and storage of sodium hypochlorite. Consequently, systems using ozone for oxidation or disinfection are required to monitor once a month at a distribution system entry point for bromate, but systems without ozone are not required to perform this monitoring. Systems that use ozone and that also add sodium hypochlorite will need to closely monitor the quality of these sodium hypochlorite products for bromate content.

Chlorite

Similar to bromate, chlorite monitoring is required only for systems using chlorine dioxide as an oxidant or disinfectant. Chlorite is a degradation product of chlorine dioxide. Chlorate is also a degradation product of chlorine dioxide but is not currently regulated. Chlorite and chlorate are potential degradation products of sodium hypochlorite, but systems using sodium hypochlorite are not required to monitor for chlorite unless they also use chlorine dioxide.

Monitoring for chlorite is more complicated than for bromate because chlorine dioxide will degrade and chlorite formation will increase over time. Therefore, chlorite monitoring requirements include daily monitoring at the distribution system entry point and monthly samples at three locations in the distribution system (first customer, average residence time, maximum residence time). Unlike the health risks for bromate, TTHM, and HAA5, the risk for chlorite requires compliance based on the average of the three chlorite-monitoring locations each month. The Stage 1 DBPR MCL for chlorite is 1.0 mg/L.

Residual Disinfectants

The maximum residual disinfectant level (MRDL) for combined or total chlorine is 4.0 mg/L as Cl_2. These values are based on the same data used to monitor minimum free and combined chlorine levels in the distribution system as required by the SWTR, using the same monitoring locations used for the TCR. Chlorine dioxide

Table 3-3 Target pH values for coagulation when TOC removal rates are not sufficient

Alkalinity	pH Value
0–60	5.5
>60–120	6.3
>120–240	7.0
>240	7.5

residual also has an MRDL of 0.8 mg/L as ClO_2, based on daily samples at the treatment plant.

 WATCH THE VIDEO
Disinfection Byproducts: DBPR Stage One & Stage Two Rules
(www.awwa.org/wsovideoclips)

Surface Water Treatment Rule (SWTR)

IESWTR and LT1ESWTR

The goal of the IESWTR is to limit human exposure to harmful organisms, including *Cryptosporidium*, by promoting achievement of particle and turbidity removal targets for surface water treatment systems. Among the IESWTR requirements that apply to surface water treatment plants are the following:

- Combined filter effluent turbidity must be ≤0.3 ntu for 95 percent of samples collected each month, including none with >1 ntu. Compliance is based on combined filter effluent samples collected at 4-hour intervals during the entire month).

- The utility must monitor each individual filter for turbidity at 15-minute intervals and must report results, including a filter profile (graphical representation of filter performance), if either of the following two conditions are met: (1) turbidity in any filter for two consecutive 15-minute intervals exceeds 1 ntu or (2) turbidity during the first 4 hours of a given filter run exceeds 0.5 ntu for two consecutive 15-minute samples. Results must be reported within 10 days of the end of the month.

- Any newly constructed finished water reservoirs must include covers to keep out dust, debris, birds, etc.

- The utility must complete a sanitary survey every 3 years. Existing surveys conducted after December 1995 can be used if they meet minimum requirements. Variances can be granted to decrease frequency to 5 years. The IESWTR explicitly requires that sanitary surveys include efforts to evaluate and control *Cryptosporidium*, in addition to other target organisms.

- Systems where the average of quarterly TTHM or HAA5 values exceeds 64 and 48 µg/L, respectively, need to complete disinfection profiling and benchmarking. *Profiling* involves determination of *C × T* values (concentration of disinfectant × contact time) for each segment of the treatment plant (see later discussion). *Benchmarking* involves determining the lowest monthly average during 12-month monitoring of *Giardia* and virus inactivation. This procedure is required for any systems that are considering a major change to their disinfection practice. Consultation with the primacy agency is also required before any disinfection change.

- Turbidity monitoring records must be maintained for a minimum of 3 years.

Facilities in compliance with these requirements, chiefly the turbidity monitoring provisions, are designated by the IESWTR to have provided 2-log virus removal, 2.5-log *Giardia* removal, and 2-log *Cryptosporidium* removal. Literature and other information cited in the IESWTR final rule indicate that these credits are conservative, and most facilities meeting these requirements are probably achieving far greater levels of virus, *Giardia*, and *Cryptosporidium* removal than the minimum credits previously cited. The level of *Cryptosporidium* protection cited is sufficient to meet all requirements of the IESWTR, but the rule requires a total of 3.0

C × T value
The product of the residual disinfectant concentration C, in milligrams per liter, and the corresponding disinfectant contact time T, in minutes. Minimum C × T values are specified by the Surface Water Treatment Rule as a means of ensuring adequate kill or inactivation of pathogenic microorganisms in water.

credits for *Giardia* and 4.0 credits for viruses. The additional credits (0.5-log for *Giardia* and 2-log for viruses) are required to be achieved by disinfection with free chlorine, chloramines, ozone, or chlorine dioxide by meeting $C \times T$ requirements described later in this chapter.

Provisions of the IESWTR apply to large systems (>10,000 persons) using surface water sources. However, similar provisions are applied to smaller surface water systems (<10,000 persons), as outlined in the LT1ESWTR. The objectives of the LT1ESWTR and IESWTR are identical, though some of the compliance deadlines and other regulatory provisions are slightly different based on greater financial and personnel resources for larger systems.

Sanitary Surveys

Sanitary surveys are a requirement of the Interim Enhanced Surface Water Treatment Rule (IESWTR). A sanitary survey is "an onsite review of the water source, facilities, equipment, operation, and maintenance of the public water system for the purpose of evaluating the adequacy of such source, facilities, equipment, operation, and maintenance for producing and distributing safe drinking water." Surveys are usually performed by the state primacy agency and are required of all surface water systems and groundwater systems under the direct influence of surface water.

These surveys are typically divided into eight main sections, although some state primacy groups may have more:

1. Water sources
2. Water treatment process
3. Water supply pumps and pumping facilities
4. Storage facilities
5. Distribution systems
6. Monitoring, reporting, and data verification
7. Water system management and operations
8. Operator compliance with state requirements

Sanitary surveys are required on a periodic basis usually every 3 years. Surveys may be comprehensive or focused according to the regulatory agency requirements.

C × T Requirements

Every water system that uses surface water as a source must meet treatment technique requirements for the removal and/or inactivation of *Giardia*, viruses, *Legionella*, and other bacteria. Because these pathogens are not easily identified in the laboratory on a routine basis, USEPA has set quality goals in lieu of MCLs in this instance. Meeting SWTR treatment technique goals demonstrates all or part of the required microbial protection, as previously noted, but additional protection is required through the use of approved disinfection treatment chemicals. The effectiveness of disinfection depends on the type of disinfectant chemical used, the residual concentration, the amount of time the disinfectant is in contact with the water, the water temperature, and, when chlorine is used, the pH of the water.

According to USEPA, a combination of the residual concentration, C, of a disinfectant (in milligrams per liter) multiplied by the contact time, T (in minutes), can be used as a measure of the disinfectant's effectiveness in killing or inactivating microorganisms. For water plant operators, this means that high residuals held for a short amount of time or low residuals held for a long period of time will produce similar results. Water plants are required to provide this computation daily, and the measure must always be higher than the required minimum value.

Long-Term 2 Enhanced SWTR (LT2ESWTR)

The LT2ESWTR supplements the SWTR requirements contained in the IESWTR for large surface water systems (>10,000 persons) and the LT1ESWTR for small systems (<10,000 persons). Details of the rules can be reviewed in the *Federal Register* or at the USEPA website (http://water.epa.gov/drink/index.cfm). One of the key elements of the LT2ESWTR was the use of *Cryptosporidium* monitoring results to classify surface water sources into one of four USEPA-defined risk levels called "bins." Facilities in the lowest bin (bin 1) are required to maintain compliance with the current IESWTR. Facilities in higher bins (bins 2 to 4) are required to either (1) provide additional *Cryptosporidium* protection from new facilities or programs not currently in use at a facility or (2) demonstrate greater *Cryptosporidium* protection capabilities of existing facilities and programs using a group of USEPA-approved treatment technologies, watershed programs, and demonstration studies, referred to collectively as the *Microbial Toolbox*.

Implementation of the LT2ESWTR was phased over many years according to system size. Four separate size categories were established (schedules 1–4, with 4 being the smallest at <10,000 population) for implementing the rule. The rule for schedule-4 systems allows filtered supplies to perform initial monitoring for fecal coliform to determine if *Cryptosporidium* monitoring is required.

Filter Backwash Recycle Rule (FBRR)

The FBRR currently applies to systems of all sizes and is intended to help utilities minimize potential health risks associated with recycle, particularly associated with respect to *Giardia* and *Cryptosporidium*. Other contaminants of concern in the recycle stream include suspended solids (turbidity), dissolved metals (especially iron and manganese), and dissolved organic carbon. Plants that control recycle will also help minimize operational problems.

Prior to the FBRR, no USEPA regulation governed recycle. Regulations within the United States regarding recycle had been established by the states, if at all. State regulatory approaches varied from a requirement of equalization of two backwashes in Illinois to 80 percent solids removal prior to recycle and maintaining recycle flows at less than 10 percent of raw water flow in California. Virginia discourages recycling.

Key components of the FBRR include (1) recycle must reenter the treatment process *prior* to primary coagulant addition, (2) direct filtration plants must report their recycle practices to the state and may need to treat their recycle streams, and (3) a self-assessment must be done at those plants that use direct recycle (i.e., no separate equalization and/or treatment of recycle stream) and that operate fewer than 20 filters. The goal of the self-assessment is to determine if the design capacity of the plant is exceeded due to recycle practices.

WATCH THE VIDEO
FBRR (www.awwa.org/wsovideoclips)

Ground Water Rule (GWR)

USEPA promulgated the final GWR in October 2006 to reduce the risk of exposure to fecal contamination that may be present in PWSs that use groundwater sources. The rule establishes a risk-targeted strategy to identify groundwater systems that are at high risk for fecal contamination. The GWR also specifies when corrective action (which may include disinfection) is required to protect consumers who receive water from groundwater systems from bacteria and viruses.

A sanitary survey is required, by the state primacy agency, at regular intervals depending on the condition of the water system as determined in the initial survey. Systems found to be at high risk for fecal contamination are required to provide 4-log inactivation of viruses. Increased monitoring for fecal contamination indicators may be required by the regulatory authority.

Total Coliform Rule (TCR) and Revised Total Coliform Rule (RTCR)

The objective of the TCR is to promote routine surveillance of distribution system water quality to search for fecal matter and/or disease-causing bacteria. All points in a distribution system cannot be monitored, and complete absence of fecal matter and disease-causing bacteria cannot be guaranteed. The TCR is an attempt to persuade water utilities to implement monitoring programs sufficient to verify that public health is being protected as much as possible, as well as allowing utilities to identify any potential contamination problems in their distribution system. The rule requires monthly sampling at each distribution sampling point.

The TCR, and the RTCR that was finalized in 2013, impact all PWSs. The RTCR requires PWSs that are vulnerable to microbial contamination to identify and fix problems. The RTCR also established criteria for systems to qualify for and stay on reduced monitoring, thereby providing incentives for improved water system operation.

The RTCR rule established an MCLG and an MCL for *Escherichia coli* (*E. coli*) and eliminated the MCLG and MCL for total coliform, replacing it with a treatment technique for coliform that requires assessment and corrective action. The rule establishes an MCLG and an MCL of zero for *E. coli*, a more specific indicator of fecal contamination and potentially harmful pathogens than total coliform. USEPA removed the MCLG and MCL of zero for total coliform. Many of the organisms detected by total coliform methods are not of fecal origin and do not have any direct public health implication.

Under the treatment technique for coliform, total coliform serves as an indicator of a potential pathway of contamination into the distribution system. A PWS that exceeds a specified frequency of total coliform occurrence must conduct an assessment to determine if any sanitary defects exist and, if found, correct them. In addition, a PWS that incurs an *E. coli* MCL violation must conduct an assessment and correct any sanitary defects found.

The RTCR also changed monitoring frequencies. It links monitoring frequency to water quality and system performance and provides criteria that well-operated small systems must meet to qualify and stay on reduced monitoring. It also requires increased monitoring for high-risk small systems with unacceptable compliance history and establishes some new monitoring requirements for seasonal systems such as state and national parks.

The revised rule eliminated monthly public notification requirements based only on the presence of total coliforms. Total coliforms in the distribution system may indicate a potential pathway for contamination but in and of themselves do not indicate a health threat. Instead, the rule requires public notification when an *E. coli* MCL violation occurs, indicating a potential health threat, or when a PWS fails to conduct the required assessment and corrective action.

The rule requires that PWSs collect total coliform samples at sites representative of water quality throughout the distribution system according to a written plan approved by the state or primacy agency. Samples are collected at regular intervals monthly. Positive total coliform samples must be tested for *E. coli*. If any positive total coliform sample is also positive for *E. coli* the state must be notified by the end of the day on which the result was received. Repeat samples are required within 24 hours of any total coliform–positive routine sample. Three repeat samples are required, one at the site of the positive sample and one within five service taps both upstream and downstream of the positive site. Any positive total coliform samples must be tested for *E. coli*. Any positive *E. coli* (EC+) samples must be reported by the end of the day. Any positive total coliform (TC+) samples require another set of repeat samples.

A Level 1 or Level 2 sanitary assessment and corrective action is triggered to occur within 30 days if there is indication of coliform contamination. A Level 1 assessment by the PWS is triggered if more than 5 percent of the routine/repeat monthly samples (if at least 40 are required) are total coliform positive or a repeat sample is not taken for a total coliform positive result. A Level 2 assessment conducted by the state or its representative is triggered if the PWS has an *E. coli* violation or repeated Level 1 assessment triggers.

Major violations of the RTCR are MCL violations and treatment technique violations. A PWS will receive an *E. coli* MCL violation when there is any combination of an EC+ sample result with a routine/repeat TC+ or EC+ sample result, as follows:

E. coli MCL Violation Occurs With the
Following Sample Result Combination

Routine	Repeat
EC+	TC+
EC+	Any missing sample
EC+	EC+
TC+	EC+
TC+	TC+ (but no *E. coli* analysis)

A PWS will receive a treatment technique violation given any of the following conditions:

- Failure to conduct a Level 1 or Level 2 assessment within 30 days of a trigger.
- Failure to correct all sanitary defects from a Level 1 or Level 2 assessment within 30 days of a trigger or in accordance with the state-approved time frame.
- Failure of a seasonal system to complete state-approved start-up procedures prior to serving water to the public.

Lead and Copper Rule (LCR)

The objective of the LCR is to control corrosiveness of the finished water in drinking water distribution systems to limit the amount of lead (Pb) and copper (Cu) that may be leached from certain metal pipes and fittings in the distribution system. Of particular concern are pipes and fittings connecting the household tap to the distribution system service line at individual homes or businesses, especially because water can remain stagnant in these service lines for long periods of time, increasing the potential to leach Pb, Cu, and other metals. Although the utility is not responsible for maintaining and/or replacing these household connections, they are responsible for controlling pH and corrosiveness of the water delivered to the consumers.

Details of the LCR include the following:

- The LCR became effective December 7, 1992.
- The action level for Pb is 0.015 mg/L and for Cu is 1.3 mg/L.
- A utility is in compliance at each sampling event (frequency discussed below) when <10 percent of the distribution system samples are above the action levels for Pb and Cu (i.e., 90th percentile value for sampling event must be below action level).
- Utilities found not to be in compliance must modify water treatment until they are in compliance. The term *action level* is used rather than *MCL* because noncompliance (i.e., exceeding an action level) triggers a need for modifications in treatment.

After identifying sampling locations and determining initial tap water Pb and Cu levels at each of these locations, utilities must also monitor other water quality parameters (WQPs) at these same locations as needed to monitor and evaluate corrosion control characteristics of treated water. The only exemptions from analysis of these WQPs are systems serving less than 50,000 people for which Pb and Cu levels in initial samples are below action levels.

Pb, Cu, and WQPs are initially collected at 6-month intervals; this frequency can be reduced if action levels are not exceeded and optimal water treatment is maintained. Systems that are in noncompliance and are performing additional corrosion-control activities must continue to monitor at 6-month intervals, plus they must collect WQPs from distribution system entry points every 2 weeks.

Each utility must complete a survey and evaluate materials that comprise their distribution system, in addition to using other available information, to target homes that are at high risk for Pb/Cu contamination.

Revisions to the LCR were enacted in 2007. These clarifications to the existing rule were made in seven areas:

- Minimum number of samples required
- Definitions for compliance and monitoring periods
- Reduced monitoring criteria
- Consumer notice of lead tap water monitoring results
- Advanced notification and approval of long-term treatment changes
- Public education requirements
- Reevaluation of lead service lines

Consult your local regulatory agency for those revisions that are applicable to your system.

Phase I, II, and V Contaminants

The Phase I, II, and V regulations were finalized in 1989, 1992, and 1995, respectively, and include various inorganic and organic contaminants. Sampling and reporting frequency vary with constituent, though sampling is typically required once every 3 years after the initial sampling period. Variances or waivers are possible for a number of constituents based on analytical results and/or a vulnerability assessment.

Public Notification Rule

USEPA has implemented a regulation called the *Public Notification Rule*. This rule is separate from the Consumer Confidence Report (CCR) Rule. The Public Notification Rule includes requirements for reporting certain water quality monitoring violations and other water quality incidents, as well as requirements for the timing, distribution, and language of the public notices. For example, the Public Notification Rule includes requirements that some incidents be reported within 24 hours, others within 30 days, and others included as part of the annual CCR. Some of these reporting requirements are more stringent than those currently required by USEPA. The regulation also includes requirements regarding how notices are to be distributed/broadcast (i.e., TV, radio, newspaper, hand delivery, regular mail, etc.), the format of the notices, the wording of certain items in the notice, and the need to include information in languages other than English.

Public notification according to the rule might include the following:

- Templates, or model notices, to be available for adaptation for certain potential incidents.
- Consolidated and updated lists of phone numbers and contacts for government (local, county, state), regulatory agencies, hospitals, radio and TV, newspapers, etc., that should be contacted per requirements of the Public Notification Rule
- Checklists and flow diagrams outlining activities that would need to be completed for certain potential events outlined in the regulation
- Identification of key personnel and what their roles and responsibilities would be to respond as required by the regulation
- A plan to periodically review and update all lists, templates, and other aspects of a response plan every year or when/if the Public Notification Rule is modified by future federal or state regulations

Unregulated Contaminant Monitoring Rule (UCMR)

The 1996 amendments to the SDWA require USEPA to establish criteria for a monitoring program for currently unregulated contaminants to generate data that USEPA can use to evaluate and prioritize contaminants that could potentially be regulated in the future. USEPA has developed three cycles of the UCMR:

1. UCMR1 in 1999
2. UCMR2 in 2007
3. UCMR3 in 2012

Failing to (1) perform required sampling and analysis, (2) use the appropriate analytical procedures, or (3) report these results is a violation of the UCMR. However, the numerical results of these analytical efforts cannot result in a violation because none of the constituents in the UCMR are currently regulated (i.e., no MCLs, action levels, or other standards apply).

Although the UCMR contaminants have no standards associated with them, the data from this monitoring will need to be reported in the annual CCR. Therefore, the CCR will need to address implications of any constituents found above detection limits. Reporting UCMR results in the CCR would also fulfill the notification requirements for "unregulated contaminants" included in the recently promulgated Public Notification Rule.

Note that the UCMR is an ongoing part of the regulatory development process that will be repeated every 5 years. Utilities will be performing similar mandatory sampling for a new list of constituents every 5 years.

UCMR3 was signed by USEPA Administrator Lisa P. Jackson on April 16, 2012. As finalized, UCMR3 requires monitoring for 30 contaminants using USEPA and/or consensus organization analytical methods during 2013–2015. Together, USEPA, states, laboratories, and PWSs will participate in UCMR3.

Operator Certification

Amendments to the 1996 SDWA required USEPA to develop national guidance for operator certification. The final rule was published on February 5, 1999, and became effective on February 5, 2001. State operator certification programs were required to address nine baseline standards, including operator qualifications, certification renewal, and program review. Indirect impacts of the rule on most water utilities include availability of Drinking Water State Revolving Fund (DWSRF) money and perhaps some slight modifications in paperwork/record-keeping requirements.

Arsenic MCL

The MCL for arsenic was reduced from 50 µg/L to 10 µg/L in the *Federal Register* published on January 22, 2001. This was the second time USEPA has established an MCL that was higher than the technically feasible level (3 µg/L), with the first being the uranium rule in 2000. The original SDWA required the MCL to be set as close to the health goal (zero for arsenic and all other suspected carcinogens) as technically feasible. Amendments to the SDWA allowed USEPA the discretion to set the MCL above the technically feasible level.

The final rule, including the revised MCL, became effective 3 years after the rule was published.

Radionuclides Rule

The Radionuclide Rule was published in December 2000. In the final rule, USEPA maintained the gross alpha MCL at 15 pCi/L MCL, 4 mrem/yr for beta emitters, 4 mrem/yr for photon emitters, and 5 pCi/L for combined radium 226 and 228 isotopes, and an MCL for uranium of 30 µg/L.

Analytical Methods

Each of the individual USEPA regulations contains its own information regarding analytical methods approved for compliance monitoring. These and other approved analytical methods are compiled in a final rule titled "Analytical Methods for Chemical and Microbiological Contaminants and Revisions to Laboratory Certification Requirements," published December 1, 1999. These analytical methods were approved for compliance monitoring effective January 3, 2000. The USEPA-approved methods include analytical procedures developed by USEPA, plus procedures developed by others that USEPA endorses, including specific procedures developed by the American Society for Testing and Materials (ASTM) and some specific procedures included in *Standard Methods for the Examination of Water and Wastewater*, published jointly by the American Public Health Association (APHA), AWWA, and the Water Environment Federation (WEF).

Currently, only approved analytical methods can be used for compliance monitoring. In the future, USEPA hopes to implement a performance-based measurement system that will allow utilities to use alternative screening methods instead of requiring only USEPA-approved reference methods. The 1996 SDWA amendments require USEPA to review new analytical methods that may be used for the screening and analysis of regulated contaminants. After this review, USEPA may approve methods that may be more accurate or cost-effective than established methods for compliance monitoring. These screening methods are expected to provide flexibility in compliance monitoring and may be better and/or faster than existing analytical methods.

The approval of new drinking water analytical methods can be announced through an expedited process in the *Federal Register*. This allows laboratories and water systems more timely access to new alternative testing methods than the traditional rule-making process. If alternate test procedures perform the same as or better than the approved methods, they can be considered for approval using the expedited process.

Study Questions

1. Water being served to the public for a population greater than 3,300 must not have a disinfectant residual entering the distribution system below _____ for more than 4 hours.
   a. 0.1 mg/L
   b. 0.2 mg/L
   c. 0.3 mg/L
   d. 0.4 mg/L

2. What is the action level for copper?
   a. 0.5 mg/L
   b. 1.0 mg/L
   c. 1.3 mg/L
   d. 1.8 mg/L

3. Where are the sampling point(s) located for required sampling of turbidity in a community water system?
   a. At representative points within the distribution system
   b. 75% at locations representative of population distribution and 25% at the farthest points in the distribution system
   c. At point(s) where water enters the distribution system, including all filter effluents if a surface water treatment plant
   d. At effluents of all filters if a surface water treatment plant, at entry points to the distribution system, and at locations representative of population distribution

4. Under the Surface Water Treatment Rule, disinfection residuals must be collected at the same location in the distribution system as
   a. coliform samples.
   b. total trihalomethanes.
   c. disinfection by-products.
   d. alkalinity, conductivity, and pH for corrosion studies.

5. What is the MCL for haloacetic acids (HAA5)?
   a. 0.040 mg/L
   b. 0.060 mg/L
   c. 0.080 mg/L
   d. 0.100 mg/L

6. An example of _____ is determining the lowest monthly average during 12-month monitoring of *Giardia* and virus inactivation.
   a. benchmarking
   b. profiling
   c. categorizing
   d. surveying

7. What is the MCLG for substances that are suspected carcinogens?

8. Which regulation is designed to limit human exposure to harmful organisms, including *Cryptosporidium*, by promoting achievement of particle and turbidity removal targets for surface water treatment systems?

9. Which regulation is designed to reduce the risk of exposure to fecal contamination that may be present in public water systems that use groundwater sources?

10. Which regulation is designed to promote routine surveillance of distribution system water quality to search for contamination from fecal matter and/or disease-causing bacteria?

Chapter 4

Coagulation and Flocculation Process Operation

Operation of the Processes

Operation of the coagulation and flocculation processes usually consists of the following steps:

1. Consider the water characteristics affecting the selection of chemicals to be used.
2. Use jar-testing or other lab-related strategies to predict dosage/feed rates for chemicals.
3. Apply the chemicals.
4. Monitor the effectiveness of the process.

Water Characteristics Affecting Chemical Selection

The selection of chemical coagulants and coagulant aids is a continuing process of trial and evaluation. To do a thorough job of chemical selection, the following characteristics of the raw water to be treated should be considered:

- Type and concentration of contaminants
- Water temperature
- pH
- Alkalinity
- Turbidity
- Color
- Total organic carbon

The effect of each characteristic on coagulation and flocculation is briefly described in this chapter.

The effectiveness of a coagulant will change as raw-water characteristics change. The effectiveness of coagulation may also change for no apparent reason, suggesting that there are other factors, not yet understood, that affect the process.

Water Temperature

Lower-temperature water usually causes poorer coagulation and flocculation and can require that more of a chemical or longer time be used to maintain acceptable results.

pH

Extreme values of pH, either high or low, can interfere with coagulation and flocculation. The optimal pH varies depending on the coagulant used.

---

coagulation

The water treatment process that causes very small suspended particles to attract one another and form larger particles. This is accomplished by the addition of a chemical, called a coagulant, that neutralizes the electrostatic charges on the particles that cause them to repel each other.

flocculation

The water treatment process, following coagulation, that uses gentle stirring to bring suspended particles together so that they will form larger, more settleable clumps called floc.

pH

A measurement of how acidic or basic a substance is. The pH scale runs from 0 (most acidic) to 14 (most basic). The center of the range (7) indicates the substance is neutral, neither acidic nor basic.

Alkalinity

Alum and ferric sulfate interact with the chemicals that cause alkalinity in the water, thus reducing the alkalinity and forming complex aluminum or iron hydroxides that begin the coagulation process. Low alkalinity limits this reaction and results in poor coagulation; in these cases, it may be necessary to increase the alkalinity of the water.

Turbidity

The lower the turbidity, the more difficult it is to form a proper floc. Fewer particles mean fewer random collisions and hence fewer chances for floc to accumulate. It may be necessary to add a weighting agent, such as clay (bentonite), to low-turbidity water.

Color

Color is caused by organic compounds in the raw water. The organics can react with the chemical coagulants, making coagulation more difficult. Pretreatment with oxidants or adsorbents may be necessary to reduce the concentration of organics. Generally, the dosage of coagulant chemical needed to coagulate color must be increased as the color concentration increases.

Total Organic Carbon

The total organic carbon (TOC) test measures the concentration of organic carbon present in water. In most waters, the organic carbon is composed of humic substances. The Stage 1 Disinfectants/Disinfection By-products Rule specifies the percent removal of TOC required as a function of water alkalinity to comply with the rule and minimize the potential formation of disinfection by-products (DBPs).

Specific Ultraviolet Absorbance

Specific ultraviolet absorbance (SUVA) is a test that measures the absorbance of UV light at 254 nm and divides that value by the dissolved organic carbon concentration. The SUVA is an indicator of the humic content—DBP precursor. If the value is less than 3, the organic carbon is largely nonhumic. If the value is in the range of 4 to 5, the organics are largely humic in composition.

Applying Coagulant Chemicals

An operator should begin chemical selection by using the jar test with various chemicals, both singly and in combination. The jar test is used experimentally to determine the optimal conditions for the coagulation, flocculation, and sedimentation processes. This process involves trying various combinations of chemical dosage, mixing speed, and settling interval.

Jar test results are expressed in milligrams per liter, which are converted to the equivalent daily dose required for setting the chemical feeders for plant operation. Details of performing jar tests are described in Chapter 10.

Monitoring Process Effectiveness

Although jar tests provide a good indication of the results to expect, full-scale plant operation may not match these results. The adequacy of flash mixing and flocculation is not something that can be observed directly.

---

alkalinity

A measurement of water's capacity to neutralize an acid. Compare pH.

turbidity

A physical characteristic of water making the water appear cloudy. The condition is caused by the presence of suspended matter.

total organic carbon (TOC)

The amount of carbon bound in organic compounds in a water sample as determined by a standard laboratory test.

specific ultraviolet absorbance (SUVA)

A test that determines humic content by measuring the absorbance of UV light at 254 nm and divides that value by the dissolved organic carbon concentration.

The following are indicators of inadequate mixing or incorrect chemical dosage:

1. Very small floc (called *pinpoint floc*)
2. High turbidity in settled water
3. Too-frequent filter backwashing
4. Too-long filter run lengths

Operating Factors That Could Affect Floc Development

Plant operating factors that could make a difference in the proper development of floc include the following:

- Inadequate flash mixing
- Improper flocculation mixing
- Inadequate flocculation time
- Incorrect chemical dosage

Inadequate Flash Mixing

Successful coagulation is based on rapid and complete mixing. Though coagulation occurs in less than 1 second, the chamber may provide up to 30 seconds of detention time. Mixing should be turbulent enough so that the coagulant is dispersed throughout the coagulation chamber.

Some experts maintain that during the first tenth of a second the coagulant must be thoroughly mixed with every drop of water to start an efficient floc; otherwise, after that point, the efficiency of the entire process declines. If polymers are used as primary coagulants, rapid mixing is less critical, but thorough mixing remains very important in encouraging as many particle collisions as possible.

Improper Flocculation Mixing

Proper flocculation requires long, gentle mixing. Mixing energy must be high enough to bring coagulated particles constantly into contact with each other, but not so high as to break up those particles already flocculated. For this reason, flocculation basins and equipment are usually designed to provide higher mixing speeds immediately following coagulation and progressively slower speeds as the water flows through the basin.

Properly coagulated and flocculated particles will look like small snowflakes or tufts of wool suspended in very clear water. The water should not look cloudy or foggy as a result of poorly formed floc. A cloudy appearance is usually caused by an inadequate alum dosage. If the water does appear cloudy, or if it displays any of the four signs of inadequate mixing listed earlier, the speed of the flocculators may be incorrect. Under some circumstances, floc can become as large as a quarter, but at that point it may be too buoyant and will not settle well. Some systems have found that pinhead-size floc is about optimal for settling efficiency.

Inadequate Flocculation Time

It takes time to develop heavy floc particles. Although some systems can be operated with as little as 10 minutes of detention time, others may require up to 1 hour. The average is about 30 minutes. Because short-circuiting can be a major problem, a minimum of three flocculation basins in series is recommended.

Operating Problems

Common operating problems encountered during the coagulation–flocculation process include the following:

- Low water temperature
- Weak floc
- Slow floc formation

Low Water Temperature

Raw-water temperature that approaches the freezing point interferes with the coagulation and flocculation processes. As the water temperature decreases, the viscosity of the water increases, which slows the rate of floc settling.

The colder temperature also slows chemical reaction rates, although this effect on coagulation is relatively insignificant. Moreover, cold-water floc has a tendency to penetrate the filters, indicating that floc strength has decreased.

The problems caused by low water temperature can best be overcome with the following techniques:

- Operate the coagulation process as near as possible to the best pH value for the water temperature.
- Increase the coagulant dosage as the water temperature decreases. This increases the number of particles available to collide and also reduces the effect of changes in pH resulting from the drop in temperature.
- Add weighting agents, such as clay, to increase floc particle density and add other coagulant aids to increase floc strength and encourage rapid settling.

Weak Floc

Weak floc is often not noticed until it has an adverse effect on filtration. A weak floc does not adhere well to the filter media; instead, it is broken up and carried deeper into the filter until it finally breaks through, causing increased turbidity in the filter effluent. Weak floc is often the result of inadequate mixing in the rapid-mix or flocculation basins. This can be checked by varying the mixing speeds and taking samples from various points to see if the floc settles any better. Jar tests should also be used to determine if other combinations of coagulants and coagulant aids produce a better floc.

Slow Floc Formation

Slow or inadequate floc formation is often a problem in water with low turbidity. If there are fewer particles, then there are fewer random collisions for floc particles to form and grow. One way to correct this problem is to recycle some previously settled sludge from the sedimentation basin to add turbidity. This is the same principle used in the solids-contact clarification process. Another way to improve floc formation is to increase turbidity artificially by adding a weighting agent.

If alum or ferric sulfate is used as the coagulant, slow floc formation could also be a result of inadequate alkalinity in the raw water. If this is the case, then alkalinity in the water can be increased by adding lime or soda ash.

 WATCH THE VIDEO
CFS: Coagulation and Flocculation (www.awwa.org/wsovideoclips)

Dosage Control

The success of coagulant and coagulant-aid chemical addition can be assured only if operators make frequent checks (make rounds) on the application of the chemicals and the results they produce. Typically, this means that the operator must do the following:

- Measure and record the actual amount of chemical being fed into the mixing process at regular intervals.
- Take samples of the water at points prior to and after chemical addition and measure the quality (e.g., pH, alkalinity, and turbidity) to determine if the expected results are being met.
- Make adjustments in feed rates as necessary.

Operators sometimes devise charts or tables to help them calculate the amounts of chemicals needed per unit of time. These tables can be referenced to determine if rate changes to the feeders are necessary. Table 4-1 is an example of a liquid alum feed table.

Safety Precautions

The following are some special safety precautions applicable to the coagulation–flocculation process:

- Most dry chemicals can irritate eyes, skin, and mucous membranes. Dry-chemical feeders should be equipped with dust-control equipment, and protective clothing, goggles, and a respirator should be worn when the chemicals are handled.
- Liquid chemicals, particularly polymers, can create dangerous slick areas if spilled, so spills should be cleaned up promptly.
- Dry alum and quicklime, when mixed together, create tremendous heat; if the temperature should reach 1,100°F (593°C), highly explosive hydrogen gas will be released. These and other chemicals should be stored and used in a manner that will prevent improper mixing.

Table 4-1 The amount of liquid alum (in mL/min) needed to achieve dry-basis dosage at various pumpages

Flow rate in MGD → mg/L dry-basis alum desired ↓	1 mL/min needed	2 mL/min needed	3 mL/min needed	4 mL/min needed	5 mL/min needed	6 mL/min needed	7 mL/min needed	8 mL/min needed	9 mL/min needed
12	49	98	147	196	246	295	344	393	442
14	57	115	172	229	286	344	401	458	516
16	65	131	196	262	327	393	458	524	589
18	74	147	221	295	368	442	516	589	663
20	82	164	246	327	409	491	573	655	737
22	90	180	270	360	450	540	630	720	810
24	98	196	295	393	491	589	687	786	884

Record Keeping

Records should be maintained of past raw-water quality and the coagulants and dosages that work best for that water. Notes should also be kept of general observations relating to the operation of the coagulation–flocculation process. This is particularly important for surface water supplies because water quality often varies in each season or in relation to natural events, such as heavy rains, snowmelt discharge, or droughts. Past experience during similar water quality conditions is often a good guide to how chemical addition or equipment should be adjusted to obtain optimal treatment. Records of seasonal jar-test details should be kept. Figure 4-1 is a suggested record-keeping form for the coagulation–flocculation process.

Type of Coagulant _____ Date Started _____									
	Results (by Date)								
Item	Date	Date	Date	Date	Date	Date	Date	Date	Date
Coagulant Dosage									
Raw Water									
Temperature (°C)									
pH									
Alkalinity (mg/L as CaCO₃)									
Turbidity (ntu)									
Taste and Odor									
Color (cu)									
Suspended Solids (mg/L)									
Algae Content									
Coagulated Water									
Filterability (Volume/Time)									
Zeta Potential (mV)									
Settled-Water Turbidity									
Filtered Water									
Turbidity (ntu)									
Color (cu)									
Taste and Odor									
Algae Content									
Residual Coagulant (mg/L)									

Figure 4-1 Record-keeping form for the coagulation–flocculation process

Study Questions

1. What is the optimal pH range for the removal of particulate matter when using alum as a coagulant?
   a. 4.5 to 5.7
   b. 5.8 to 6.5
   c. 6.5 to 7.2
   d. 7.3 to 8.1

2. Which forces will pull particles together once they have been destabilized in the coagulation–flocculation process?
   a. van der Waals forces
   b. Zeta potential
   c. Ionic forces
   d. Quantum forces

3. Which of the following is a common mistake that operators make in regard to flocculation units?
   a. Excessive flocculation time
   b. Lack of food-grade NSF-approved grease on the flocculator bearings
   c. Keeping the mixing energy the same in all flocculation units
   d. Too short a flocculation time

4. Ferric sulfate has which advantage over aluminum sulfate (alum)?
   a. Less staining characteristics
   b. Less cost
   c. More dense floc
   d. Not as corrosive

5. The lower the _____, the more difficult it is to form a proper floc.
   a. temperature
   b. turbidity
   c. pH
   d. alkalinity

6. Jar test results are expressed in
   a. milligrams per liter.
   b. gallons per day.
   c. whatever unit the treatment system has decided to use.
   d. parts per million.

7. List four characteristics of raw water that must be considered in selecting a treatment chemical.

8. Raw-water temperature that approaches the freezing point has what effect on coagulation and flocculation processes?

9. How much time is required to develop heavy floc particles during the flocculation process?

Chapter 5
Sedimentation and Clarifiers

S and, grit, chemical precipitates, pollutants, floc, and other solids are kept in suspension in water as long as the water is flowing with sufficient velocity and turbulence. Sedimentation removes these solids by reducing the velocity and turbulence. Efficient solids removal by sedimentation greatly reduces the load on filtration and other treatment processes. Sedimentation basins should be designed and operated such that they apply a continually low-turbidity water to the filters regardless of the incoming raw-water turbidity.

Process Description

Sedimentation, which is also called *clarification*, is the removal of settleable solids by gravity. The process takes place in a rectangular, square, or round tank called a sedimentation (or settling) basin (or tank).

In the conventional water treatment process, sedimentation is typically used as a step between flocculation and filtration. Sedimentation is also used to remove the large amounts of chemical precipitates formed during the lime–soda ash softening process (see Chapter 12).

Performance goals for the sedimentation process should be instituted at treatment plants. Operators take samples of the influent and effluent water of the basin at regular intervals and calculate the percent removal of each contaminant or additive (e.g., chlorine) to determine the efficiency of this treatment process. When performance goals are not met, operator intervention is necessary. The following might represent a common goal for turbidity removal in a conventional water treatment plant sedimentation basin:

- Effluent turbidity of 1 ntu or less when the 95th percentile raw-water turbidity is 10 ntu or less
- Effluent turbidity of 2 ntu or less when the 95th percentile raw-water turbidity is more than 10 ntu

Basins designed for efficient sedimentation allow the water to flow very slowly, with a minimum of turbulence at the entry and exit points and with as little short-circuiting of flow as possible. Sludge, the residue of solids and water, accumulates at the bottom of the basin and must then be pumped out of the basin for disposal or reuse.

sedimentation
The water treatment process that involves reducing the velocity of water in basins so that the suspended material can settle out by gravity.

sedimentation basin
A basin or tank in which water is retained to allow settleable matter, such as floc, to settle by gravity. Also called a settling basin, settling tank, or sedimentation tank.

influent
Water flowing into a basin.

effluent
Water flowing from a basin.

sludge
The accumulated solids separated from water during treatment.

Sedimentation Facilities

This section describes different types of sedimentation basins and their associated equipment and features.

Types of Basins

Although there are many variations in design, sedimentation basins can generally be classified as either rectangular or center-feed types. Figure 5-1 shows overhead views of the flow patterns in different types of sedimentation basins. Figure 5-1A shows a rectangular settling tank, and Figures 5-1B through 5-1E show circular and square tanks. The operating principles of these basins are described next.

Conventional Rectangular Basins

Rectangular basins are usually constructed of concrete or steel and designed so that the flow is parallel to the basin's length. This type of flow is called rectilinear flow. The basins must be designed to keep the flow distributed evenly across the width of the basin to minimize the formation of currents and eddies that would keep the suspended matter from settling. The basins are often constructed with a bottom that slopes slightly downward toward the inlet end to make sludge removal easier. Figure 5-2 shows uncovered rectangular sedimentation basins at a large treatment plant.

Conventional Center-Feed Basins

Basins can also be constructed as either round or square, so that the water flows radially from the center to the outside. This type of flow is called radial flow. It is important that these basins also be designed to keep the velocity and flow distribution as uniform as possible. Their bottoms are generally conical and slope downward toward the center of the basin to facilitate sludge removal.

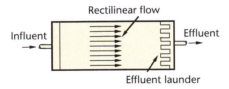

A. Rectangular settling tank, rectilinear flow

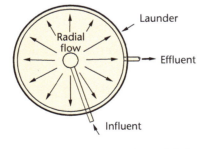

B. Center-feed settling tank, radial flow

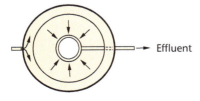

C. Peripheral-feed settling tank, radial flow

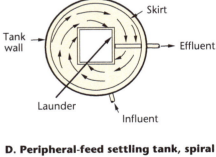

D. Peripheral-feed settling tank, spiral flow

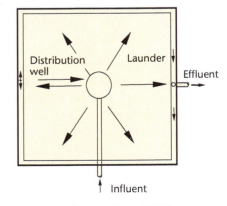

E. Square settling tank, radial flow

Figure 5-1 Overhead views of flow patterns in sedimentation basins

rectilinear flow
Uniform flow in a horizontal direction.

radial flow
Flow that moves across a basin from the center to the outside edges or vice versa.

Figure 5-2 Typical uncovered sedimentation basins

Peripheral-Feed Basins Peripheral-feed basins are designed to feed incoming water from around the outer edge and collect it at the center. This type of basin also has radial flow. The design is otherwise similar to that of a center-feed basin.

Spiral-Flow Basins Spiral-flow basins have one or more points around the outer edge where water is admitted at an angle. This design causes the flow to circle around the basin and ultimately leave the basin at a center collector (launder).

Basin Zones

All basin types have four zones, each with its own function. As illustrated in Figure 5-3, these zones are as follows:

1. The *influent zone* decreases the velocity of the incoming water and distributes the flow evenly across the basin.
2. The *settling zone* provides the calm (quiescent) area necessary for the suspended material to settle.
3. The *effluent zone* provides a smooth transition from the settling zone to the effluent flow area. It is important that currents or eddies do not develop in this zone because they could stir up settled solids and carry them into the effluent.
4. The *sludge zone* receives the settled solids and keeps them separated from other particles in the settling zone.

These zones are not actually as well defined as Figure 5-3 illustrates. There is normally a varying gradation of one zone into another. The settling zone is particularly affected by the other three zones, based primarily on how the basin is designed and operated.

 WATCH THE VIDEO
CFS: Sedimentation Clarifiers (www.awwa.org/wsovideoclips)

Parts of a Sedimentation Basin

Equipment used in conventional settling basins varies depending on the design and manufacturer. Figures 5-4 and 5-5 show the parts of typical rectangular and

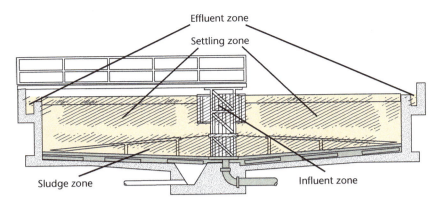

Figure 5-3 Zones in a sedimentation basin

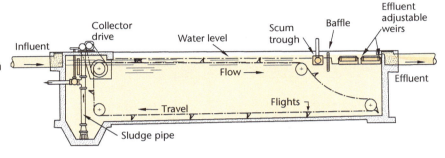

Figure 5-4 A typical rectangular sedimentation basin (with a continuous chain collector sludge removal system)

Courtesy of US Filter Envirex Products.

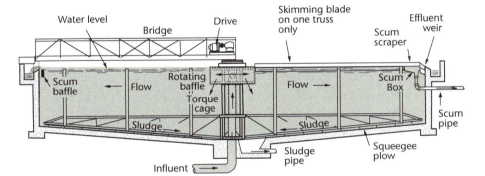

Figure 5-5 A typical circular sedimentation basin

Courtesy of US Filter Envirex Products.

circular basins, respectively. The inlet distributes the influent evenly across (or around) the basin so the water will flow uniformly. A baffle installed downstream of the inlet reduces the velocity of the incoming water and helps produce calm, nonturbulent flow conditions for the settling zone. The water flows underneath the baffle and into the main part of the basin.

The effluent launder (also called the effluent trough) collects the settled water as it leaves the basin and channels it to the effluent pipeline, which carries the water to the next treatment process. Launders can be made of fiberglass or steel, or they may be cast concrete as a part of the tank.

The launder is equipped with an effluent (overflow) weir, which is a steel, plastic, or fiberglass plate designed to distribute the overflow evenly to all points of the launder. One of the most common types of effluent weirs is the V-notch, as illustrated in Figure 5-6. In some designs, launders can also receive the flow of water from beneath the water surface through holes in the launder wall.

 WATCH THE VIDEO
CFS: Sedimentation Basin Zones (www.awwa.org/wsovideoclips)

Figure 5-6 V-notch weir
Courtesy of Fisher Scientific.

Shallow-Depth Sedimentation

Shallow-depth sedimentation basins are designed to shorten the detention time required for sedimentation; this means the basins are smaller. Shallow basins and plate and tube settlers are two shallow-depth designs.

Shallow Basins

Basins are occasionally designed to have a fairly shallow depth to reduce the time necessary for floc to settle to the bottom. Some rectangular sedimentation basins have two or three levels; the flow of water at the inlet to the basin divides into parallel flows, one over each level. These basins are designed on the principle that surface area is more important than depth. For shallow basins to work properly, it is important that coagulant doses and flash mixing be carefully controlled.

Plate and Tube Settlers

Several types of shallow-depth settling units are constructed of multiple, individual modules. These modules are either plates or tubes of fiberglass, steel, or other suitable material. They are spaced only a short distance apart and tilted at an angle with respect to horizontal (Figures 5-7 and 5-8). If the angle is greater than 50–60 degrees, they will be self-cleaning; in other words, the sediment will settle until it hits the plate or the tube bottom and will then slide to the bottom

Figure 5-7 Tube settlers
Courtesy of Wheelabrator Engineered Systems—Microfloc.

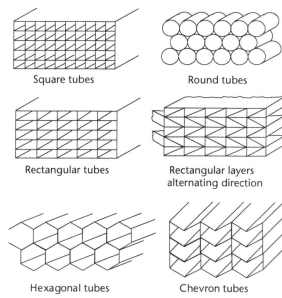

Square tubes

Round tubes

Rectangular tubes

Rectangular layers alternating direction

Hexagonal tubes

Chevron tubes

Figure 5-8 Various formats for tube modules
Courtesy of *Water Quality and Treatment.* 5th ed. 1999.

of the basin. An angle as small as 7 degrees is used when sludge is removed from the tubes or plates using periodic backflushing, possibly in conjunction with filter backwashing. A typical separation distance between the inclined surfaces of tube or plate settlers is 2 in. (50 mm); the inclined length is 3–6 ft (1–2 m).

The direction of flow through tube or plate settlers varies by the manufacturer's design. With countercurrent settling (Figure 5-9), the suspension is fed to the lower end and flows up the channels. If the angle of inclination is great enough, the solids slide down the surface, counter to the flow of the liquid. Concurrent settlers are designed so that the suspension is fed to the upper end and leaves at the bottom. In crossflow settlers, the flow is horizontal between the surfaces.

Tube and plate settlers are prefabricated in modules that can either be incorporated into new construction or be used to retrofit old basins so as to increase their settling efficiency. Their advantages include lightweight construction, structural rigidity, and the ability to settle a given flow rate in a much smaller basin size. As with all installations, however, the addition of tube or plate settlers to a treatment plant is not advisable unless a thorough engineering evaluation is made of the plant design to ensure they will operate properly. Figure 5-10 illustrates the installation of tube and plate settlers in rectangular and circular basins.

Another design for shallow-depth sedimentation uses inclined plates (Lamella® plates), as illustrated in Figure 5-11. This design incorporates parallel plates installed at a 45-degree angle. In this case, the water and sludge both flow downward. The clarified water is then returned to the top of the unit by small tubes.

Sludge Removal

As solids settle to the bottom of a sedimentation basin, a sludge layer develops. If this sludge is not removed before the layer gets too thick, the solids can become resuspended or tastes and odors can develop as a result of decomposing organic matter (Figure 5-12). Methods of removal are discussed in Chapter 17.

WATCH THE VIDEO
WW Treatment: Preliminary (www.awwa.org/wsovideoclips)

Direction of flow

To sludge collection

Figure 5-9 Countercurrent flow in tubes

Courtesy of Wheelabrator Engineered Systems—Microfloc.

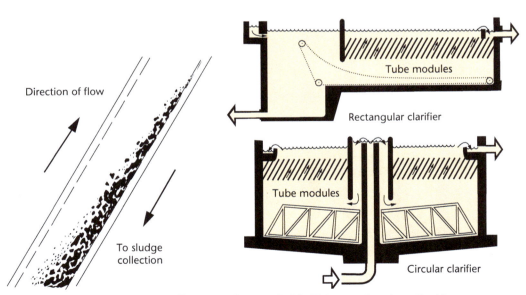

Tube modules

Rectangular clarifier

Tube modules

Circular clarifier

Figure 5-10 Tube settlers installed in sedimentation basins

Courtesy of Wheelabrator Engineered Systems—Microfloc.

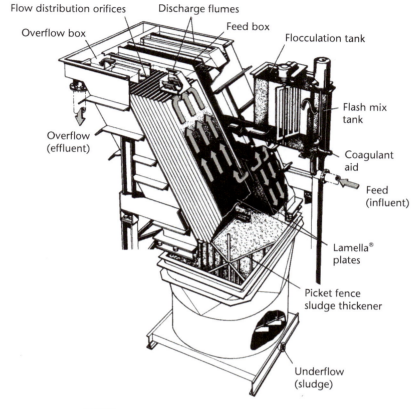

Flow distribution orifices

Discharge flumes

Overflow box

Feed box

Flocculation tank

Flash mix tank

Coagulant aid

Feed (influent)

Overflow (effluent)

Lamella® plates

Picket fence sludge thickener

Underflow (sludge)

Figure 5-11 Lamella® plates

Courtesy of Parkson Corporation.

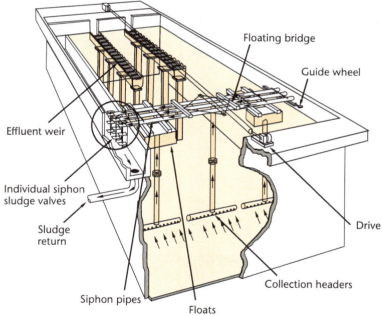

Floating bridge

Guide wheel

Effluent weir

Individual siphon sludge valves

Sludge return

Drive

Siphon pipes

Floats

Collection headers

(a)

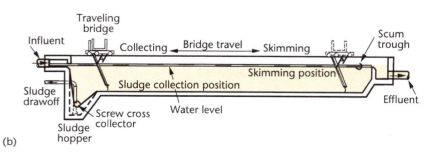

Traveling bridge

Influent

Collecting

Bridge travel

Skimming

Scum trough

Skimming position

Sludge collection position

Sludge drawoff

Screw cross collector

Water level

Effluent

Sludge hopper

(b)

Figure 5-12 Two mechanical sludge removal systems used in rectangular basins. (a) Floating-bridge siphon collector. (b) Traveling-bridge collector.

Courtesy of F. B. Leopold Company, Inc.

Other Clarification Processes

Other clarification processes, including the use of solids-contact basins, dissolved-air flotation, and contact clarifiers, are discussed next.

Solids-Contact Basins

Solids-contact basins (also called *upflow clarifiers*, *solids-contact clarifiers*, or *sludge-blanket clarifiers*) are generally circular and contain equipment for mixing, flow circulation, and sludge scraping. A wide variety of designs are available from different manufacturers. All are divided by baffles into two distinct zones: mixing and settling. Coagulation and flocculation take place in the mixing zone, where the raw water and coagulant chemicals are combined and slowly agitated. A typical unit is illustrated in Figure 5-13.

Near the bottom of the basin, the flow is directed upward into the settling zone, which is separated from the reaction flocculation area by baffles. A point exists at which the upflow velocity can no longer support the floc particles. This point defines the top of the sludge blanket. The water has essentially been "filtered" through this sludge blanket and the clarified water flows upward into the effluent troughs. The floc particles in the sludge blanket contact other particles and grow larger until they settle to the bottom. A portion of the sludge is recycled to the mixing zone, and the remainder settles in a concentration area for periodic disposal through the blow-off system. This must be done to maintain an almost constant level of solids in the unit.

The advantage of this type of unit is that the chemical reactions in the mixing area occur more quickly and completely because of the recycled materials from the sludge blanket. This allows much shorter detention times than with conventional basins. One problem with these clarifiers is that a quick change in water temperature or a quick increase in flow can cause currents that upset the sludge blanket.

Dissolved-Air Flotation

Dissolved-air flotation (DAF) can be used in some installations in place of sedimentation. It is a process in which gas bubbles are generated so that they will attach to solid particles, causing them to rise to the surface rather than settle to the bottom.

solids-contact basin

A basin in which the coagulation, flocculation, and sedimentation processes are combined. The water flows upward through the basin. It is used primarily in the lime softening of water. It can also be called an upflow clarifier or sludge-blanket clarifier.

dissolved-air flotation (DAF)

A clarification process in which gas bubbles are generated in a basin so that they will attach to solid particles causing them to rise to the surface. The sludge that accumulates on the surface is then periodically removed by flooding or mechanical scraping.

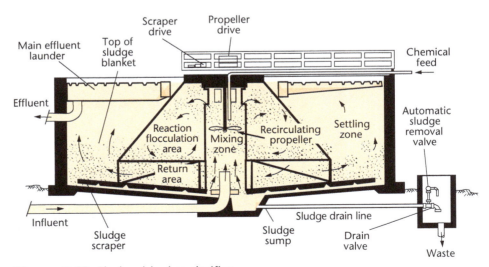

Figure 5-13 Sludge-blanket clarifier
Courtesy of US Filter-General Filter Products.

DAF units consist of a flotation basin, saturation unit, air compressor, and recycle pumping. In operation, about 5 to 10 percent of the clarified water is recycled by the pump to the saturator, where high-pressure air is brought into contact with the water. When this water is reintroduced to the influent of the flotation basin, the gases in the water—which are supersaturated—gently come out of solution. The effect is similar to the bubbles in a carbonated beverage that rise to the top when the bottle is opened. There are several variations to this process; some are more appropriate than others for potable water treatment.

Flotation can be used in either a rectangular or a circular tank. The sludge that accumulates on the surface of the flotation tank is called *float*, and it can be removed continuously or intermittently either by flooding the basin to overflow the float or by mechanical scraping. The most widely used method is to have rubber scrapers that travel over the tank surface and push the float over a ramp, called a *beach*, to a trough for disposal.

Flotation is widely used for industrial water clarification and wastewater clarification. For potable water treatment, the process is particularly good for algae removal because algae naturally tend to float.

Each of the different DAF systems is named for the method of producing the gas bubbles.

WATCH THE VIDEO
Dissolved Air Flotation (www.awwa.org/wsovideoclips)

Electrolytic Flotation

Electrolytic flotation uses very small bubbles of hydrogen and oxygen generated by passing a direct current between two electrodes.

Dispersed-Air Flotation

There are two different types of dispersed-air processes: foam flotation and froth flotation. These processes are not suitable for potable water treatment because high turbulence and undesirable chemicals are required to produce the froth.

Contact Clarifiers

A number of unique proprietary clarification processes have been developed in recent years and are gradually gaining acceptance. A few of them are mentioned here as examples. Additional information can be obtained from the various companies.

Pulsator Clarifier

The Pulsator, developed and marketed by Infilco Degremont (Richmond, Virginia), uses a unique pulsating hydraulic system to maintain a homogeneous sludge layer. When flow is not introduced into the clarifier, the sludge blanket settles. Pulsation of the sludge blanket about every 40 to 50 seconds acts to maintain a uniform layer and to reduce the potential for short-circuiting of the flow through the sludge blanket. The Pulsator is often used to treat highly colored, low-turbidity water, in which the formation of settleable floc is usually difficult.

Superpulsator Clarifier

The Superpulsator (Figure 5-14), also developed and marketed by Infilco Degremont, combines the hydraulic pulsation system of the Pulsator with a series of inclined plates to help maintain high solids concentrations at increased upflow rates. The units can therefore be operated at rates two to three times greater than

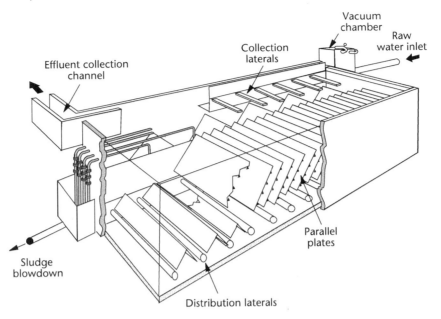

Figure 5-14
Superpulsator flat-bottom
clarifier with lateral flow
distribution

Courtesy of Infilco Degremont
Inc., Richmond, Virginia.

Figure 5-15 Super-P
installation at
Steubenville, Ohio

Courtesy of MWH Americas.

those of the Pulsator. The plates are inclined at an angle of 60 degrees from the horizontal and spaced from 12 to 20 in. (305 to 510 mm) apart.

Superpulsator (Super-P) units, such as the one engineered by MWH Americas for Steubenville, Ohio (Figure 5-15), are being used in water treatment plants that have river sources subject to rapid changes in turbidity or that are subject to organic spills. Pilot-scale testing of the Steubenville Super-P, conducted for two, 2-week seasonal periods, showed that settled water of very high quality was produced when the pilot was operated at 3 gpm/ft². Subsequent full-scale testing after construction demonstrated that the units are a cost-effective alternative to conventional treatment systems with much larger footprints.

This type of unit is expected to operate under the difficult conditions experienced when oil spills require high dosage of activated carbon for organics control. The sludge layer can be subjected to high concentrations of powdered activated carbon, thus providing the needed water/carbon contact without fear of carryover of fines to the filters.

Trident Contact Adsorption Clarifier

The Trident treatment system, marketed by WesTech, combines coagulation, flocculation, and clarification processes in a single upflow clarifier. The raw water, with coagulant added, enters at the bottom of the unit and passes through a bed, consisting of a plastic medium, that floats on top of the water. The solids adhere to the medium, with relatively high removal rates.

When the solids accumulation results in excessive head loss, or when effluent quality becomes unacceptable, the medium is cleaned by upflow hydraulic flushing with raw water and air. Trident operating times typically average 4–8 hours between backwashes.

Regulations

There are no federal regulations that specifically apply to the design of sedimentation systems. However, the requirements of the Surface Water Treatment Rule (SWTR) discussed in Chapter 3 can have a significant impact on the design or operation of these facilities. The SWTR basically requires that surface water systems maintain specified $C \times T$ values (the product of disinfectant concentration and contact time) before the treated water reaches the first customer. The contact time of raw water with chlorine (or other disinfectant) while the water is in the sedimentation basin provides a major portion of the contact time required for most treatment plants.

Many older sedimentation basins were not designed with serious consideration given to avoiding dead water spaces in corners or to minimizing short-circuiting of flow. The SWTR requirements may make it necessary for some water systems to alter their sedimentation basins to achieve maximum detention. This can be done by installing additional baffles. An improvement in detention time can best be obtained by improving inlet and outlet baffles and improving flow patterns.

Operators who are considering adding baffles to sedimentation basins should obtain professional design assistance to ensure that the improvement will maximize detention time without causing operating problems.

Operation of the Process

This section discusses certain key features associated with the operation of the sedimentation process.

Sedimentation Facility Operation

The primary function of the sedimentation process is to prepare water for effective and efficient filtration. An effective sedimentation process also removes substantial amounts of organic compounds that cause color and that can be precursors for the formation of disinfection by-products. Because effective sedimentation is closely linked with proper coagulation and flocculation, the operator must ensure that the best possible floc is being formed.

Conventional Basins

The minimum detention time (i.e., the theoretical time it takes a particle of water to pass through the basin) is usually designed to be 4 hours for surface water systems. The actual detention time for a basin is often shorter than the calculated period because currents and improper design features may cause the water to short-circuit through the basin. For this reason, weir overflow rates are often better indicators of the operation of sedimentation basins.

The flow rate over outlet weirs is commonly designed not to exceed 20,000 gpd per foot (250,000 L/d per meter) of weir. For light alum floc, it may have to be reduced to 14,400 gpd per foot (180,000 L/d per meter). Raw-water sources and treatment facilities are not all the same, so an operator must experiment to determine the most effective overflow rates. The most effective rate may vary at different times of the year because of changes in water quality. If water demand makes it necessary to exceed the optimal overflow rate, adjustments will have to be made elsewhere in the plant operation to compensate for the poorer-quality settled water.

A minimum of two sedimentation units should be provided, so that one can be taken out of service for cleaning or repair without disrupting plant operation. The operator must ensure that the total flow is divided evenly among the basins and that any changes in flow rate are made smoothly. This will prevent the basins from receiving surges that cause eddy currents or break up floc. The flow rate over the effluent weirs should be uniform and evenly distributed along the weir length. If it is not, the weirs may have to be adjusted.

High-Rate Tube and Plate Settlers

High-rate settlers were developed to increase the rate of conventional settlers for application in site areas (e.g., package plants). Because the overflow rates used for tube and plate settlers are two to three times those for conventional basins, the floc must have good settling characteristics. Ensuring these characteristics often requires special attention to the flocculation process, including the addition of a coagulant aid.

At times, the floc may bridge across the upper edge of the tube opening, resulting in a buildup of solids several inches (centimeters) thick. To dislodge the accumulation, the water level of the basin is usually lowered. If this is not possible, a gentle current of water must be directed across the top of the tubes with a hose or a permanent perforated header pipe mounted along the length of the settler unit. Sludge withdrawal is likely to be required more frequently for a system with tube or plate settlers than for one with conventional basins.

Solids-Contact Basins

The optimal detention time for solids-contact basins varies with equipment design and raw-water quality. The upflow rates used for turbidity removal are typically about 1 gpm/ft^2 (0.7 mm/sec), which results in detention times of only 1 or 2 hours. Proper coagulation–flocculation and control of the solids concentration in the slurry or sludge blanket are essential for good results. Weir loadings normally should not exceed 10 gpm per foot (124 L/min per meter) of weir length.

To maintain a good sludge blanket, coagulant aids or weighting agents may be necessary. The solids concentration should be determined at least twice a day—more frequently if the water quality is always changing. Monitoring the solids concentration is done by taking samples from the taps provided on the basin and conducting settling tests as prescribed by the manufacturer. Because solids-contact basins are often used in water softening, their operation is further discussed in Chapter 12.

Actiflo® Process

The Actiflo process is an ultra-high-rate proprietary process that combines microsand-enhanced flocculation and lamellar settling to treat surface water. The process removes suspended solids by conventional coagulation–flocculation and lamellar settling. The process is followed by filtration to meet SWTR standards.

A typical process flow incorporates a rapid-mix basin where polyelectrolyte and sand are introduced and mixed with the incoming water. The flocculation process is conventional. The sedimentation process uses lamellar settling to separate microsand-weighted flocs and microsand-weighted ballast. Hydrocyclone technology is used to separate sludge solids from the microsand. Sludge solids are further processed using conventional dewatering.

Monitoring the Process

The primary test used to indicate proper sedimentation is the test for turbidity. The turbidity of samples taken from the raw water and the outlet of each basin should be tested at least every 2 hours. More frequent testing may be necessary if water quality is changing rapidly. By comparing these turbidities, an indication of the efficiency of the sedimentation process can be obtained. For example, if the raw-water turbidity is 50 ntu and effluent turbidity is 40 ntu, very little turbidity is being removed. This indicates that better coagulation–flocculation is needed or that short-circuiting is occurring. Visual examination of water samples from the effluent can also indicate if floc is being carried over onto the filters.

The turbidity of the settled water should be kept below 1-2 ntu. If it is above 3 ntu, the process should be checked and the operation improved. Turbidity test procedures are discussed in *Water Quality*, also part of this series. Lime softening plants may differ and have more relaxed goals for turbidity removal.

The raw-water temperature should be measured and recorded at least daily. As water becomes colder, it is more viscous, thus presenting more resistance to the settling particles. To compensate for this, the surface overflow rates may have to be reduced.

Operating Problems

The most common operating problems in sedimentation facilities are poorly formed floc, short-circuiting, density currents, wind effects, and algae growth.

Poorly Formed Floc

Poorly formed floc is characterized by small or loosely held particles that do not settle properly and are carried out of the settling basin. This problem is the result of inadequate rapid mixing, improper coagulants or dosages, or improper flocculation. Jar tests can usually provide the information necessary to find the specific problem. The solution may be to increase the mixing energy, use a coagulant aid, or install additional baffles in the flocculation basin.

Short-Circuiting

If a basin is not properly designed, water bypasses the normal flow path through the basin and reaches the outlet in less than the normal detention time. This occurs to some extent in every basin. It can be a serious problem in some installations, causing floc to be carried out of the basin as a result of shortened sedimentation time.

The major cause of short-circuiting is poor inlet baffling. If the influent enters the basin and hits a solid baffle, strong currents will result. A perforated baffle can successfully distribute inlet water without causing strong currents (Figure 5-16). If short-circuiting is suspected, tracer studies are the best method of determining the extent of the problem. (Publications detailing the SWTR requirements of the Safe Drinking Water Act can provide more information on tracer studies.)

Figure 5-16 Perforated baffles

Figure 5-17 Barrier around an uncovered basin

Density Currents

Currents that disrupt the sedimentation process in a basin occur when the influent contains more suspended solids, and thus has a greater density, than the water in the basin. They can also occur when the influent is colder than the water in the basin. In either case, the influent sinks to the bottom of the basin, where it can create upswells of sludge and short-circuits. If a system encounters this problem, an engineering study should be conducted to determine the best solution. The problem can often be lessened by modifying the effluent weirs.

Wind Effects

Wind can create currents in open basins, thus causing short-circuiting. If wind is a problem, a barrier should be constructed around uncovered basins to lessen the wind's effect and keep debris out of the water (Figure 5-17).

Algae and Slime Growth

Growth of algae and slime on the basin walls, which can cause taste-and-odor problems, is a problem that often occurs in open, outdoor basins. If the algae and slime detach from the walls, they can clog weirs or filters.

The growths can be controlled by coating the walls with a mixture of 10 g of copper sulfate ($CuSO_4$) and 10 g of lime ($Ca[OH]_2$) per liter of water. The basin should be drained and the mixture applied to the problem areas with a brush.

Waste Disposal

Regardless of the type of basin used, sludge is probably the most troublesome operating problem. The collection and disposal methods depend in part on the nature and volume of the sludge formed. This determination is primarily governed by the raw-water quality and the types of coagulants used. For example, different methods of handling and disposal are typically used for alum sludge than for lime sludge. Methods of handling lime sludge are discussed in Chapter 12.

In the past, sedimentation basin sludge was typically discharged into lakes and streams without treatment. This is no longer allowed under current environmental laws because the sludge can form deposits in the receiving water that are harmful to aquatic life and can produce objectionable tastes and odors.

Alum Sludge

Alum sludge is the most common form of sludge resulting from the sedimentation process, because alum is the coagulant most often used to remove turbidity. Alum sludge is a gelatinous, viscous material, typically containing only 0.1 to 2 percent solid material by weight. However, it is extremely hard to handle and dewater because much of the water is chemically bound to the aluminum hydroxide floc.

Alum sludge is commonly pumped into specially designed lagoons. When a lagoon is full, the sludge is diverted to another one, and the filled lagoon is allowed to dry until the sludge can be removed for final disposal. Water decanted from the top can be returned to the treatment plant. The process can take a year or longer, and the sludge will still contain over 90 percent of the original water, making it difficult to handle and unfit for disposal in a sanitary landfill. Therefore, it usually must be placed on land owned by the water utility. Because of the large land requirements, lagooning may not be possible for large treatment plants or where land is not available. In areas where freezing temperatures are common, the freezing and thawing of sludge can hasten the dewatering process.

Sand drying beds can accomplish more efficient dewatering of sludge than lagoons. The sludge is spread in layers over the sand, which overlies gravel and drain tiles. However, land requirements, difficulties with sludge removal, and poor performance during rainy periods are disadvantages of this method that must be considered.

If there is not much land available, dewatering can be improved using mechanical equipment. Vacuum filters, centrifuges, and filter presses can successfully dewater alum sludge to at least 20 percent solids by weight. The sludge can then be placed in most landfills or used as a soil conditioner. Mechanical dewatering usually has a higher operation and maintenance cost than lagoons and sand drying beds.

Regardless of the mechanical equipment used, the sludge must be pretreated to make the subsequent dewatering more effective. This usually requires a sludge thickener, which is a circular tank much like a clarifier, equipped with a stirring mechanism (Figure 5-18). This mechanism breaks apart the floc particles in the sludge, allowing the water to escape and the solids to settle. A polymer is usually added to the sludge as it enters the thickener to enhance settling. A thickener can concentrate sludge up to about 5 percent solids by weight.

Because of the difficulties with dewatering and disposing of the alum sludge, many operators reduce alum dosages by effectively using coagulant aids, such as polymers. This greatly reduces the volume of alum sludge that must be handled.

Other Sludges

Ferric salt coagulants also produce a sludge that is difficult to dewater. The sludge is also difficult to thicken because it has a low density, which causes it to settle slowly.

Disposal to a Waste Treatment Plant

A cost-effective method for disposing of treatment waste is to pump it to a sanitary sewer, for ultimate disposal by the local wastewater treatment plant. The amount

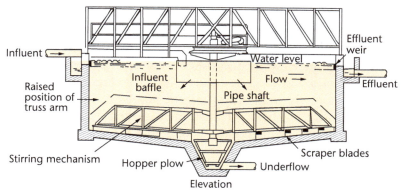

Figure 5-18 Sludge thickener

of alum or ferric chloride sludge in proportion to the sanitary waste sludge is usually small enough that it does not disrupt the normal processing of sludge by the wastewater treatment plant. Some municipalities or authorities operate water and wastewater treatment plants and may have a means of handling their own wastes.

Wastewater treatment authorities that will accept water treatment plant waste usually charge a fee to cover their increased operating costs. In addition, they usually have a number of restrictions, including the quantity that can be disposed of each day and a ban on any disposals during rainstorms or other times when the waste treatment facility may be operating near capacity. A study of the sewer system must be made to ensure that it can adequately handle this additional flow along with its normal domestic flow. Lime sludge cannot be discharged to a sanitary sewer system because lime deposits might build up and block the sewers.

Filter Backwash Water

The handling and disposal of filter backwash water are discussed in greater detail in Chapter 3. The Filter Backwash Recycle Rule (FBRR) establishes guidance on how to treat filter backwash. In general, the backwash water can be combined with the sedimentation basin sludge prior to disposal, it can be treated separately, or it can be recycled by being added to the raw water entering the plant. If recycling, plants will need to perform an audit of the waste streams as prescribed in the FBRR.

Equipment Maintenance

Manually cleaned conventional basins should be inspected at the completion of each sludge removal. Basins with mechanical sludge removal should be drained at least once a year for a general inspection. Slimes that build up on walls and appurtenances harbor *Legionella* and other microbiological growths that can accumulate and overwhelm the disinfection system. The slimes need to be power-blasted with fire hoses using pressurized clean water.

In addition to inspection of the operation of mechanical equipment, the inlet system should be examined closely to ensure that all openings are clear of obstructions that could cause unequal flow distribution. The weirs should also be inspected to ensure they are level and not blocked by debris or chemical deposits. Uneven flow over the weirs could cause uneven flow in the basin and eddies that would disturb the settling zone.

The baffles should also be checked for deterioration that will cause short-circuiting, and any algae or slime accumulation on basin walls, weirs, and baffles should be removed.

Solids-contact basins need to be drained at least annually so that the condition of the baffles and mixing equipment can be checked. Sludge-pumping lines and equipment should be inspected routinely to ensure that they are not becoming plugged. In addition, the lines should be flushed occasionally to prevent a buildup of solids.

Safety Precautions

Open sedimentation basins should be equipped with guardrails that will prevent falls into the basins. Walkways and bridges connected to basins should also have guardrails. Life rings or poles should be kept near the basins for rescue purposes.

If basins are not covered, care must be exercised during periods of rain, ice, or snow because the walkways become very slippery. Caution is also needed when drained basins are being cleaned or inspected because growths of aquatic organisms and sludge deposits can make the surfaces slippery.

The moving parts of all machinery should be equipped with guards to prevent the machinery from catching legs, fingers, or clothing. Guards should never be removed while the machinery is operating.

Record Keeping

Good records are invaluable in helping the operator to solve problems and produce high-quality water. As raw-water quality or other conditions change, the operator can review past records to help determine what adjustments are needed. Because sedimentation is closely linked with coagulation–flocculation, both records should be maintained together. Sedimentation records should include the following:

- Surface and weir overflow rates, calculated by using the flow rate through each basin
- Turbidity results for raw water and effluent from each basin plotted as percentage or log removal
- Chemical analysis results of disinfectant if used, and iron and manganese if appropriate
- Quantity of sludge pumped or cleaned from each basin
- Types of operating problems and corrective actions taken

Study Questions

1. If enteric disease-causing protozoans have been found in the effluent of a water plant, which of the following is the most probable solution?
   a. Where possible, use powdered activated carbon (PAC) throughout the water plant; backwashing filters will remove the PAC.
   b. Use PAC only in the sedimentation basin; backwashing the filters will remove the PAC.
   c. Use the multibarrier approach—coagulation, flocculation, sedimentation, and filtration.
   d. Superchlorinate the water plant.

2. What is the major cause of short-circuiting in a sedimentation basin?
   a. Open basins that are subject to algal growths and thick slime growths on the side of the basin
   b. Basins without a wind break
   c. Poor inlet baffling
   d. Density currents

3. Conventional sedimentation has a _____ removal of *Cryptosporidium* oocysts.
   a. less than 0.5-log
   b. 0.5-log
   c. 1.0-log
   d. 2.0-log

4. Dissolved-air flotation is particularly good for removing
   a. sulfides.
   b. inorganics.
   c. manganese and iron.
   d. algae.

5. As solids settle to the bottom of a sedimentation basin,
    a. tank pressure is reduced.
    b. a sludge layer develops.
    c. movement of sediment accelerates.
    d. water pH decreases.

6. What is the maximum weir loading for light alum floc in solids-contact basins?

7. List four components of a typical rectangular sedimentation basin.

8. Which regulation basically requires that surface water systems maintain specified $C \times T$ values before the treated water reaches the first customer?

9. What is the primary test used to indicate proper sedimentation?

10. Which type of sludge cannot be discharged to a sanitary sewer system because its deposits might build up and block the sewers?

Chapter 6
Filtration

Equipment Associated With Gravity Filters

This section describes equipment associated with gravity filters, including filter tanks, filter media, underdrain systems, wash-water troughs, and filter bed agitation equipment.

Filter Tanks

Filter tanks are generally rectangular and constructed of concrete. However, prefabricated tanks and units for package treatment plants are often made of steel. Several filter tanks are usually constructed side by side on either side of a central pipe gallery to minimize piping. Figure 6-1 illustrates this arrangement (it shows only one row of filters).

Filter Media

Sand or other filter media must be prepared specifically for filtration use. The original design of the conventional sand filters used for many years placed three to five layers of graded gravel between the sand and the underdrain system. The gravel bed served the dual purposes of preventing sand from entering the underdrains and of helping distribute the backwash water evenly across the bed. The total gravel bed may be from 6 to 18 in. (150 to 450 mm) thick, depending on the type of underdrain system.

Many types of filter underdrain systems are now available that do not require a gravel layer. They are designed to support the sand or other media directly.

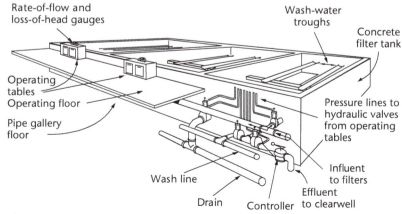

Figure 6-1 Filter tank construction

---

filter tank

The concrete or steel basin that contains the filter media, gravel support bed, underdrain, and wash-water troughs.

filter media

Granular material through which material is collected and stored when water passes through it.

Additional details on the requirements for filtering materials are covered in AWWA Standard B100, *Filtering Material*.

Filter Sand

The sand used for filtration is specifically manufactured for filtration use. If the sand is too fine, it will resist the flow of water and require frequent backwashing. If too coarse, it may not effectively remove turbidity. Included in the specifications for filter sand are the grain size, size distribution, shape, density, hardness, and porosity. Hardness is particularly important because soft sand will break down quickly during the agitation of backwashing.

Grain size has important effects on the efficiency of filtration and on backwashing requirements. Proper size is determined by sieve analysis that uses the American Society for Testing and Materials (ASTM) Standard Test C136 (most recent edition). In the United States, the sand is described in terms of the effective size (ES) and the uniformity coefficient (UC). The ES is that size for which 10 percent of the grains are smaller by weight. The UC is a measure of the size range of the sand.

A conventional rapid sand filter typically uses sand that has a fairly uniform grain size of 0.10–0.15 in. (2.5–3.8 mm) in diameter, in a bed 24–30 in. (0.6–0.75 m) deep. This sand is much coarser than that used for slow sand filters. The sand ordinarily used in conventional rapid sand filters has an ES of 0.11–0.13 in. (2.8–3.3 mm) and a UC varying from 1.6 to 1.75.

Anthracite

Crushed and graded anthracite is often used as a filter medium along with sand. It is much lighter than sand, so it always stays on top of the sand during backwashing. It is usually graded to a size slightly larger than sand.

Granular activated carbon (GAC) can be used instead of anthracite as a filter medium, in which case it plays a dual role by also adsorbing organic compounds in the water. Its principal applications have been for taste-and-odor control, but it can also be used for removing organic compounds suspected of being carcinogenic or causing adverse health effects. Experience has shown that where the organic loading is not too heavy, tastes and odors can successfully be removed for periods of 1–5 years. When the effectiveness of the GAC has ceased, it must be removed and regenerated, or replaced with a new medium.

When GAC is used in the retrofit of a conventional rapid sand filter, 15–30 in. (0.38– 0.76 m) of GAC is placed over several inches (centimeters) of sand. The effectiveness of GAC as an adsorber in a filter is limited because the water is in contact with it for only a few minutes while passing through the filter.

The use of GAC in filters is covered in more detail in Chapter 14.

Garnet Sand

Garnet sand, or ilmenite, is used as the bottom medium in multimedia filters. It is graded to be finer than sand, but because of its greater density, it tends to stay at the bottom of the filter.

 WATCH THE VIDEO
Filtration: Rapid Sand (www.awwa.org/wsovideoclips)

Underdrain Systems

Filter underdrains serve two functions: (1) they collect the filtered water uniformly across the bottom of the filter, keeping the filtration rate uniform across the filter surface, and (2) they distribute the backwash water evenly, allowing the sand and gravel beds to expand but not be unduly disturbed by the backwashing. The common types of underdrain systems are

- pipe lateral collectors,
- perforated tile bottoms,
- Wheeler filter bottoms,
- porous plate bottoms, and
- false-floor underdrain systems.

Pipe lateral, perforated tile, and Wheeler filter bottom systems all require a gravel support bed to prevent the sand or anthracite from flowing into the underdrains and to distribute the backwash water evenly. More new systems are now available that allow fine media to be placed directly on the filter bottom so that a gravel layer will not be required.

Pipe Lateral Collectors

Pipe lateral collectors are the first and oldest type of underdrain system; many installations are still in use (Figure 6-2). These systems have a central manifold pipe with perforated lateral pipes on each side. The pipes can be cast iron, asbestos cement, or polyvinyl chloride (PVC). Small holes are usually spaced along the underside of the pipes, where they are least likely to become plugged with sand. This hole placement will also force the backwash water against the floor where it will be evenly distributed and not disrupt the media.

In some cases, the holes are fitted with brass inserts to prevent plugging caused by corrosion. Plugged holes in the underdrain piping can cause one area to be "dead" during backwashing and can increase velocity through the holes in other areas. This can eventually lead to a serious disruption of the filter media if the problem is not corrected.

Perforated Tile Bottoms

Perforated tile bottoms have been installed in many water filtration plants (Figure 6-3). They consist of perforated, vitrified clay blocks with channels inside to

> **pipe lateral collector**
> A filter underdrain system using a main pipe (header) with several smaller perforated pipes (laterals) branching from it on both sides.

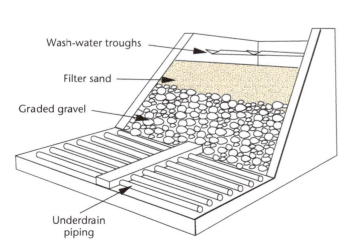

Figure 6-2 Pipe lateral collector under a conventional rapid sand system

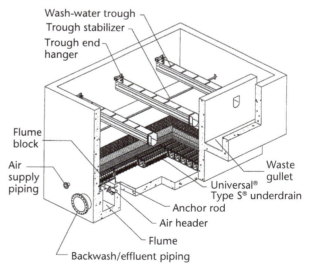

Figure 6-3 Conventional rapid sand filter with perforated tile underdrain system

Courtesy of F. B. Leopold Company, Inc.

carry and distribute the water. Figure 6-4 shows details of the same type of filter underdrain blocks that are currently available; the blocks are made of high-density polyethylene.

Wheeler Filter Bottoms

Wheeler filter bottoms have also been widely used and consist of conical concrete depressions filled with porcelain spheres (Figure 6-5). Each cone has an opening in the bottom and contains 14 spheres ranging in diameter from 1⅜ to 3 in. (35 to 76 cm). The spheres are arranged to lessen the velocity of the wash water and distribute it evenly.

Porous Plate Bottoms

Porous plates can be used to make up an entire filter bottom. The plates are made of a ceramically bonded aluminum oxide and are supported over the bottom of the filter tank by long bolts, steel beams, concrete piers, or fiberglass supports. Typical

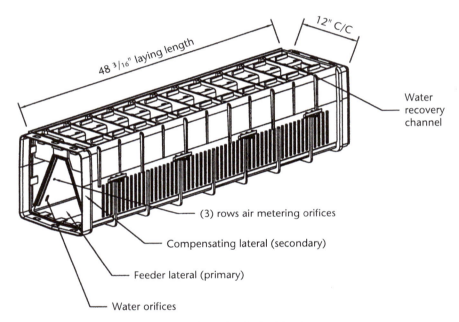

Figure 6-4 A section of a perforated underdrain block

Courtesy of F. B. Leopold Company, Inc.

Wheeler filter bottom

A patented filter underdrain system using small porcelain spheres of various sizes in conical depressions.

Figure 6-5 A Wheeler bottom

Courtesy of Honeywell, Industrial Measurement and Control.

installations are illustrated in Figures 6-6 and 6-7. Porous plates are not recommended for filtering hard water because a calcium buildup can form in the plates.

False-Floor Underdrain Systems

False-floor underdrain systems are constructed by placing a concrete or steel plate 1–2 ft (0.3–0.6 m) above the filter tank bottom, creating an underdrain plenum below the false floor (Figure 6-8). Various types of nozzles are available for installation in the floor. These nozzles have coarse openings if a gravel layer is to be used or very fine openings if a fine medium is to be placed directly adjacent to them (Figures 6-9 and 6-10).

Other Systems

Several other types of underdrain systems are offered by various manufacturers. Each has some advantages and disadvantages that should be carefully investigated before a new system is installed.

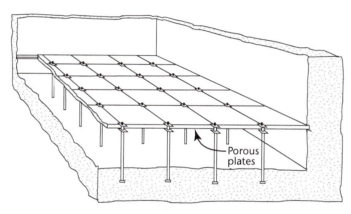
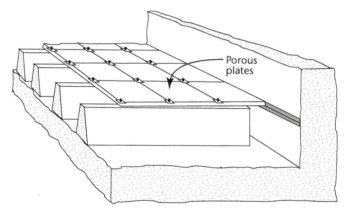

A. Supported by long bolts

B. Supported by concrete piers

Figure 6-6 Porous plate underdrain system
Courtesy of Christy Refractories Co.

Figure 6-7 Porous plate installation showing contour gaskets between plates
Courtesy of Christy Refractories Co.

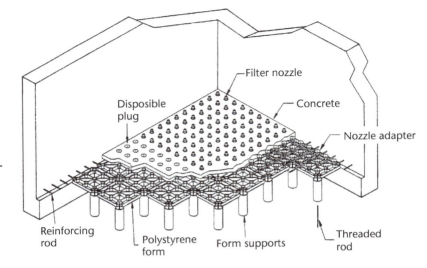

Figure 6-8 Typical false-floor filter underdrain system constructed of poured concrete

Courtesy of ONDEO Degremont Inc., Richmond, Virginia.

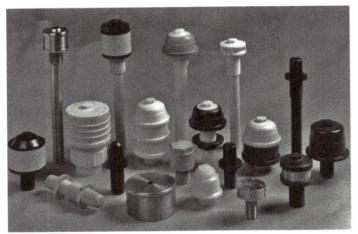

Figure 6-9 Typical distribution nozzles used on false-floor underdrain systems

Courtesy of Orthos Liquid Systems, Inc.

Figure 6-10 Typical distribution nozzle installed on a thin plate

Courtesy of Orthos Liquid Systems, Inc.

Wash-Water Troughs

All filters require a wash-water trough placed over the filter media to collect the backwash water during a wash and also carry it to waste. Proper placement of the troughs is very important. The media bed must be properly expanded during a wash, but no appreciable quantity of media should be wasted by carrying it over into the troughs. The troughs must also be placed so that equal flow is maintained in all parts of the filter bed and so that the wash will be uniform across the bed. The troughs are usually made of concrete, fiberglass, or steel.

Filter Bed Agitation

Most filters require some type of surface agitation to help release suspended matter trapped in the upper layers of filter media. Normally, this matter is not completely removed by backwashing alone. An additional system is added to provide adequate agitation of the top few inches (centimeters) of media.

Surface wash systems consist of nozzles attached to either a fixed-pipe or rotary-pipe arrangement installed just above the filter media (Figure 6-11).

wash-water trough

A trough placed above the filter media to collect the backwash water and carry it to the drainage system.

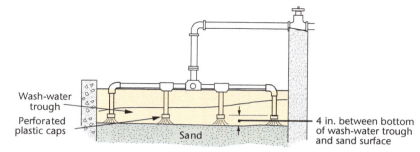

A. Fixed-nozzle

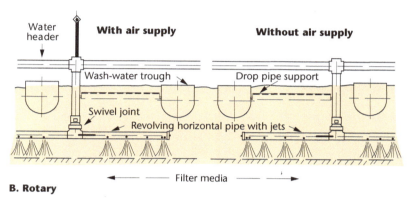

B. Rotary

Figure 6-11 Filter bed agitators using surface wash systems
Courtesy of the Roberts Filter Group.

Water or an air–water mixture is pumped through the nozzles, producing high-velocity jets.

Air-scouring systems are sometimes used in place of surface wash systems to ensure complete cleaning of the media. Air is usually applied through the back-wash piping or separate outlets located at the bottom of the filter. The air is applied at the beginning of the wash cycle, before the rising water reaches the lip of the wash-water troughs. Air scouring tends to disrupt the media considerably, so it must be followed by a fluidization wash period without air to allow mixed-media beds to restratify.

When air is used with mixed-media beds, provision must be made to ensure uniform distribution of the air across the filter bottom to prevent undue disruption of the media. The length and rate of air application must also be closely controlled; otherwise, undue amounts of media will be lost over the wash-water troughs. For these reasons, air scouring is not always the best choice for some operations. Air scouring works particularly well with monomedium beds because disruption of the bed is not a problem for them.

In a typical air-scour sequence, air is applied at the beginning of the backwash cycle *before* backwash water is introduced. Typically air alone will be applied for a period of 2–3 minutes. This portion of the air scour tends to scour quite thoroughly the top 12–18 in. (0.3-0.45 m) of filter media. Following the air-alone portion of the wash cycle, water is introduced at a rate of 5–7 gpm/ft² (3.4–4.8 mm/sec), which is called a *concurrent air/water wash*. During the concurrent backwash, pockets of air form within the media and collapse, causing an enhanced scouring condition called *collapse pulsing*. This process allows the entire column of filter media to be scoured and is essential for plants using deep-bed filtration to ensure that their filter beds are adequately maintained.

air scouring
The practice of admitting air through the underdrain system to ensure complete cleaning of media during filter backwash. Normally an alternative to using a surface wash system.

Filter Control Equipment

Rapid sand and multimedia filters usually require the following control equipment:

- A rate-of-flow controller
- A loss-of-нead indicator
- Online turbidimeters

Rate-of-Flow Controllers

Modulating control valves are usually installed on the effluent discharge of each filter. The controller maintains a fairly constant flow through the filter so that flow surges do not occur. Without a controller, the surges would force suspended particles through the filter. The controller typically consists of a flow-measuring device, a throttling valve, and a means to set the throttling valve automatically or manually to maintain a fixed flow rate. A typical rate-of-flow controller is illustrated in Figure 6-12.

Loss-of-Head Indicators

Indicators are required to monitor the status of resistance to flow in the filter as suspended matter builds up in the media. The head loss should be continuously measured to help determine when the filter should be backwashed. A filter run should not continue if terminal head loss is reached. In the simplest form of measurement, a clear plastic hose can be connected to the filter influent and another to the filter effluent. The difference in height between the water levels of the two hoses represents the head loss through the filter.

The head loss can also be measured by devices that use air pressure data (Figure 6-13) or by electronic equipment. Recorders are commonly installed on loss-of-head indicators for maintaining a record of operation. The recorder also provides a visible indication of the rate at which loss of head is increasing, so that a projection can be made of when it will be necessary to backwash a filter.

Online Turbidimeters

After a filter has been operating for a while, the suspended material will begin to break through the filter bed. This will cause turbidity to increase in the filtered water. If the filter effluent is continuously monitored, the filter can be backwashed

Figure 6-12 Rate-of-flow controller
Courtesy of Honeywell, Industrial Measurement and Control.

rate-of-flow controller

A control valve used to maintain a fairly constant flow through the filter.

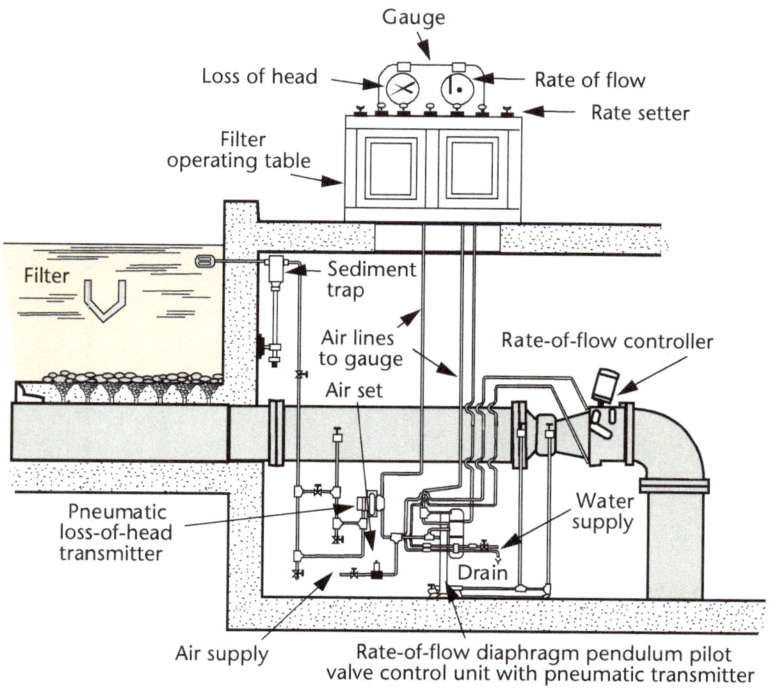

Figure 6-13 Configuration to measure head loss based on air pressure data

Courtesy of Hach Company, USA.

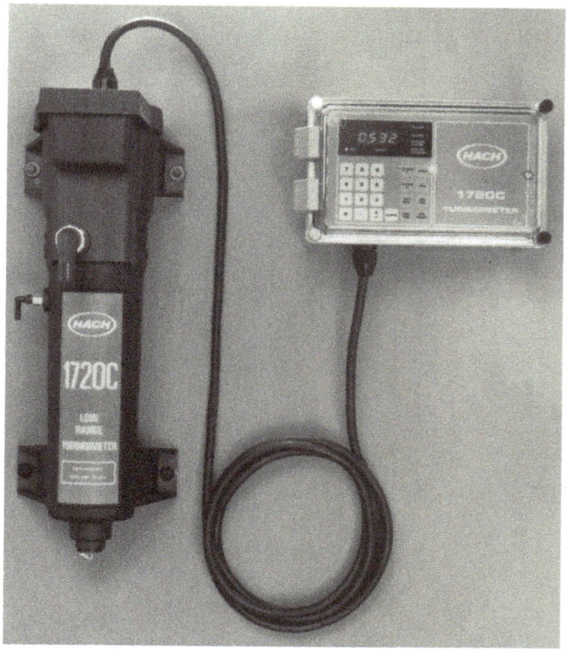

Figure 6-14 Online turbidimeter

Courtesy of Hach Company, USA.

as soon as the breakthrough starts, preventing excessive turbidity from passing into the distribution system. The Interim Enhanced Surface Water Treatment Rule (IESWTR) requires that all filters in a surface water treatment plant have online **turbidimeters**, and that they monitor the filter while it is in operation. A typical online turbidimeter is illustrated in Figure 6-14. If it is necessary to analyze the size and density of particles in the filter effluent at levels below the range of a turbidimeter, then a particle counter must be used.

WATCH THE VIDEO
Turbidity and Particle Counting: Turbidimeters
(www.awwa.org/wsovideoclips)

Operation of Gravity Filters

Filtration involves three procedures: filtering, backwashing, and filtering to waste. Figure 6-15 indicates the key valves used for filtration and their positions during the filtration procedures. The goal of filter operations is to produce low-turbidity water during an acceptable unit filter run volume (UFRV) period of operation. The UFRV is defined as the amount (gallons) of water produced per square foot of filter surface area. A properly designed and operated filter should be able to produce a minimum UFRV of 5,000.

Filter Operation Methods

The flow rate through filters can be controlled by a rate-of-flow controller, or it may proceed at a variable declining rate. Regardless of the control method used, as filtration progresses, suspended matter builds up within the filter bed and is at risk

turbidimeter

A device that measures turbidity, the amount of suspended particulate matter in the water.

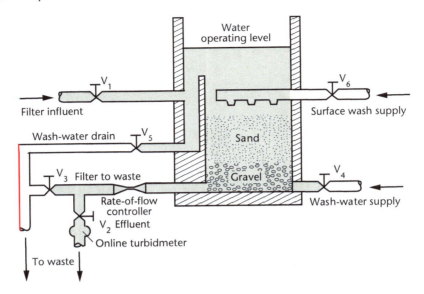

Valve position during filtration

Valve	Filtering	Backwashing	Filtering to waste
V_1 — Influent	Open	Closed	Open
V_2 — Effluent	Open	Closed	Closed
V_3 — Filter to waste	Closed	Closed	Open
V_4 — Wash-water supply	Closed	Open	Closed
V_5 — Wash-water drain	Closed	Open	Closed
V_6 — Surface wash supply	Closed	Open	Closed

Figure 6-15 Key valves used in filter operation

of passing particulates into the clearwell. Before that time, the filter must be cleaned by backwashing. Backwashing is typically done when the filter begins to show unacceptable particle count or turbidity trending, and before terminal head loss is met.

Controlled Rate of Flow

During filtering, water is applied to the filter to maintain a constant depth of 4–5 ft (1–1.5 m) over the media. The media bed is initially clean, and head loss is very low. The filtration rate can be kept at the desired level by using a rate-of-flow controller. The controller is also important in preventing harmful surges that can disturb the media and force floc through the filter.

The controller is used to maintain a constant desired filtration rate—usually 2 gpm/ft² (1.4 mm/sec) for rapid sand filters and 4–6 gpm/ft² (2.8–4 mm/sec) for high-rate filters. At the beginning of the filter run, the flow controller valve is almost closed, which produces the necessary head loss and maintains the desired flow rate.

As filtration continues and suspended material builds up in the bed, head loss increases. To compensate for this increase, the controller valve is gradually opened. When the valve is fully opened, the filter run must be ended because further head loss cannot be compensated for and the filter rate will drop sharply. Rate-of-flow controllers must be carefully maintained because a malfunctioning controller can damage the filter bed and degrade water quality by allowing sudden changes in the filtration rate.

Variable Declining-Rate Filtration

Another commonly used control method is variable declining-rate filtration. As shown in Figure 6-16, the filtration rate is not kept constant for this method. The rate for a particular filter starts high and gradually decreases as the filter gets dirty and head loss increases. One advantage of this system is that it does not require

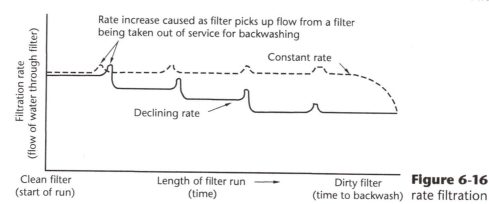

Figure 6-16 Declining-rate filtration

a rate-of-flow controller or constant attention by the operator. In addition, harmful rate changes cannot occur.

With this method, each filter accepts the proportion of total flow that its bed condition can handle. As a filter gets dirty, the flow through it decreases. Flow then redistributes to cleaner filters, and total plant capacity does not decrease. To prevent excessive flow rates from occurring in clean filters, a flow-restricting orifice plate is placed in the effluent line of each filter.

Backwashing Procedures

The Importance of Careful Backwashing

Backwashing is a critical step in the filtration process. Inadequate backwashing causes most of the operating problems associated with filtration. For a filter to operate efficiently, it must be cleaned efficiently before the next filter run begins. In addition, properly backwashed filters require far less maintenance.

Treated water is always used for backwashing so that the bed will not be contaminated. This treated water can be delivered from elevated storage tanks designed for this purpose or pumped in from the clearwell.

During filtration, the voids between the media grains fill with filtered material (**floc**). The grains also become coated with the floc and become very sticky, making the filter bed difficult to clean. To clean the filter bed, the media grains must be agitated violently and rubbed against each other to dislodge the sticky coating. Therefore, the backwash rate must be high enough to completely suspend the filter media in the water.

The backwash causes the filter bed to expand, as shown in Figure 6-17. The expansion, however, should not be so large that the media flow into the wash-water troughs. Because normal backwash rates are not sufficient to clean the media thoroughly, auxiliary scour (surface wash) or air-scour equipment is recommended to provide the extra agitation needed before backwashing begins. Auxiliary scouring is a must for high-rate filters because the filtered material penetrates much deeper into the bed.

A common mistake that operators make in filtration plants is backwashing the filters for too long or too vigorously, thus stripping the bed of its ripening. When this happens, the filter will need extra time to reripen and may produce water of inferior quality for a longer time than is tolerable. Maintaining proper backwashing conditions is especially important in a plant that cannot filter to waste. Filter inspection techniques encompassing solids retention analyses and spent filter backwash water analyses are good tools for preventing overly cleaned beds.

The properly cleaned filter bed is a result of a combination of applied science and art. A filter must be backwashed in similar fashion by all of the operators in the plant. Differences in backwash procedures from operator to operator can

floc

Collections of smaller particles (such as silt, organic matter, and microorganisms) that have come together (agglomerated) into larger, more settleable particles as a result of the coagulation–flocculation process.

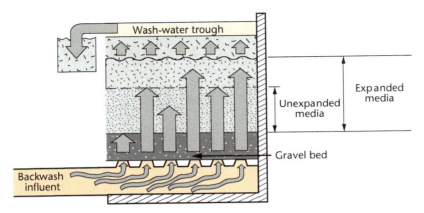

Figure 6-17 Filter bed expansion

lead to filter failure. For this reason, water treatment plants typically have written standard operating procedures for this important task.

Factors That Determine Backwash Frequency

Head loss, filter effluent turbidity (and particle counts if available), and UFRV must all be considered in deciding when a filter needs to be backwashed. Usually, a filter should be backwashed when any one or more of the following occur:

- Head loss is so high that the filter no longer produces water at the desired rate and may risk air binding. This condition is known as *terminal head loss*.
- Floc starts to break through the filter bed, causing the filter effluent turbidity (or particle counts) to increase. Visual trending is a good tool for detecting this problem.
- A filter has produced an acceptable UFRV and is still producing quality effluent. The UFRV is one that has been previously agreed on by the plant staff. Operators usually want to achieve a UFRV between 5,000 and 10,000. Higher UFRVs are dangerous and may lead to catastrophic floc breakthrough. Triggers for backwash, in order of importance, might be:
  - increasing effluent turbidity or particle counts,
  - head loss approaching terminality, and
  - UFRV approaching unreasonable limits.

A filter must be backwashed when the effluent turbidity trend graph shows that the trend begins to increase, as illustrated in Figure 6-18. The turbidity should never be allowed to increase to 0.3 ntu before backwashing. Most surface systems must now maintain a turbidity of less than 0.1 ntu for water entering the distribution system, according to the IESWTR. In fact, filtration tests have shown that microorganisms start passing through the filter rapidly once breakthrough begins, even though the turbidity may be well below 0.3 ntu.

Length of Filter Runs

If high-quality water is applied to the filter, then filter runs, based on head loss or effluent turbidity, can be very long. The length of time that a filter operates is variable because the UFRV dictates the length of run. Since the filter may operate at different flow rates during its run, operating it for the same length of time doesn't make sense. A 100 ft² filter operating at 3 gpm/ft² will produce 300 gpm. If it is operated for an entire day, it will produce 0.432 million gallons. That same filter

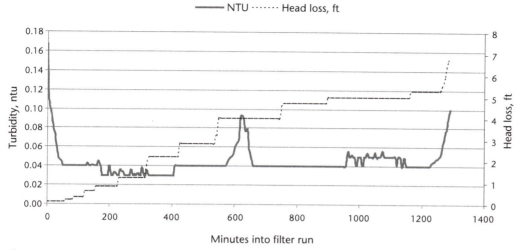

Figure 6-18 Typical filter run showing head loss and turbidity output

could produce the same amount of water in a shorter time if it were operated more aggressively. If the water quality from the filters is meeting the stated goals, then time is not a factor. In some groundwater softening plants, filter runs may last days. Surface water treatment plants, however, risk floc breakthrough if the UFRV is too great. Therefore, a maximum UFRV is advised to minimize the risk.

However, short filter runs decrease finished-water production because of the time the filters are out of production and the increased quantity of backwash water needed. In general, the amount of backwash water used should not exceed 4 percent of the amount of water treated for rapid sand filters, and 6 percent for dual-media and multimedia filters. Filter runs in most treatment plants will produce a minimum UFRV of 5,000.

Backwash Sequence

Although no two water treatment plants can use the same procedures, a typical backwash sequence might be as follows:

1. The water in the filter is drained down to a level about 6 in. (150 mm) above the media.

2. The surface washers are turned on and allowed to operate for 1–2 minutes. This allows the high-velocity water jets to break up any surface layers of filtered material. Alternatively, air scouring might be used here.

3. The backwash valve is partly opened to allow the bed to expand to just above the level of the washers. This provides violent scrubbing of the top portion of the media, which has the greatest accumulation of filtered material. Intense scrubbing is particularly important for rapid sand filters because the top 8 in. (200 mm) of media removes most of the suspended solids.

4. After a few minutes, the backwash valve is fully opened to allow a filter bed expansion of 20–30 percent. The actual amount of expansion that is best for a filter depends on how much agitation is needed to suspend the coarsest grains of media in the bed. With multimedia filters, the bed must be expanded so that the surface washers can scrub the area between the anthracite and sand layers (known as the *interface*), where most of the filtered material has penetrated. A backwash rate of 15–20 gpm/ft^2 (10–14 mm/sec) is usually sufficient to provide the expansion needed.

5. The surface washers are usually turned off about 1 minute before the back-wash flow is stopped. This allows the bed to restratify into layers, which is particularly important for multimedia filters.

6. The expanded bed is washed for 5–15 minutes, depending on how dirty the filter is. The clarity of the wash water as it passes into the wash-water troughs can be used as an indicator of when to stop washing. AWWA suggests a turbidity of 10 ntu.

If surface wash equipment is not available, a two-stage wash should be used. The initial wash velocity should be just enough to expand the top portion of the bed slightly, usually about 10 gpm/ft² (6.8 mm/sec). Although not as effective as surface washing, this method will provide some scrubbing action to clean the surface media grains. After the surface has been cleaned, the full backwash rate is applied.

Turning on the backwash too quickly can severely damage the underdrain system, as well as heave the gravel bed and media to a point where they will not restratify. The time from starting backwash flow to reaching the desired backwash flow rate should be 30–40 seconds. To prevent accidents, the backwash valve controls should be set to open slowly.

 WATCH THE VIDEO
Filtration: Filter Backwash (www.awwa.org/wsovideoclips)

Disposal of Backwash Water

To avoid water pollution, backwash water must not be returned directly to streams or lakes. The water is usually routed to a lagoon or basin for settling. After settling, the water may be recycled to the treatment plant, as shown in Figure 6-19. The settled solids are combined with the sludge from the sedimentation basins and disposed of as discussed in Chapter 5.

Because backwash water does not usually contain a very high concentration of suspended solids, some treatment plant operators have found that they can send the wash water directly to the intake well to be blended with the incoming raw water.

The Cleveland Division of Water recycles backwash water through simple sedimentation and then blends it with raw source water. They report that the blend is kept to less than 5 percent of the total instantaneous flow. Testing has shown that there is less than a 10 percent increase in blended water total organic carbon, iron and manganese, and turbidity during recycle events.

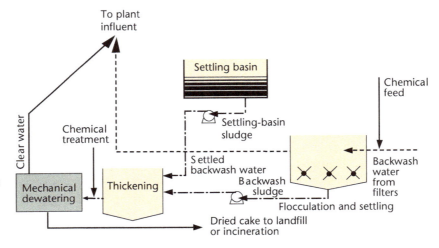

Figure 6-19 Disposal of backwash water

Courtesy of US Filter-Microfloc® Products.

Optimizing Filter Effluent Quality

Filtering to Waste

Once backwashing is complete, the water applied to the filter should be filtered to waste until the turbidity of the effluent drops to an acceptable level. As shown in Figure 6-18, the effluent turbidity remains high for a certain time period at the beginning of the filter run. Depending on the type of filter and treatment processes used before filtering, this period may be from 2 to 20 minutes. This is probably due, in part, to filtered material that remains in the bed after backwashing. In addition, the initial high filtration rates continue to wash fine material through the filter until the media grains become sticky and more effectively adsorb the suspended material.

If filtering to waste cannot be done, then filter resting can be used. Also, a slower filtration rate can be used for the first 15–30 minutes of each filter run to minimize breakthrough of filtered material. The material that passes through the filter at this time can include large quantities of microorganisms, so it is extremely important that breakthrough be prevented.

Resting Filters

After filters are backwashed, they should be rested if not needed for service. Allowing a filter to remain offline for even a half hour before putting it into service has been shown to produce a shorter ripening period in some water plants. Operators should not allow filters to rest for more than a few days because that would allow the water in the filter to stagnate and perhaps allow for unwanted microbiological growth.

Where water treatment plants have multiple filters and do not treat water at a rate that is constantly near treatment capacity, operators can easily test this method to see if positive results can be had by filter resting. Comparisons of initial turbidity output of nonrested versus rested filters should help to determine if this filter ripening technique should be implemented.

Filter Aids

As water passes through the filter, the floc can be torn apart, resulting in very small particles that can penetrate the filter. In such instances, turbidity breakthrough occurs well before the terminal head loss is reached, resulting in short filter runs and greater use of backwash water. This scenario is more likely with high-rate filters because their loadings, in terms of suspended material and the filter rate, are much higher than with rapid sand filters.

To help solve this problem, polymers (also called *polyelectrolytes*) can be used as filter aids. Polymers are water-soluble, organic compounds that come in either dry or liquid form. The dry form is more difficult to use because it does not dissolve easily. Mixing and feeding equipment, as illustrated in Figure 6-20, is required.

Polymers have high molecular weights and may be nonionic, cationic, or anionic. They are also used as primary coagulants or coagulant aids. Nonionic or slightly anionic polymers are normally used as filter aids.

When used as a filter aid, the polymer strengthens the bonds between the filtered particles and coats the media grains to improve adsorption. The floc then holds together better, adheres to the media better, and resists the shearing forces exerted by the water flowing through the filter. For best results, the polymer should be added to the water just upstream of the filters. In order to select the

polymer
High-molecular weight, synthetic organic compound that is used to aid in the coagulation and flocculation process.

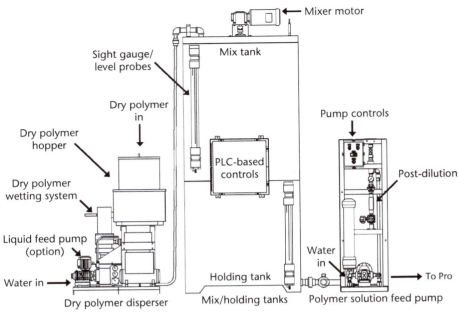

Figure 6-20 Filter aid feed system for use with dry polymers

Courtesy of US Filter-Microfloc® Products.

proper polymer for a particular treatment plant, several kinds should be tested and their performance and relative cost compared.

The required dosage is normally less than 0.1 mg/L, but the proper dosage must be determined through actual use. Continuous turbidity monitoring of the filtered water is essential to ensure that the proper dose is being applied. Figure 6-21A illustrates the effect of an inadequate polymer dosage: turbidity breakthrough occurs well before the terminal head loss is reached. Figure 6-21B illustrates the effect of an excessive polymer dosage. In this case, too much polymer makes the filtered material stick in the upper few inches (centimeters) of the filtered bed, creating a rapid head loss. The optimal dosage causes maximum head loss to be reached just before turbidity breakthrough occurs, as illustrated in Figure 6-21C.

Monitoring Filter Operation

Head Loss Monitors

Head loss must be monitored and recorded continuously to provide the operator with the most important information concerning filter operation. If head loss is

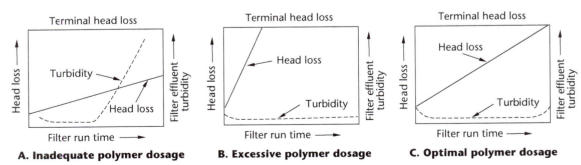

A. Inadequate polymer dosage   **B. Excessive polymer dosage**   **C. Optimal polymer dosage**

Figure 6-21 Effect of polymer dosage on turbidity and head loss

Courtesy of Hach Company, USA.

allowed to become excessive, the operating efficiency of the filter will decrease, effluent turbidity may increase, and turbidity will be driven deeper than desired into the bed, where it may be hard to remove.

Effluent Turbidity Monitors

The turbidity of filter effluent should be monitored with an online turbidimeter (Figure 6-14) to meet the requirements of the IESWTR (Figure 6-22). If continuous monitors are used, they should be checked periodically for accuracy against a laboratory unit. A sudden increase in effluent turbidity indicates filter breakthrough.

Filterability Tests and Zeta Potential Tests

Filterability and zeta potential tests can be used to monitor the coagulation process to ensure the optimal coagulant dosage. If the direct filtration process is being used, one of these types of tests is essential to properly control the chemical dosage in the short time span between application of the chemicals and the point when the water reaches the filters.

Visual Inspection

Filter beds should regularly be visually inspected by observing the backwash and the filter's surface. Media loss is usually apparent from the undue amounts of media grains flowing into the backwash troughs. The top of the media surface should also be measured periodically to see if media grains are being lost. A serious decline in the media level may indicate an excessive wash-water or air-scour rate.

A bed in good condition and with even backwash distribution should appear very uniform, with the media moving laterally on the surface. Violent upswelling

Figure 6-22 Nephelometric turbidimeter
Courtesy of Hach Company, USA.

zeta potential

A measurement (in millivolts) of the particle charge strength surrounding colloidal solids. The more negative the number, the stronger the particle charge and the repelling force between particles.

or boils of water indicate problems. If some areas do not appear to clear up as quickly as others, an uneven distribution of backwash flow is indicated. If backwash is uneven, the problem will not fix itself and will probably get worse.

When the filter is drained, its surface should appear smooth. Cracks, mudballs, or ridges also indicate problems with backwashing. Backwash problems could mean that some of the underdrain system is blocked or broken, depending on the type of system. This indicates that the media will have to be removed for repair; at the same time, the media will probably have to be regraded or new media purchased for replacement. This is a difficult and expensive task, and expert advice should be sought before these repairs are attempted.

Mudball Volume Determination

The volume of mudballs in the filter sand can be determined easily without expensive equipment. See Chapter 2 for a description of the procedure.

Filter Operating Problems

Most filtration problems occur in the following major areas:

1. Chemical treatment before the filter
2. Control of filter flow rate
3. Backwashing the filter

If these three procedures are not performed effectively, the quality of filtered water will suffer and additional maintenance problems will occur.

Chemical Treatment Before the Filter

The importance of proper coagulation and flocculation and the advantages of using a filter aid have already been discussed. Successful adjustments to these processes are possible only if the filtration process is closely monitored. Because many raw-water characteristics, such as turbidity and temperature, are not constant, dosage changes during filtration are usually necessary. Consequently, instruments that continuously record turbidity, head loss, and flow rate are very important.

If short filter runs occur because of turbidity breakthrough, perhaps more coagulant, better mixing, or less filter aid is needed. If short runs are caused by a rapid buildup of head loss, perhaps less coagulant or less filter aid is required. It is the operator's job to recognize these types of problems and choose the proper solution.

Control of Filter Flow Rate

Increases or rapid fluctuations in the flow rate can force previously deposited filtered material through the media. The dirtier the filter, the more problems that rate fluctuations can cause. These fluctuations may be caused by an increase in total plant flow, a malfunctioning rate-of-flow controller, a flow increase when a filter is taken out of service for backwashing, or operator error.

Obviously, as demand increases, filter rates may have to be increased to meet the demand, particularly if there is inadequate treated-water storage. If an increase is necessary, it should be made gradually over a 10-minute period to minimize the impact on the filters. Filter aids can also reduce the harmful effects of a rate increase.

When a filter is backwashed, the filters remaining in operation must pick up the nonoperating filter's load. This can create an abrupt surge through the filters, particularly if rate-of-flow controllers are used. This problem can be avoided if a clean filter is kept in reserve. When a filter is taken out of service for backwashing, the clean filter is placed in service to pick up the extra load. This will probably not work for plants with fewer than four filters, because all filters are typically needed to keep up with demand.

In plants that operate for only part of a day, the filtered material remaining on the filters can be shaken loose by the momentary surge that occurs when filtration is again started. These problems can be avoided if the filters are backwashed before they are placed into service.

If rate-of-flow controllers are used, they should be well maintained so that they function smoothly. Malfunctioning controllers will "hunt" for the proper valve position, causing harmful rate fluctuations.

Results of Ineffective Backwashing

Effective backwashing is essential for the consistent production of high-quality water. Ineffective backwashing can cause many problems. Most filtration problems can be traced back to lack of standard operating procedures or poor operator training.

Mudball Formation   During filtration, grains of filter media become covered with sticky floc material. Unless backwashing removes this material, the grains clump together and form mudballs. As the mudballs become larger, they can sink into the filter bed during backwashing and clog those areas where they settle. These areas then become inactive, causing higher-than-optimal filtration rates in the remaining active areas and unequal distribution of backwash water (Figure 6-23).

Additional problems, such as cracking and separation of the media from the filter walls, may also result. Mudballs are usually seen on the surface of the filter after backwashing, particularly if the problem is severe, as shown in Figure 6-24.

A periodic check for mudballs should be made. Mudballs can be prevented by adequate backwash flow rates and surface agitation. Filter agitation is essential for dual-media and multimedia filters because mudballs can form deep within the bed.

Filter Bed Shrinkage   Bed shrinkage, or compaction, can result from ineffective backwashing. Clean media grains rest directly against each other, with little compaction even at terminal head loss. However, dirty media grains are kept apart by the layers of soft filtered material. As the head loss increases, the bed compresses

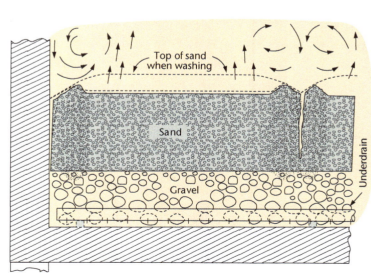

Figure 6-23 Clogged filter bed

Figure 6-24 Mudballs on the filter surface

and shrinks, resulting in cracks (Figure 6-25) and separation of the media from the filter walls. The water then passes rapidly through the cracks and receives little or no filtration.

Gravel Displacement If the backwash valve is opened too quickly, the supporting gravel bed can be washed into the overlying filter media. Gravel can also wash into the filter media over a period of time if a section of the underdrain system is clogged, causing unequal distribution of the backwash flow. Eventually, the increased velocities displace the gravel and create a sand boil, as shown in Figure 6-26. When this occurs, there is little or no media over the gravel at the boil, so water passing

Figure 6-25 Cracks in the filter bed

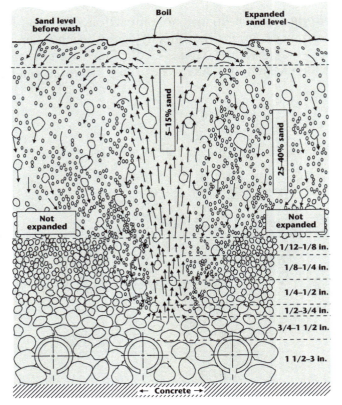

Figure 6-26 A sand boil in a filter bed

sand boil

The violent washing action in a filter caused by uneven distribution of backwash water.

through the filter at that point receives very little filtration. In addition, media may wash down at this point and start flowing into the underdrain system.

Because some gravel movement always occurs, filters should be probed at least once a year to locate the gravel bed. This can be done with a metal rod 1/4 in. (6 mm) in diameter while the filter is out of service. Probing the bed on a grid system and keeping track of the depths at which the gravel is located will reveal whether serious displacement has occurred.

If displacement has occurred, the media must be removed and the gravel re-graded or replaced. Future displacement can be minimized by placing a 3-in. (76-mm) layer of coarse garnet between the gravel and the media and by not using excessive backwash rates.

Additional Operating Problems

Air binding and media loss are common problems for gravity filters.

Air Binding   If a filter is operated so that the pressure in the bed is less than atmospheric (a condition known as *negative head*), then the air dissolved in the water will come out of solution and form bubbles within the filter bed. This process, known as air binding, creates resistance to flow through the filter and leads to very short filter runs.

Upon backwashing, the release of the trapped air causes violent agitation, which can cause a loss of media into the wash-water troughs. Negative head typically occurs in filters with less than 5 ft (1.5 m) of water depth over the unexpanded filter bed. If a filter has been designed for a shallower water depth, filter runs may have to be terminated at a head loss of about 4.5 ft (1.4 m) to prevent negative head.

Air binding can also occur when cold water that is supersaturated with air warms up, typically in the spring. Unfortunately, not much can be done in this case except to keep the water at maximum levels over the filter bed and to backwash frequently.

Media Loss   Some media are always lost during backwashing. This is especially true if surface washers are used. However, if considerable quantities of media are being lost, backwashing procedures should be examined. Because the bed is usually completely fluidized at 20 percent expansion, further expansion may not be needed. Turning the surface washers off about 1–2 minutes before the end of the main backwash will also help reduce media loss. If serious loss continues, the only solution is to raise the level of the wash-water troughs. Attempting to replace lost media with several inches of new media can be detrimental to the filter if the new media are not chosen carefully. What is normally lost in the backwash procedures are smaller filter media particles. It is nearly impossible to replace the lost media with an equal amount of the same-sized media. Typically, a composite of smaller- and larger-diameter filter material is added. The addition of the extra larger media can alter the filter configuration to the point that the filter will not produce low-turbidity water or operate at the same UFRV or head losses.

Pressure Filtration

Pressure filters have been used extensively to filter water for swimming pools, and more small water systems are now using them because installation and operating costs are low.

air binding
A condition that occurs in filters when air comes out of solution as a result of pressure decreases and temperature increases. The air clogs the voids between the media grains, which causes increased head loss through the filter and shorter filter runs.

Sand or Mixed-Media Pressure Filters

The operating principle of pressure filters is similar to that of gravity filters, but the filtration process takes place in a cylindrical steel tank. Another difference is that water entering the filter is under pressure, so it is forced through the filter. As with gravity filters, sand or a combination of media is used. Filtration rates are about the same as for gravity filters.

If these filters will not be backwashed very often, the valves may be manually operated. Systems with automatic backwashing that use mechanically operated valves are also available. Because the water is under pressure, air binding will not occur.

The principal disadvantage of pressure filters is that the filter bed cannot be observed during operation or backwashing. The operator consequently cannot observe the backwash process or the condition of the filter bed. Horizontal and vertical pressure filter tanks are shown in Figures 6-27 and 6-28.

Diatomaceous Earth Filters

Diatomaceous earth is the skeletal remains of microscopic aquatic plants called *diatoms*. Prehistoric deposits of diatoms are mined and processed to produce the diatomaceous earth medium. This material is mostly silica and inert so it is suitable to use to filter water. Diatomaceous earth (DE) filters were developed by the US military during World War II to remove the organisms causing amoebic dysentery. Today, they are used extensively to filter swimming pool water and, to some extent, by small public water systems.

The process uses DE to strain particulates from water and normally does not use coagulant chemicals. A thin layer of DE coats filter leaves or septum, which is a plastic or wire mesh covering a hollow collection channel. The DE coating (precoat) is formed by a recirculating slurry of DE. Only previously filtered water should be used for applying the precoat. Figure 6-29 shows the components of a DE filter. Figure 6-30 shows the basic operation of a DE filter, and Figure 6-31 shows a precoat tank used in the process.

The precoat normally has a thickness of about ⅛ in. (3 mm). Untreated water is applied to the precoat filter with a small amount of DE added; this water is called *body feed*. The body feed must be added continuously while the filter is in operation to prevent possible cracking and clogging of the precoat.

> **diatomaceous earth filter**
>
> A pressure filter using a medium made from diatoms. The water is forced through the diatomaceous earth by pumping.

Figure 6-27 Horizontal pressure filter tanks
Courtesy of ONDEO Degremont.

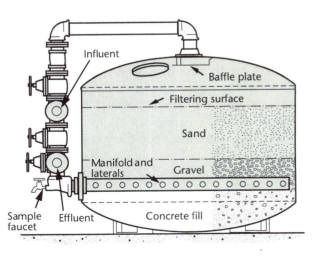

Figure 6-28 Vertical sand pressure filter tank
Courtesy of Celite Corporation.

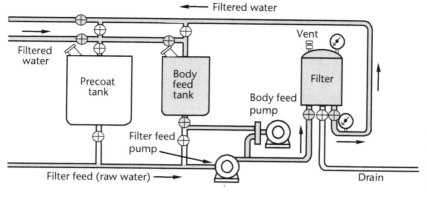

Figure 6-29
Components of a
diatomaceous earth filter
Courtesy of Celite Corporation.

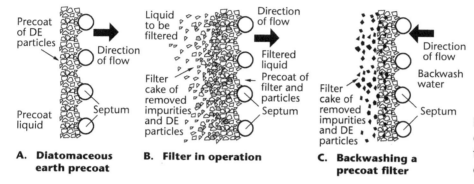

A. Diatomaceous earth precoat

B. Filter in operation

C. Backwashing a precoat filter

Figure 6-30 Operation of a diatomaceous earth filter
Courtesy of Celite Corporation.

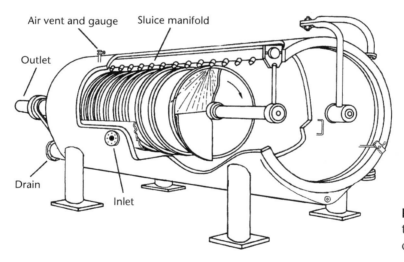

Figure 6-31 Precoat tank of rating leaf type
Courtesy of Celite Corporation.

Particulates in the water are removed on the filter surface, and this causes pressure to build (head loss). When the terminal head loss is reached, the filter is taken out of service for cleaning. When the filter run is completed, the filter cake is washed from the septum by a reversal of the water flow, and it is usually discharged to a lagoon for disposal. The process is then repeated starting with the filter precoat with fresh DE.

DE filtration is used for raw water with a maximum turbidity of 10 ntu. The filter loading rate can be between 0.5 and 2.0 gpm/ft² (0.3–1.4 mm/sec). DE filtration is very effective for removing *Giardia* and *Cryptosporidium*. Dissolved substances and color are not effectively removed by DE filtration.

DE filters may be installed in either horizontal or vertical tanks and can be operated as either pressure or vacuum filters. A vacuum filter has a pump on the filter discharge that pulls the water through the filter. Regardless of the design, the components and the operation are similar.

The use of DE filters for potable water treatment has been limited because of the difficulty in maintaining an effective filter cake at all times. If there is a defect in the cake at any time, microorganisms could easily pass into the water system. If DE filters are to be used for treating surface water, then the filtered water must be monitored continuously.

Regulations

The Surface Water Treatment Rule (SWTR), enacted by the US Environmental Protection Agency (USEPA) in June 1989, affected the operation of essentially every public water system in the United States that uses surface water as a source. It also imposed new requirements on systems using groundwater that might become contaminated by surface water. These systems are described as having "groundwater under the direct influence of surface water," or GWUDI.

Amendments to the Safe Drinking Water Act (SDWA) in 1996 require USEPA to develop rules that balance risks. To strengthen protection against microbial contaminants, especially *Cryptosporidium parvum*, and at the same time reduce potential health risks from disinfection by-products, USEPA promulgated the Enhanced Surface Water Treatment Rule (ESWTR), the Stage 1 Disinfectants/Disinfection By-products Rule, and the Filter Backwash Rule (FBR).

These rules have significant impact on the way that water treatment plants filter water. The ESWTR includes the following requirements:

- A maximum contaminant level goal of zero for *Cryptosporidium*
- 2-log *Cryptosporidium* removal requirements for systems that filter
- Strengthened combined filter effluent turbidity performance standards
- Individual filter turbidity monitoring provisions

The FBR requires the following:

- Recycle streams (e.g., spent filter backwash water) must be returned to the point of primary coagulation in the system.
- Selected systems must perform self-assessment studies.

Safety Precautions

Guardrails should be installed to prevent falls into the filters. Operators should work in pairs if at all possible, and life rings or poles should be located near the filter area to allow quick rescue in case of an accident.

Pipe galleries are ideal places for accidents. They should be well ventilated, well lit, and provided with good drainage to reduce the possibility of slipping. Any ramps or stairs in this area should be equipped with nonskid treads. Painting pipes also makes them easier to see and maintain.

Polymers used for filter aids are particularly slippery, so spills should be cleaned up immediately. If the containers or tanks are leaking, they should be repaired or discarded.

Record Keeping

Good record keeping can identify problems and indicate the proper steps to be followed. The type of filtration records maintained will depend on the treatment process being used, but all records should include the following:

- Rate of flow, in million gallons per day (or megaliters per day)
- Head loss, in feet (or meters)
- Length of filter runs, in hours
- UFRV, in gallons per square foot
- Backwash water rate, in gallons per minute (or liters per minute)
- Volume of wash water used, in gallons (or liters)
- Volume of water filtered, in gallons (or liters)
- Length of backwash, in minutes
- Length of surface wash, in minutes
- Filter aid dosage, in grains per gallon (milligrams per liter)

Figures 6-32 and 6-33 are examples of typical filtration record-keeping forms.

Daily Filter Record
Filter Plant No. 2

Filter No. _____   Prev. Run _____ Hours   Date: ____ / ____ / 20___

Oper.	Time	Rate of Flow, mgd	Loss of Head, ft	Surface Wash, min.	Rate Water Wash, mgd	Water Wash, min.	Polymer Feed On	Polymer Feed Off	Wash Water Used	Water Filtered, mil gal
	12 mid									
	1 a.m.									
	2									
	3									
	4									
	5									
	6									
	7									
	8									
	9									
	10									
	11									
	Noon									
	1									
	2									
	3									
	4									
	5									
	6									
	7									
	8									
	9									
	10									
	11									
	12 mid									

Figure 6-32 Typical daily filter record

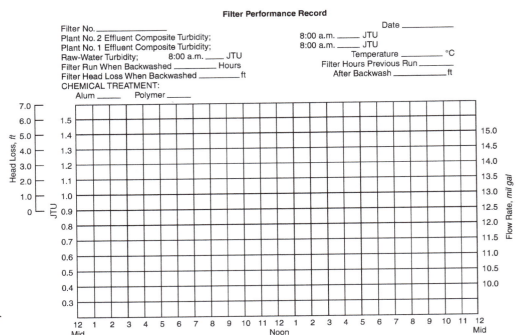

Figure 6-33 Typical filter performance record

Study Questions

1. If filter run times between backwashes are long (e.g., 1 week) because high-quality (low-turbidity) water is being applied to the filters, which problem could still arise?
   a. Mudball formation
   b. Air binding and formation of mudballs
   c. Extended backwashing due to media becoming too compacted
   d. Floc breakthrough

2. Gravel displacement in a filter bed from backwash rates with too high of a velocity could eventually cause
   a. compaction of the filter media.
   b. loss of media into the backwash troughs.
   c. a sand boil.
   d. bed shrinkage.

3. Which conventional treatment step is eliminated by direct filtration?
   a. Oxidation
   b. Aeration
   c. Flocculation
   d. Sedimentation

4. Which of the following is the layer of solids and biological growth that forms on the top of a slow sand filter?
   a. Biosolids film
   b. Bio-carbonated scale layer
   c. Schmutzdecke
   d. Saprophytic layer

5. Filter tanks are generally
   a. spherical and constructed of steel.
   b. rectangular and constructed of concrete.
   c. rectangular and constructed of steel.
   d. cylindrical and constructed of concrete.

6. In a typical air-scour sequence, air is applied _____ backwash water is introduced.
   a. before
   b. while
   c. after
   d. both before and after

7. What is the filtration flow rate through a high-rate mixed media filter?

8. Which system is the first and oldest type of underdrain system, having a central manifold pipe with perforated lateral pipes on each side?

9. Which regulation requires that all filters in a surface water treatment plant have online turbidimeters, and that they monitor the filter while it is in operation?

10. Prehistoric deposits of the skeletal remains of microscopic aquatic plants are the source of what useful filtering element?

Chapter 7
Chlorine Disinfection

Gas Chlorination Facilities

Chlorine gas, Cl_2, is about 2.5 times as dense as air. It has a pungent, noxious odor and a greenish-yellow color, although it is visible only at a very high concentration. The gas is very irritating to the eyes, nasal passages, and respiratory tract, and it can kill a person in a few breaths at concentrations as low as 0.1 percent (1,000 ppm) by volume. Its odor can be detected at concentrations above 0.3 ppm.

Chlorine liquid is created by compressing chlorine gas. The liquid, which is about 99.5 percent pure chlorine, is amber in color and about 1.5 times as dense as water. It can be purchased in cylinders, containers, tank trucks, and railroad cars (Figures 7-1 through 7-4).

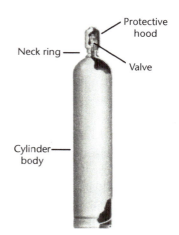

Figure 7-1 Chlorine cylinder
Courtesy of the Chlorine Institute.

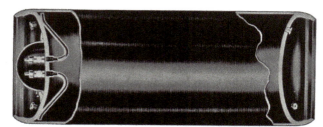

Figure 7-2 Chlorine ton container
Courtesy of the Chlorine Institute.

Figure 7-3 Chlorine ton container truck
Courtesy of the Chlorine Institute.

Figure 7-4 Chlorine tank car
Courtesy of the Chlorine Institute.

Liquid chlorine changes easily to a gas at room temperatures and pressures. One volume of liquid chlorine will expand to about 460 volumes of gas. Dry chlorine gas will not corrode steel or other metals, but it is extremely corrosive to most metals in the presence of moisture.

Chlorine will not burn. But, like oxygen, *it will support combustion*; that is, it takes the place of oxygen in the burning of combustible materials. Chlorine is not explosive, but it will react violently with greases, turpentine, ammonia, hydrocarbons, metal filings, and other flammable materials. Chlorine will not conduct electricity, but the gas can be very corrosive to exposed electrical equipment. Because of the inherent hazards involved, chlorine requires special care in storage and handling.

Handling and Storing Chlorine Gas

Safe handling and storage of chlorine are vitally important to the operator and to the communities immediately surrounding a treatment plant. An error or accident in chlorine handling can cause serious injuries or even fatalities.

The containers commonly used to supply chlorine in smaller water treatment plants are 150-lb (68-kg) cylinders. Larger plants find it more economical to use ton containers. Some very large plants are equipped to draw chlorine directly from tank cars.

The decision of whether to use cylinders or ton containers should be based on cost and capacity. The cost per pound (kilogram) of chlorine in cylinders is usually substantially more than that of chlorine in ton containers. If a plant's needs for chlorine are lower than 50 lb/d (23 kg/d), cylinders should usually be selected. For systems that use large amounts, ton containers will probably be more economical.

Cylinders

Chlorine cylinders hold 150 lb (68 kg) of chlorine and have a total filled weight of 250–285 lb (110–130 kg). They are about 10.5 in. (270 mm) in diameter and 56 in. (1.42 m) high. As illustrated in Figure 7-1, each cylinder is equipped with a hood that protects the cylinder valve from damage during shipping and handling. The hood should be properly screwed in place whenever a cylinder is handled and should be removed only during use.

Cylinders are usually delivered by truck. Each cylinder should be unloaded to a dock at truck-bed height if possible. If a hydraulic tailgate is used, the cylinders should be secure to keep them from falling. Cylinders must never be dropped, including "empty" cylinders, which actually still contain some chlorine.

The easiest and safest way to move cylinders in the plant is with a hand truck. As shown in Figure 7-5, the hand truck should be equipped with a restraining chain that fastens snugly around the cylinder about two-thirds of the way up. Slings should never be used to lift cylinders, and a cylinder should never be lifted by the protective hood because the hood is not designed to support the weight of the cylinder. Cylinders should not be rolled to move them about a plant. Tipping the cylinders over and standing them up can lead to employee injury. In addition, the rolled cylinders might strike something that could break off the valve.

Cylinders can be stored indoors or outdoors. If cylinders are stored indoors, the building should be fire resistant, have multiple exits with outward-opening doors, and be adequately ventilated. Outdoor storage areas must be fenced and protected from direct sunlight, and they should be protected from vehicles or falling objects that might strike the cylinders. If standing water accumulates in an outdoor storage area, the cylinders should be stored on elevated racks. Avoiding contact with water will help minimize cylinder corrosion.

Some operators find it convenient to hang "full" or "empty" identification tags on cylinders in storage, so that the status of the chlorine inventory can be

chlorine cylinder

A container that holds 150 lb (68 kg) of chlorine and has a total filled weight of 250–285 lb (110–130 kg).

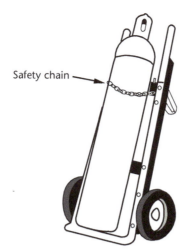

Figure 7-5 Chlorine cylinder
Courtesy of the Chlorine Institute.

quickly determined. Other plants maintain separate storage areas for full and empty cylinders, but all cylinders, full or empty, should receive the same high level of care. In addition, protective hoods should be placed on empty and full cylinders in storage. Even when a cylinder no longer has sufficient chlorine for plant use, a small amount of gas remains and could escape if the cylinder or valve were damaged. Both full and empty cylinders should always be stored upright and secured with a chain to prevent them from tipping over.

Ton Containers

The **ton container** is a reusable, welded tank that holds 2,000 lb (910 kg) of chlorine. Containers weigh about 3,700 lb (1,700 kg) when full and are generally 30 in. (0.76 m) in diameter and 80 in. (2.03 m) long. As shown in Figure 7-2, the ends are concave. The containers are crimped around the perimeter of the ends, forming good gripping edges for the hoists used to lift and move them. The ton container is designed to rest horizontally both in shipping and in use. It is equipped with two valves that provide the option of withdrawing either liquid or gaseous chlorine. The upper valve will draw gas, and the lower valve will draw liquid.

Handling the heavy containers is, by necessity, far more mechanized than handling cylinders. Containers are loaded or unloaded by a lifting beam in combination with a manual or motor-operated hoist mounted on a monorail that has a capacity of at least 2 tons (1,815 kg) (Figure 7-6). To prevent accidental rolling, containers are stored on trunnions, as illustrated in Figure 7-7. The trunnions allow the container to be rotated so that it can be positioned correctly for connection to the chlorine supply line.

Ton containers can be stored indoors or outdoors and require the same precautions as chlorine cylinders. The bowl-shaped hood that covers the two valve assemblies when the tank is delivered should be replaced each time the container is handled, as well as right after it has been emptied.

The chlorine storage area should provide space for a 30- to 60-day supply of chlorine. Some systems feed chlorine directly from this storage area. When ton containers are used, the chlorination feed equipment is usually housed in a separate room (Figure 7-8).

WATCH THE VIDEO
Chlorine Safety: Overview (www.awwa.org/wsovideoclips)

> **ton container**
> A reusable, welded tank that holds 2,000 lb (910 kg) of chlorine. Containers weigh about 3,700 lb (1,700 kg) when full and are generally 30 in. (0.76 m) in diameter and 80 in. (2.03 m) long.

Figure 7-6 Lifting beam with motorized hoist for ton containers

Figure 7-7 Ton containers stored on trunnions

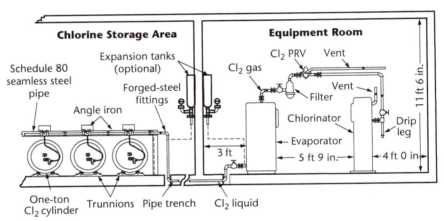

Figure 7-8 Chlorination feed equipment located in a separate room

Source: *Handbook of Chlorination and Alternative Disinfectants.* 4th ed. by Geo. Clifford White, copyright © 1998. Reprinted by permission of John Wiley & Sons, Inc.

Feeding Chlorine Gas

Chlorine feeding begins where the cylinder or ton container connects to the manifold that leads to the chlorinator. The feed system ends at the point where the chlorine solution mixes into the water being disinfected. The system is composed of the following main components:

- Weighing scale
- Valves and piping
- Chlorinator
- Injector or diffuser

Weighing Scales

It is important that an accurate record be kept of the amount of chlorine used and the amount of chlorine remaining in a cylinder or container. A simple way to do this is to place the cylinders or containers on weigh scales. The scales can be calibrated to display either the amount used or the amount remaining. By recording weight readings at regular intervals, the operator can develop a record of chlorine-use rates. Figure 7-9 shows a common type of two-cylinder scale. Figure 7-10 shows a portable beam scale. Figure 7-11 shows a combination trunnion and scale for a ton container; this scale operates hydraulically and has a dial readout.

Valves and Piping

Chlorine cylinders and ton containers are equipped with valves, as shown in Figures 7-12 and 7-13. The valves must comply with standards set by the Chlorine Institute.

It is standard practice for an auxiliary tank valve to be connected directly to the cylinder or container valve, as illustrated in Figure 7-14. The connection is made with either a union-type or a yoke-type connector. The auxiliary valve can be used to close off all downstream piping, thus minimizing gas leakage during

Figure 7-9 Two-cylinder scale
Courtesy of US Filter/Wallace & Tiernan.

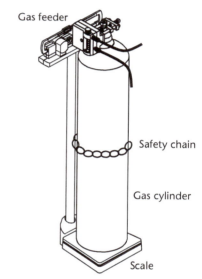

Figure 7-10 Portable beam scale
Courtesy of Severn Trent Services.

Figure 7-11 Combination trunnion and scale for a ton container
Courtesy of Force Flow.

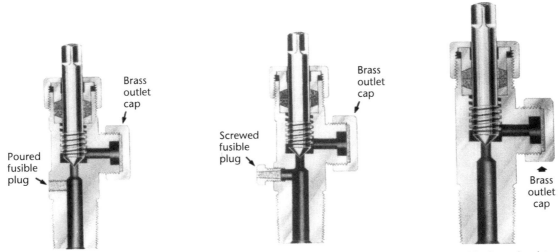

Figure 7-12 Standard cylinder valves: poured-type fusible plug (left) and screw-type fusible plug (right)

Courtesy of the Chlorine Institute.

Figure 7-13 Standard ton container valve

Courtesy of the Chlorine Institute.

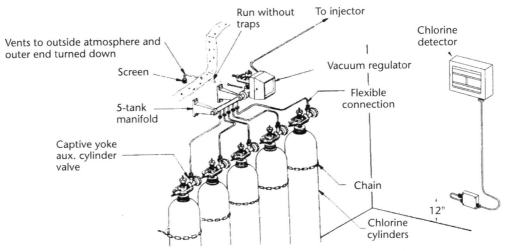

Figure 7-14 Auxiliary tank valve connected directly to container valve

Courtesy of US Filter/Wallace & Tiernan.

container changes. The auxiliary tank valve will also serve as an emergency shut-off if the container valve fails. If a direct-mounted chlorinator is used, an auxiliary tank valve is not required (Figure 7-15).

The diagram in Figure 7-14 is of a typical valve assembly. The figure shows that the assembly is connected to the chlorine-supply piping by flexible tubing, which is usually 3/8-in. (10-mm) copper rated at 500 psig (3,500 kPa).

When more than one container is connected, a manifold must be used, as shown in Figure 7-14. The manifold channels the flow of chlorine from two or more containers into the chlorine-supply piping. The manifold and supply piping must meet the specifications of the Chlorine Institute. Manifolds may have from 2 to 10 connecting points. Each point is a union nut suitable for receiving flexible connections. Notice in Figure 7-14 that the header valve is connected at the manifold discharge end, providing another shutoff point. Additional valves are used along the chlorine supply line for shutoff and isolation in the event of a leak.

Chlorinators

The **chlorinator** can be a simple direct-mounted unit on a cylinder or ton container, as shown in Figure 7-15. This type of chlorinator feeds chlorine gas directly to the water being treated. A free-standing cabinet-type chlorinator is illustrated in Figure 7-16. Cabinet-type chlorinators, which operate on the same principle as cylinder-mounted units, have a sturdier mounting, and are capable of higher feed rates. Schematic diagrams of two typical chlorinators are shown in Figures 7-17 and 7-18.

The purpose of the chlorinator is to meter chlorine gas safely and accurately from the cylinder or container and then accurately deliver the set dosage. To do this, a chlorinator is equipped with pressure and vacuum regulators that are actuated by diaphragms and orifices for reducing the gas pressure. The reduced pressure allows a uniform gas flow, accurately metered by the rotameter (feed rate indicator). In addition, a vacuum is maintained in the line to the injector for safety

Figure 7-15 Direct-mounted chlorinator

Courtesy of Severn Trent Services.

Figure 7-16 Free-standing chlorinator cabinet

Courtesy of Severn Trent Services.

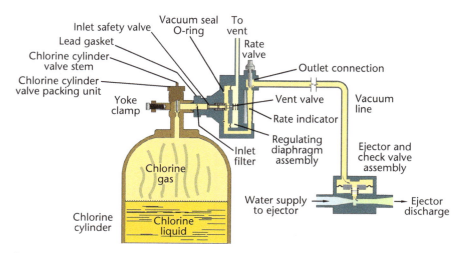

Figure 7-17 Schematic of direct-mounted gas chlorinator

Courtesy of Severn Trent Services.

chlorinator
Any device that is used to add chlorine to water.

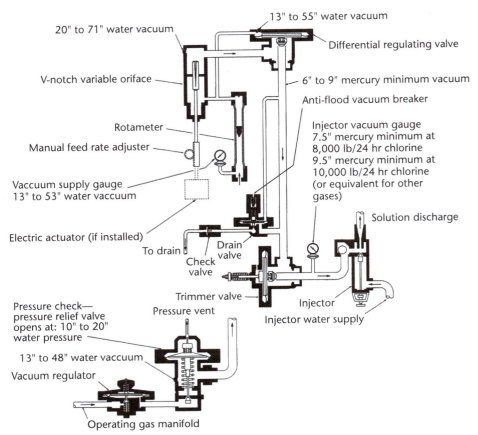

Figure 7-18 Schematic of cabinet-style chlorinator
Courtesy of US Filter/Wallace & Tiernan.

purposes. If a leak develops in the vacuum line, air will enter the atmospheric vent, causing the vacuum relief valve to close and stopping the flow of chlorine gas. To vary the chlorine dosage, the operator manually adjusts the setting of the rotameter.

It is normally required that each treatment plant have at least one standby chlorinator ready for immediate use in the event that the primary chlorinator should fail. Automatic switchover equipment is also strongly recommended.

Injectors

An **injector** (or ejector) is located within or downstream of the chlorinator, as illustrated in Figure 7-17. It is a venturi device that pulls chlorine gas into a passing stream of dilution water, forming a strong solution of chlorine and water. The injector also creates the vacuum needed to operate the chlorinator. The highly corrosive chlorine solution (pH of about 2–4) is carried to the point of application in a corrosion-resistant pipeline. The type of pipe typically used is polyvinyl chloride (PVC), fiberglass, or steel pipe lined with PVC or rubber. A strainer should be installed on the water line upstream of the injector. This strainer prevents any grit, rust, or other material from entering and blocking the injector or causing wear of the injector throat.

Diffusers

A **diffuser** is one or more short lengths of pipe, usually perforated, that quickly and uniformly disperses the chlorine solution into the main flow of water. There

injector

The portion of a chlorination system that feeds the chlorine solution into a pipe under pressure.

diffuser

A section of a perforated pipe or porous plates used to inject a gas, such as carbon dioxide or air, under pressure into water.

are two types of diffusers: those used in pipelines and those used in open channels and tanks. A properly designed and operated diffuser is necessary for the complete mixing needed for effective disinfection.

The diffuser used in pipelines less than 3 ft (0.9 m) in diameter is simply a pipe protruding into the center of the pipeline. Figure 7-19 shows a diffuser made from Schedule 80 PVC, and Figure 7-20 shows how the turbulence of the flowing water completely mixes the chlorine solution throughout the water. Complete mixing should occur downstream at a distance of 10 pipe diameters.

Figure 7-21 shows a perforated diffuser for use in larger pipelines. A similar design is used to introduce chlorine solution into a tank or open channel, as shown in Figure 7-22. (During normal operations, the diffuser would be completely submerged, but in the figure, the water level has been dropped, for illustrative purposes only, to show the chlorine solution passing out of each perforation.)

Gas Chlorination Auxiliary Equipment

A variety of auxiliary equipment is used for chlorination. The following discussion describes the functions of the more commonly used items.

Booster Pumps

A booster pump (Figure 7-23) is usually needed to provide the water pressure necessary to make the injector operate properly. The booster pump is usually a low-head, high-capacity centrifugal type. It must be sized to overcome the pressure

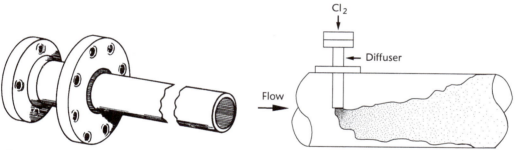

Figure 7-19 Diffuser made from Schedule 80 PVC

Figure 7-20 Chlorine solution mixing in a large-diameter pipeline

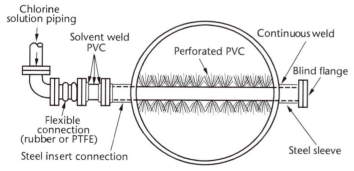

Figure 7-21 Perforated diffuser for pipelines larger than 3 ft (0.9 m) in diameter

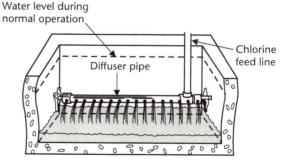

Figure 7-22 Open-channel diffuser

Source: *Handbook of Chlorination and Alternative Disinfectants.* 4th ed. by Geo. Clifford White, copyright © 1998. Reprinted by permission of John Wiley & Sons, Inc.

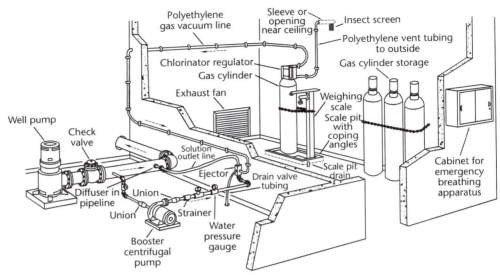

Figure 7-23 Typical chlorinator deep-well installation showing booster pump
Courtesy of Severn Trent Services.

in the line that carries the main flow of water being treated, and it must be rugged enough to withstand continuous use.

Automatic Controls

If a chlorination system is to be manually operated, adjustments must be made each time the flow rate or the chlorine demand changes. For constant or near-constant flow rate situations, a manual system is suitable.

However, when flow rate or chlorine demand is continually changing, the operator is required to change the rotameter settings frequently. In these situations, automatic controls are valuable. Although many automatic control arrangements are possible, there are two common types: flow proportional control and residual flow control.

Flow Proportional Control If chlorine demand rarely changes and it is necessary to compensate only for changes in the pumping rate, flow proportional control works well. It will automatically increase or decrease the chlorine feed rate as the water flow rate increases or decreases. The required equipment includes a flowmeter for the treated water, a transmitter to sense the flow rate and send a signal to the chlorinator, and a receiver at the chlorinator. The receiver responds to the transmitted signal by opening or closing the chlorine flow rate valve.

Residual Flow Control If the chlorine demand of the water changes periodically, it is necessary to make corresponding changes in the rate of feed to provide adequate disinfection. Residual flow control, also called *compound loop control*, automatically maintains a constant chlorine residual, regardless of chlorine demand or flow rate changes. The system uses an automatic chlorine residual analyzer (Figure 7-24) in addition to the signal from a meter measuring the flow rate. The analyzer uses an electrode to determine the chlorine residual in the treated water. Signals from the residual analyzer and flow element are sent to a receiver in the chlorinator, where they are combined to adjust the chlorine feed rate to maintain a constant residual in the treated water.

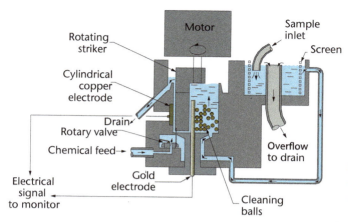

Figure 7-24 Automatic chlorine residual analyzer
Courtesy of Severn Trent Services.

Evaporators

An evaporator is a heating device used to convert liquid chlorine to chlorine gas. Ton containers are equipped with valves that will draw either liquid or gas. At 70°F (21°C), the maximum gas withdrawal rate from a ton container is 400 lb/d (180 kg/d). If higher withdrawal is required, the liquid feed connection is used and connected to an evaporator. The evaporator accelerates the evaporation of liquid chlorine to gas, so that withdrawal rates up to 9,600 lb/d (4,400 kg/d) can be obtained.

An evaporator (Figure 7-25) is a water bath heated by electric immersion heaters to a temperature of 170–180°F (77–82°C). The pipes carrying the liquid chlorine pass through the water bath, and liquid chlorine is converted to gas by the heat.

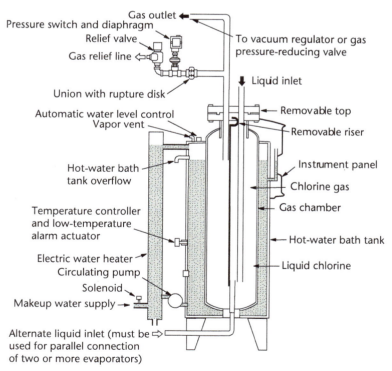

Figure 7-25 Chlorine evaporator
Courtesy of US Filter/Wallace & Tiernan.

Automatic Switchover Systems

For many small water systems, it is either impossible or uneconomical to have an operator available to monitor operation of the chlorination system at all times. An automatic switchover system provides switchover to a new chlorine supply when the online supply runs out. The switchover is either pressure or vacuum activated. The vacuum type of installation is shown in Figure 7-26. The automatic changeover mechanism has two inlets and one outlet. As the online supply is exhausted, the vacuum increases, causing the changeover mechanism to close on the exhausted supply and open the new chlorine supply. The unit can also send a signal to notify operating personnel that the one tank is empty and should be replaced. Figure 7-27 shows a typical installation. This system is ideal for remote locations to ensure uninterrupted chlorine feeding.

Chlorine Alarms

Chlorinators are often equipped with a vacuum switch that triggers an alarm when it senses an abnormally low or high vacuum. A low-vacuum condition can

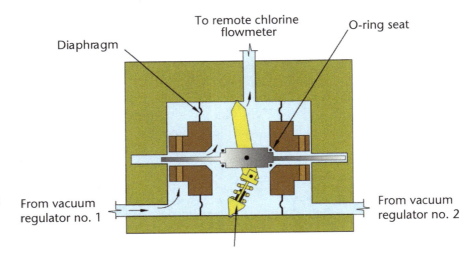

Figure 7-26 Automatic switchover unit

Courtesy of Severn Trent Services.

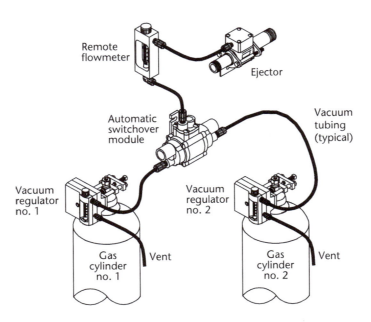

Figure 7-27 Typical installation of switchover system

Courtesy of Severn Trent Services.

mean an injector failure, vacuum line break, or booster pump failure. A high-vacuum condition can be caused by a plugged chlorine supply line or by empty chlorine tanks.

Safety Equipment

Safety in and around the gas chlorination process is important to prevent serious accidents and equipment damage. Certain items of equipment, such as the following, are essential for the safe operation of a chlorination facility:

- Chlorine detectors
- Self-contained breathing apparatus
- Emergency repair kits

Chlorination safety is discussed in detail later in this chapter.

Hypochlorination Facilities

Hypochlorination is a chlorination method increasingly used by water treatment plants because of its relative safety (as compared to gaseous chlorine) and its ease of use. Sodium hypochlorite is fed as a liquid, and many operators prefer to feed liquids rather than gases.

Hypochlorite Compounds

The two most commonly used compounds are calcium hypochlorite and sodium hypochlorite. Table 7-1 lists the properties of both compounds.

Calcium Hypochlorite

Calcium hypochlorite, $Ca(OCl)_2$, is a dry, white or yellow-white, granular material. It is also available in compressed tablets. It normally contains 65 percent available chlorine by weight. This means that when 1 lb (0.5 kg) of the powder is added to water, only 0.65 lb (0.3 kg) of pure chlorine is being added. Conversely, if 1 lb (0.5 kg) of chlorine is added, 1.5 lb (0.7 kg) of calcium hypochlorite must be added (Table 7-2).

Calcium hypochlorite requires special storage to avoid contact with organic material. Its reaction with any organic substances can generate enough heat and oxygen to start and support a fire. When calcium hypochlorite is mixed with water, heat is given off. To provide adequate dissipation of the heat, the dry chemical should be added to the water; the water should *not* be added to the chemical.

Calcium hypochlorite is used mostly for disinfection of new and repaired water mains, water storage tanks, and smaller water volumes, such as swimming pools.

Table 7-1 Properties of hypochlorites

Property	Sodium Hypochlorite	Calcium Hypochlorite
Symbol	NaOCl	$Ca(OCl)_2$
Form	Liquid	Dry granules, powder, or tablets
Strength	Up to 15% available chlorine	65–70% available chlorine, depending on form

hypochlorination
Chlorination using solutions of calcium hypochlorite or sodium hypochlorite.

Table 7-2 Chlorine content of common disinfectants

Compound	Percentage Cl	Amount of Compound Needed to Yield 1 lb of Pure Cl
Chlorine gas or liquid (Cl_2)	100	1 lb (0.454 kg)
Sodium hypochlorite (NaOCl)*	15 12.5 5 1	0.8 gal (3 L) 1.0 gal (3.8 L) 2.4 gal (9.1 L) 12.0 gal (45.4 L)
Calcium hypochlorite [$Ca(OCl)_2$]	65	1.54 lb (0.7 kg)

*Sodium hypochlorite is available in four standard concentrations of available chlorine. Ordinary household bleach contains 5% chlorine.

Sodium Hypochlorite

Sodium hypochlorite, NaOCl, is a clear, light-yellow liquid commonly used for bleach. Ordinary household bleach contains 5–6 percent available chlorine. Industrial bleaches are stronger, containing from 9 to 15 percent.

The sodium hypochlorite solution is alkaline, with a pH of 9–11, depending on the available chlorine content. For common strengths, Table 7-2 shows the amount of solution needed to supply 1 lb (0.5 kg) of pure chlorine. Large systems can purchase the liquid chemical in carboys, drums, and railroad tank cars. Very small water systems often purchase it in 1-gal (3.8-L) plastic jugs.

There is no fire hazard in storing sodium hypochlorite, but the chemical is quite corrosive and should be kept away from equipment susceptible to corrosion damage. At its maximum strength of 12–15 percent, sodium hypochlorite solution can lose 2–4 percent of its available chlorine content per month at room temperature. It is therefore recommended that it not be stored for more than 15–25 days. Instability of the chemical increases with increasing temperature, solution strength, and exposure to sunlight. Sodium hypochlorite solutions of 6 percent are more stable. Water plant design specifications usually call for the dilution of the 12 percent stock chemical to 6 percent, unless special storage facilities are built that keep out sunlight and heat.

Common Equipment

Disinfecting facilities using calcium hypochlorite should be equipped with a cool, dry storage area to stockpile the compound in the shipping containers. A variable-speed chemical feed pump (hypochlorinator), such as a diaphragm pump, is all that is required for feeding the chemical to the water. A mix tank and a day tank (Figure 7-28) are also required. After calcium hypochlorite is mixed with water, impurities and undissolved chemicals settle to the bottom of the mix tank. The clear solution is then transferred to the day tank for feeding. This prevents any of the solids from reaching and plugging the hypochlorinator or rupturing the diaphragm.

Because sodium hypochlorite is a liquid, it is simpler to use than calcium hypochlorite. It is fed neat (as the 12 percent stock chemical) or at the 6 percent strength, usually with peristaltic pump equipment that uses a quality-grade tubing. Redundant pumps are needed and operators must get used to changing the tubing on a frequent basis to prevent failure.

Off-gassing is a major issue with the equipment used for storing and feeding sodium hypochlorites. Equipment failure and damage are common occurrences when hypochlorite feed systems are poorly designed or poorly maintained.

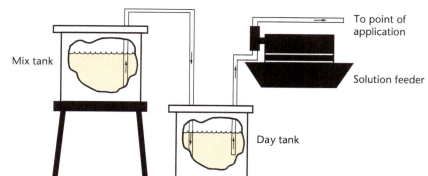

Mix tank

To point of application

Solution feeder

Day tank

Figure 7-28 Mix tank and day tank

Operation of the Chlorination Process

Successful operation of the chlorination process requires an understanding of how each of the system components operates. In addition, the operator must be aware of the safety procedures that must be followed when handling cylinders and when dealing with leaks and equipment breakdowns.

Using Cylinders and Ton Containers

Cylinders

Cylinders should always be stored and used in an upright position. In this position, the cylinders will deliver chlorine gas continuously at a maximum rate of about 42 lb/d (19 kg/d). If the gas is withdrawn at a faster rate, the drop in pressure in the cylinder will cause a drop in cylinder temperature, and frost will form on the outside of the cylinder. This frost will reduce the withdrawal rate because a cooler temperature retards the vaporization of liquid chlorine. Under extreme conditions of frosting, the valve may freeze and completely stop the flow of gas.

Moderate frosting of a cylinder can be reduced by improving air circulation, which is easily accomplished by placing a fan in the chlorinator room. Heat should never be applied directly to a chlorine cylinder. If the heat is excessive, the cylinder's fusible plug could melt or pressure could increase to the point that the valve fails. If the problem of frosting continues, multiple cylinders must be used to reduce the withdrawal rate from any single cylinder.

Ton Containers

Ton containers are transported, hoisted, stored, and used in the horizontal position. When they are to be used, they must be positioned so that the two valves are oriented with one directly above the other. In this position, the top valve delivers chlorine gas, and the bottom delivers liquid chlorine.

A single ton container can deliver chlorine gas at rates up to 400 lb/d (180 kg/d) against a backpressure of 35 psig (240 kPa) at room temperature without frosting. Liquid chlorine can be delivered at rates up to 9,600 lb/d (4,400 kg/d) if an evaporator is being used. If a plant's requirements exceed 400 lb/d (180 kg/d), the plant can still operate on gas withdrawal by connecting two or more containers to feed at the same time.

The exact maximum withdrawal rate of gas for vacuum systems can be determined from the following formula:

$$\left(\begin{array}{c} \text{chlorine room} \\ \text{temp. } °F \end{array} - \begin{array}{c} \text{threshold} \\ \text{temp. } °F \end{array} \right) \times \text{withdrawal factor} = \text{maximum withdrawal rate}$$

The withdrawal factor depends on the size and shape of the cylinder or container and can be obtained from chlorine suppliers or manufacturers.

Table 7-3 Values from the vapor pressure curve for liquid chlorine

Minimum Pressure, psig	(kPa)	Threshold Temperature, °F	(°C)
9	(62)	–10	(–23)
14	(97)	0	(–18)
21	(145)	10	(–12)
28	(193)	20	(–7)
37	(255)	30	(–1)
47	(324)	40	(4)
59	(407)	50	(10)
71	(490)	60	(16)
86	(593)	70	(21)
102	(703)	80	(27)

The threshold temperature is the temperature at which the minimum gas pressure required to operate a gas chlorinator at the point of withdrawal (i.e., the line pressure at the point of application) is reached. The threshold temperature is determined from the vapor pressure curve of liquid chlorine. Table 7-3 is a chart of values from that curve. For example, given a room temperature of 60°F (15°C), a 150-lb (68-kg) cylinder having a withdrawal factor of 1, and a minimum gas pressure of 14 psig (97 kPa), the threshold temperature is 0°F (–17°C). The maximum withdrawal rate under these conditions is

$$(60 - 0)\ 1 = 60\ \text{lb/d (27 kg/d)}$$

 WATCH THE VIDEO
Chlorine Safety: Transporting Cylinders (www.awwa.org/wsovideoclips)

Weighing Procedures

The only reliable method of determining the amount of chlorine remaining in a cylinder or container is to weigh the unit and check its weight against the tare weight (empty weight) stamped on the shoulder of the cylinder or container. The information can be used to determine the feed rate and to decide when to change cylinders or containers. Because the pressure in a cylinder depends on the liquid chlorine temperature, not on the amount of chlorine in the container, the pressure cannot be used to determine when the cylinder is empty and must be changed.

Weighing procedures depend on the scale being used. Simple scales show the combined weight of the container and chlorine. On other scales, the operator enters the tare weight into the scale when the cylinder or container is put in use, and the scale thereafter displays only the weight of the remaining chlorine.

Connecting Cylinders and Ton Containers

When a cylinder is being connected to the chlorine supply line or manifold (using a yoke and adapter), the following procedure should be used:

1. Always wear personal respirator protection when changing cylinders or containers.
2. Never lift a cylinder by its protective hood. The hood is not designed to support the cylinder weight.

3. Secure the cylinder with a safety chain or steel strap in a solid, upright position.

4. Remove the protective hood. If the threads have become corroded, a few sound raps with a wooden or rubber mallet will usually loosen the hood so it can be unscrewed.

5. Remove the brass outlet cap and any foreign matter that may be in the valve outlet recess. Use a wire brush to clear out any pieces of the washer, being careful not to scratch the threads or gasket-bearing surface.

6. Place a new washer in the outlet recess. Do not reuse old washers.

7. Place the yoke over the valve. Insert the adapter in the outlet recess; then, fitting the adapter in the yoke slot, tighten the yoke screw. Make sure the end of the adapter seats firmly against the washer. Use only the cylinder-valve wrench provided by the chlorine supplier for all chlorine cylinder or container valve connections.

8. Install the flexible connector, sloping it back toward the chlorine cylinder so that any liquid chlorine droplets will flow back to the cylinder and not to the chlorinator unit (Figure 7-29).

The procedure for connecting ton containers to the piping or manifold is basically the same as that described for cylinders. Note that when the cylinder or ton container is being connected, both the container valve and the auxiliary valve should be closed. Ton containers should have a drip leg (liquid chlorine trap) (Figure 7-8) with a heater installed before the chlorinator. This element will vaporize any liquid chlorine initially coming from the eductor at the start of gas operation, and any liquid droplets that may flow out of the container or evaporator during normal operation.

Opening the Valve

Once the cylinder or container has been installed and the flexible connector attached, the valve can be opened and the lines checked for leakage according to the following procedure:

1. Place the valve wrench provided by the chlorine supplier on the cylinder or container valve stem. Stand behind the valve outlet. Grasp the valve firmly with one hand, and give the wrench a sharp blow in a counterclockwise direction with the palm of the other hand. Do not pull or tug at the wrench because this may bend the stem, causing it to stick or fail to close properly.

   To open a stubborn valve, follow the normal opening procedure, but use a small block of wood held in the palm of the hand when striking the wrench. If the valve continues to resist opening, return the cylinder to the supplier. Do not—under any circumstances—use a pipe wrench or an ill-fitting wrench because these wrenches will round the corners of the square-end valve stem. Avoid using wrenches longer than 6 in. (150 mm) for opening stubborn valves because they might bend or break the valve stem.

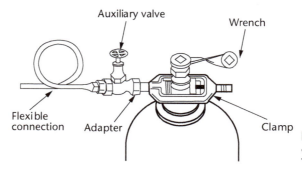

Figure 7-29 An installed yoke and auxiliary valve

2. Open the valve and close it immediately.

3. The line is now pressurized. All new joints and connections can be checked for leaks using an ammonia solution. Use only commercial 26° Bé ammonia, which can be obtained from a supplier of chemicals or chlorine. Common household ammonia is not strong enough to work properly.

   Hold an opened plastic squeeze bottle of the ammonia beneath the valve and joints, and allow the ammonia fumes to rise up around any suspected leak areas. Ammonia fumes react with chlorine gas to form a white cloud of ammonium chloride, thus making small leaks easy to locate. Do not spray, swab, or otherwise bring ammonia liquid into contact with chlorination equipment. The chlorine will combine with the ammonia and may start corrosion at that point.

4. If no leaks are found, open the cylinder valve. One complete turn will permit the maximum withdrawal rate.

5. Leave the wrench on the valve to allow for easy and rapid shutoff in an emergency. The wrench also indicates to other operators which cylinder is being used. A sign disk should be installed on the stem of the cylinder or container valve (Figure 7-30) to show the closing direction in case of emergency.

6. If the injector and chlorinator are already operating, open the auxiliary valve; the newly connected cylinder or container will start feeding chlorine to the system. Do not open the auxiliary valve until the injector is operating because the necessary vacuum will not be developed and the regulating valves will not function.

Closing the Valve

When all the chlorine has been released from the cylinder or container, as indicated by the weight reading on the scale, the unit should be disconnected and replaced according to the following procedures:

1. Close the cylinder or container valve. After about 2 minutes, close the auxiliary valve. The delay allows any remaining chlorine gas in the line to be drawn into the injector.

   To close the valve, use the wrench provided, grasping the valve in one hand and tapping the wrench in a clockwise direction with the palm of the other hand. If the valve does not close tightly on the first try, open and close it lightly several times until the proper seating is obtained. Never use a hammer or any other tool to close the cylinder valve tightly.

Figure 7-30 Sign to show which direction the valve should be turned to close it in the event of an emergency

Courtesy of Severn Trent Services.

2. Disconnect the flexible tubing from the cylinder or container, and replace the brass outlet cap on the valve immediately so that the valve parts will be protected from moisture in the air.

3. Screw the protective hood in place.

4. The outlet cap of each valve is fitted with a gasket designed to fit against the valve outlet face. If a valve leaks slightly after closing, the leak can often be stopped by drawing up the valve cap tightly.

5. The flexible copper tubing from the manifold to the container should be supported while the empty container is being replaced. Support it on another container, a wall hook, or a block to prevent any kinking or weak spots from developing in the pipe. If the flexible tubing is disconnected for any length of time, there is a danger of moisture forming in the line. Close the open end of the pipe with tape or plastic wrap and a rubber band.

Determining the Chlorine Dosage

The chlorine demand of the water being treated must be determined by performing a chlorine-demand test. By using the test result and knowing the desired residual, the operator can determine the dosage as follows:

$$Cl\ dosage = Cl\ demand + Cl\ residual,$$

where all quantities are in milligrams per liter.

Once calculated, the dosage should be converted into pounds (or kilograms) per day, and the chlorinator should be set to deliver that dosage. After the dosage rate is set, the water must be tested regularly to ensure that the proper residual is maintained.

As noted previously, temperature, pH, and contact time are important variables affecting the effectiveness of chlorination. As these factors change, the amount of residual needed will also change. Table 7-4 shows general recommendations for the minimum residuals required for effective disinfection. Notice that recommendations are included for free residual as well as combined residual. Free residual values are based on a contact time of at least 10 minutes. The combined residual concentrations require at least 60 minutes of contact time. However, systems that must comply with the Surface Water Treatment Rule must meet considerably more complex requirements for determining $C \times T$ values (i.e., concentration of disinfectant × contact time).

Additional information on chlorine dosage, demand, and residual calculations is included in Chapter 2.

Table 7-4 Recommended minimum concentrations of free chlorine residual versus combined chlorine residual

pH Value	Minimum Concentration of Free Chlorine Residual (Disinfecting Period of at Least 10 minutes), ppm	Minimum Concentration of Combined Chlorine Residual (Disinfecting Period of at Least 60 minutes), ppm
6.0–7.0	0.2	1.0
7.0–8.0	0.2	1.5
8.0–9.0	0.4	1.8
9.0–10.0	0.8	Not recommended
10.0+	0.8+ (with longer contact)	Not recommended

Chlorination Operating Problems

Operating problems related to chlorination include the following:

- Chlorine leaks
- Stiff container valves
- Hypochlorinator problems
- Tastes and odors
- Sudden change in residual
- Trihalomethane formation

Proper maintenance can prevent many problems. Most manufacturers have equipment troubleshooting guides that help locate and correct problems.

Chlorine Leaks

A major concern in the operation and maintenance of the chlorination process is the prevention of chlorine leaks. The most common place for leaks to occur is the pressurized chlorine supply line between the containers and the chlorinator. Every joint, valve, fitting, and gauge in the line is a possible point of leakage.

Some chlorine leaks are readily apparent. Others are very slow, very small, partly hidden, or otherwise difficult to locate. The usual method of detection is to open a bottle of ammonia solution, place the bottle near a suspected leak, and allow the fumes to rise around the suspected area. If there is a sizable leak, the chlorine will combine with the ammonia to form a visible white vapor. Unfortunately, this method will not indicate a very small leak. In addition, very small leaks will often not produce a noticeable odor. Small leaks can go unnoticed for weeks unless the operator periodically looks for two signs: joint discoloration and moisture.

Even the smallest leak will remove cadmium plating from chlorine tubing and fittings. The metal underneath the plating (copper, brass, or bronze) will appear reddish, and a green copper-chloride scum may appear around the edges of the area affected.

Portions of the pressure piping system (for example, the manifold) are often painted. As a result, discoloration of the metal beneath the paint will not be apparent. To locate leaks in painted piping, look for small droplets of water that may form on the underside of joints. Small, almost invisible leaks must be located early; otherwise, the corrosion they cause will often result in a sudden and massive chlorine leak after a period of time.

The best and most reliable way to find chlorine leaks is a chlorine detector, which is sensitive to leaks as small as 1 ppm chlorine in air. Such leaks are not normally detectable by the ammonia technique or by smell.

If a major leak requires shutdown of the system for repair, the tank valve should be closed, the yoke disconnected, and the injector left running with the auxiliary tank valve open, until any remaining chlorine gas is purged from the line. To prevent leaks, the operator should observe the following precautions:

- Install a new gasket every time a cylinder or container is changed.
- Each time a threaded fitting is opened, clean the threads with a wire brush. Then wrap them with polytetrafluoroethylene (PTFE) tape or use one of the following pipe joint compounds: linseed oil and graphite, linseed oil and white lead, or litharge and glycerine. If PTFE tape is used, remove any previous remnants of tape before remaking the joint.
- Replace all chlorine supply line valves annually. Refit and repack the old ones so they are ready for use the following year.

WATCH THE VIDEO
Chlorine Safety: Working with Chlorine (www.awwa.org/wsovideoclips)

Stiff Container Valves

Container valves are carefully checked before leaving the manufacturer's plant, but occasionally a valve may be stiff to turn or difficult to shut off tightly. This problem is often caused by overly tight packing. Sometimes the valve can be freed by opening and shutting it a few times. If the valve does not operate at all, set the container aside and call the supplier.

Hypochlorinator Problems

Two problems that commonly occur with hypochlorinators are clogged equipment and broken diaphragms.

Clogged Equipment

Clogging, caused by calcium carbonate ($CaCO_3$) scaling, occurs primarily in two areas of a hypochlorinator: at the pump head and in the suction and discharge hoses. Scale is most likely to form when the water used for preparing the solution has a high calcium hardness and carbonate alkalinity. Under these conditions, calcium carbonate forms and causes a scale deposit in the pump head, suction hose, and discharge hose. Scale may also form in the solution injector or diffuser.

The scale can readily be removed by pumping a dilute (5 percent) hydrochloric acid solution (also known as *muriatic acid*) through the pump head, hoses, and diffuser. The hypochlorite solution should be completely flushed out of the system with water before the acid is used.

An associated problem affecting the pump head is the accumulation of dissolved calcium hypochlorite (lime sludge). When the solution tank level is low or the suction foot valve is too near the bottom of a one-tank installation, the suction hose can draw some of the undissolved chemical up into the pump head and fill the area of the head. This congestion can result in the pump not feeding hypochlorite solution, and it can cause diaphragm rupture. To prevent these problems, it is best to use a two-tank setup when calcium hypochlorite is being used.

Broken Diaphragms

Because broken diaphragms are a common problem with hypochlorinators, it is important that the operator inspect the diaphragm regularly to ensure that it is

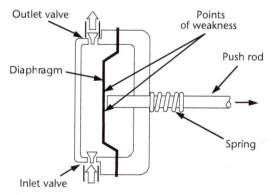

Figure 7-31 Points of weakness in a mechanically actuated diaphragm

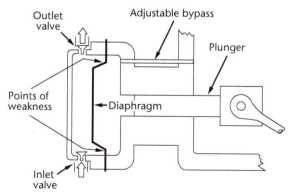

Figure 7-32 Points of weakness in a hydraulically actuated diaphragm

functioning properly. A visual inspection of the pump head may not reveal a broken diaphragm, but an outflow of solution from the diaphragm hose is a positive indication that the unit is functioning properly. Figures 7-31 and 7-32 indicate the points of diaphragm weakness in the two most common types of hypochlorinators.

Common Chlorination Issues

Tastes and Odors

The cause of "swimming pool" taste and odor in drinking water is commonly misunderstood to be too much chlorine. Oddly enough, chlorine-like taste and odor are usually caused by too little chlorine. Natural water is not pure, and the reaction of chlorine with the impurities in the water, such as organic matter, iron, manganese, nitrites, and ammonia, interferes with the formation of a free chlorine residual. Adding more chlorine to the water actually decreases the residual.

Between points 1 and 2 of Figure 7-33, added chlorine combines immediately with iron, manganese, and nitrites. These chemicals are reducing agents, and no residual can be formed until all reducing agents are completely destroyed by the chlorine. The curve segment between points 2 and 4 is predominantly combined chlorine residual. The combined residual available in this range is the weakest form of chlorine for disinfection, and the accompanying dichloramines, trichloramines, and chloroorganic compounds cause taste and odor. The decrease (shown from point 3 to point 4) results because the additional chlorine oxidizes some of the chloroorganic compounds and ammonia. The additional chlorine also changes some of the monochloramine to dichloramine and trichloramine.

As additional chlorine is added between points 3 and 4, the amount of chloramine reaches a minimum value. Beyond this minimum point, the addition of more chlorine produces free residual chlorine. The point at which this occurs is known as the *breakpoint*. By increasing the chlorine dosage beyond the breakpoint (point 4 in Figure 7-33), any taste or odor is usually eliminated or significantly reduced. Free available chlorine residual is free from taste and odor at the concentrations commonly used in disinfection.

Sudden Change in Residual

One important attribute of chlorine is that the residual present in a sample can be measured easily. Based on a chlorine residual test procedure, the strength of the disinfectant remaining in the water at any point in the distribution system can be determined. This test should take less than 5 minutes. Records of the chlorine residual kept over a period of time can help predict the type and amount of residual that will be found at various locations in the distribution system. Any sudden drop in residual is a warning of potential danger, such as a cross-connection that is allowing contaminated water into the drinking water system.

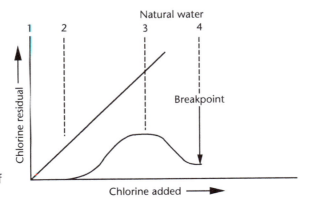

Figure 7-33 Decrease of chlorine residual

When a sudden drop in residual occurs, chlorination levels should be increased immediately to raise the residual to the desired level. This response guards against the possibility of waterborne disease while the problem is being analyzed and corrected. Samples should then be analyzed to identify the contaminants, and the distribution system should be checked to locate the source of contamination or other cause of the drop in residual.

Safety Precautions

Chlorine should never cause an accident or injury if it is used by properly trained operators in adequately equipped plants. The key factors in safely operating a chlorination system are proper safety equipment and procedures.

Proper Safety Equipment

Without adequate equipment and proper training in the use of safety equipment, the operator's life—as well as the lives and well-being of the surrounding community—is needlessly at risk.

It is essential that every chlorination facility be equipped with the following safety devices:

- Self-contained breathing equipment
- Emergency repair kits
- Adequate ventilation equipment

In addition, installation of a chlorine detector is strongly recommended at all chlorination installations, and it should *always* be installed at unattended chlorination stations.

Self-Contained Breathing Apparatus

Self-contained breathing equipment should be available wherever gas or liquid chlorine is in use. Air packs, as shown in Figure 7-34, have a positive-pressure mask with a full, wide-view face piece and a cylinder of air or oxygen carried on the operator's back. The only units that should be purchased are those that

Figure 7-34 Air pack with positive-pressure mask

have been approved by the National Institute of Occupational Safety and Health (NIOSH). Canister-type gas masks, which were used in the early days of chlorination, are no longer recommended because they are not effective against heavy chlorine concentrations.

Every operator should be familiar with the location and use of the breathing apparatus available at the treatment plant. Operating instructions are provided with each unit. Operators should study and review these instructions periodically and have regular formal training and practice sessions in the use of the equipment.

An air pack is very similar to the equipment used by scuba divers and fire departments. The tank contains a 15- to 30-minute supply of air, depending on tank size. The actual time an air supply will last also depends on an individual's pattern of breathing while under stress. The mask must fit tightly around the face. Operators with a beard or eyeglasses may find it difficult to fit the mask.

Operators should practice repairing simulated leaks while using breathing equipment, so that they will know how long they can work before the air tank is exhausted. A low-air-pressure alarm on the 30-minute air pack is activated when about 5 minutes of air is left. When the alarm sounds, the wearer should immediately leave the contaminated area to get a fresh cylinder of air.

The air pack should be located at a readily accessible point, away from the area likely to be contaminated with chlorine gas. It should not be located in the chlorine feed or storage rooms. A wall storage cabinet is usually mounted outside these rooms (Figure 7-35). The mask and air supply tanks should be inspected routinely and maintained in good condition. Spare air cylinders should be on-site for use during prolonged emergencies.

The mask and breathing apparatus should be cleaned at regular intervals and after each use. When air tanks are depleted, they should be refilled at stations where proper air compressor equipment is available. Many local fire departments have this equipment.

When stowing the equipment after use, the straps on the masks and backpacks should be extended to their limits. This allows the equipment to be fitted quickly to the next user.

WATCH THE VIDEO
Chlorine Safety: SCBA & PPE (www.awwa.org/wsovideoclips)

Figure 7-35 Storage unit for chlorine air pack

Emergency Repair Kits

Standardized emergency repair kits that meet US and Canadian specifications contain various devices and hardware for stopping leaks from chlorine shipping containers. Currently, there are three standard Chlorine Institute emergency kits:

- Emergency kit A for cylinders (Figure 7-36)
- Emergency kit B for ton containers
- Emergency kit C for tank cars and tank trucks

Each kit is designed to be used in the repair of leaks that can occur in the shipping containers. The kits contain all the equipment needed to cap a leaking valve, seal a sidewall leak, and cap a fusible plug.

Instruction booklets and other materials are available from the Chlorine Institute for training operators in the use of repair kits. It is important that each operator be trained through an established training program.

Adequate Ventilation

Because chlorine gas is 2.5 times as dense as air, it will settle and stay near the floor when leakage occurs. The rooms in which chlorine cylinders are stored and the enclosures that surround chlorinators should have sealed walls, and the doors to the rooms should open outward. The rooms should also be fitted with chlorine-resistant power exhaust fans ducted at the floor level and with fresh-air intake vents at the ceiling. In large installations, it is usually desirable to provide a fan near the ceiling to force fresh air into the chlorinator room. The ventilating equipment should be capable of completely changing the volume of air in the room every 1–4 minutes. A different required rate may be specified in the regulations of the state drinking water agency.

Exhaust fan switches should be located outside the room and should be wired to room lights so that the lights and fan go on at the same time. It is also good practice to have a window in the door or adjacent to the door, so that the operator can look into the room to detect any abnormal conditions before entering (Figure 7-37).

Figure 7-36 Emergency kit A for use with chlorine cylinders
Courtesy of the Chlorine Institute.

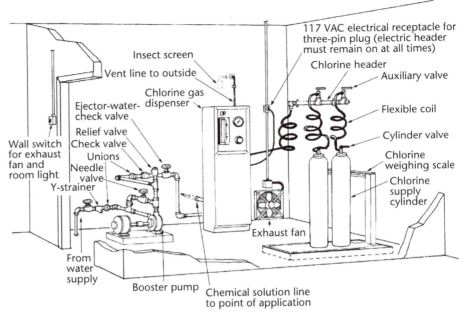

Figure 7-37 Schematic of a well-designed chlorine room

Courtesy of The Chlorine Institute.

Chlorine Detector

The installation of a chlorine detector is a wise investment for every treatment plant using chlorine gas. It detects chlorine concentrations so small that they are impossible to smell. It can give early warning so that a small leak can be stopped before becoming larger, and it can provide immediate warning of a large leak.

Various types of chlorine detectors are available. One type, shown in Figure 7-38, operates by measuring conductivity between two electrodes set in a liquid or solid electrolyte. Samples of air are drawn from the area being monitored, and any chlorine in the sample will cause an increase in the current passing between the electrodes. A sensitive circuit detects this change and activates an alarm.

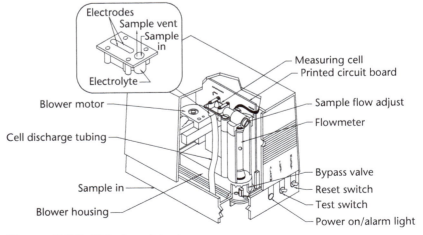

Figure 7-38 Chlorine detector

Courtesy of Severn Trent Services.

Chlorine Scrubbers

Some utilities that continue to use liquid and gas chlorine have installed chlorine scrubbers as an added measure of safety. Chlorine scrubbers are usually designed to neutralize up to 2000 lb of chlorine. Scrubbers work by evacuating the chlorine-laden gas from a chlorine storage room by means of a powerful ventilation fan through a closed duct system. The gas is then passed through a contact chamber containing a chlorine neutralizing substance, such as liquid caustic found in wet type scrubbers, or media pellets (proprietary), capable of removing 15 percent of their weight in chlorine, in dry scrubbers. Once the air has passed through the contact chamber, the scrubbed air is safely discharged to atmosphere. An example of a dry type scrubber is shown in Figure 7-39.

Proper Safety Procedures

Every facility using chlorine, regardless of size, should have established procedures for handling chlorine and chlorine leaks. These procedures should include the following:

- Safety precautions for storing and handling cylinders or containers
- Basic steps in connecting and disconnecting cylinders or containers
- Procedures to follow in case of a chlorine leak
- Emergency procedures to be taken if a chlorine leak threatens nearby residential areas
- First-aid procedures for persons exposed to chlorine

Each operator should be thoroughly familiar with all of these procedures through routine, in-plant training programs. These programs should emphasize the use of respiratory protection equipment, leak repair kits, and emergency first aid.

It is helpful to post descriptions of important procedures near the chlorination facilities. Many of the chlorine chemical and equipment manufacturers distribute wall charts on safety and handling procedures that can be used for training and display. Additional information on chlorine handling is also available from the Chlorine Institute.

Figure 7-39 Dry chlorine scrubber

Courtesy Dave Hardy, Central Utah Water Conservancy District.

Record Keeping

Records for the disinfection process should show the type and amount of disinfectant used and the bacteriological and other operational control test results. The following is a list of information that should be recorded as part of the disinfection process:

- Types of disinfectants in use
- Ordering information, including the following:
  - Manufacturer's name, address, and phone number
  - Shipper's name, address, and phone number
  - Type, size, and number of shipping containers
- Most recent costs
- Current dosage rate, in milligrams per liter or pounds per day
- Bacteriological test results
- Chlorine and other disinfectant residual test results
- Water temperature
- Raw-water pH
- Daily explanation of unusual conditions, mechanical problems, supply problems, emergencies, or unusual test results

Study Questions

1. The advantage to using the oxidant ozone is that it
   a. is easily generated using relatively little energy.
   b. is easily fed into the treatment process.
   c. is noncorrosive.
   d. has little pH effect.

2. Pretreatment with chlorine is being eliminated at many water treatment plants because it has been shown to
   a. react with floc and not much with organics, pathogens, or algae; thus it is a waste of resources and money.
   b. react with organics almost exclusively and not much with pathogens or algae; thus it is a waste of resources and money.
   c. sometimes produce disinfection by-products known to be carcinogenic.
   d. react by as much as 95% of its concentration with concrete walls and metal structures before oxidizing pathogens, organics, and algae.

3. $C \times T$ values are based on
   a. concentration of chlorine, contact time, and pH.
   b. concentration of chlorine, contact time, pH, and temperature.
   c. concentration of chlorine, contact time, pH, and water impurities.
   d. concentration of chlorine, contact time, alkalinity, pH, and temperature.

4. Which index determines the calcium carbonate deposition property of water by calculating the saturation pH, where a negative value indicates corrosive water and a positive value indicates depositing water?

   a. Baylis curve

   b. Langelier saturation index

   c. Marble test

   d. Ryzner index

5. Which type of chlorine gas feeder is most commonly used?

   a. Pressure

   b. Combination water and pressure

   c. Vacuum

   d. Combination pressure and vacuum

6. Which device(s) uniformly disperse(s) the chlorine solution into the main flow of water?

   a. Injectors

   b. Pressure regulating valve

   c. Diffusers

   d. Effluent nozzles

7. The pressure in a chlorine cylinder depends on the

   a. amount of chlorine in the cylinder.

   b. temperature of the chlorine liquid.

   c. vacuum placed on the regulator.

   d. amount of gas being withdrawn.

8. Chlorine cylinders and ton containers are equipped with valves, which must comply with standards set by which organization?

9. What is an injector?

10. What is an evaporator?

11. List four operating problems related to chlorination.

12. Compared to air, how dense is chlorine gas?

Chapter 8

Iron and Manganese Treatment

Iron and manganese are often present in groundwaters and surface waters. Iron is more prevalent in groundwaters. When manganese is found in groundwaters, it is usually accompanied by iron.

Iron and manganese found in groundwaters originate when rock strata rich in iron and manganese are exposed to acidic water devoid of oxygen from anaerobic activity. In groundwaters, iron may also be present as soluble ferrous bicarbonate in alkaline wells or as soluble ferrous sulfate in acid mine drainage waters, or in waters high in sulfur. Iron is present as suspended insoluble ferric hydroxide in groundwater exposed to air and as a product of pipe corrosion.

At times, iron and manganese in the oxidized form are present in surface waters, usually as organic complexes.

Excessive Iron and Manganese

Excessive iron and manganese in drinking water can result in aesthetic and operational problems, potentially causing customers to seek out alternate supplies of drinking water that may not be safe.

Aesthetic Problems

Iron and manganese in the concentrations that occur naturally in groundwater and surface water have no known adverse health effects. However, the aesthetic problems they can cause may be quite serious from a consumer's standpoint. Iron and manganese in raw water are generally in the soluble, reduced, divalent state. The water is clear, and the substances are not noticeable aside from the taste and odor effect that they may cause at high concentrations. When they are oxidized, iron and manganese change and discolor the water from turbid yellow to black, depending on their concentration and the presence or absence of other contaminants.

When a groundwater system pumps water directly from wells to the distribution system and uses no disinfection or other treatment, dissolved iron in the water usually first becomes oxidized when it is exposed to the oxygen in air. After a customer fills a glass, bathtub, or washing machine with water, the iron gradually oxidizes and changes color. This property results not only in water that is unpalatable for consumption, but also in stained porcelain fixtures and discolored laundry (Figure 8-1).

The reaction between the high levels of iron and the tannic acid in tea and coffee can also cause customer complaints. In some cases, the beverage will darken so that it looks like ink.

---

iron

An abundant element found naturally in the earth. As a result, dissolved iron is found in most water supplies. When the concentration of iron exceeds 0.3 mg/L, it causes red stains on plumbing fixtures and other items in contact with the water. Dissolved iron can also be present in water as a result of corrosion of cast-iron or steel pipes. This is usually the cause of red-water problems.

manganese

An abundant element found naturally in the earth. Dissolved manganese is found in many water supplies. At concentrations above 0.05 mg/L, it causes black stains on plumbing fixtures, laundry, and other items in contact with the water.

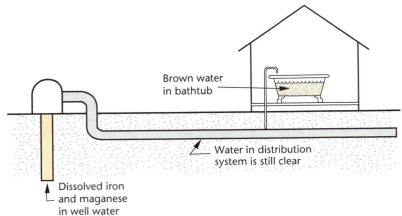

Figure 8-1 Iron and manganese oxidized after being exposed to air in a customer's plumbing fixtures

The presence of iron or manganese in the distribution system often provides a food source for bacterial growth. The bacterial slimes contribute to aesthetically objectionable tastes and odors. The presence of manganese is a problem because it creates brown spots on laundry. It is also a major problem for industries that incorporate water into their product because it will react with other chemicals to form undesirable tastes, odors, or colors. Manganese also tends to accumulate, corrode, and clog industrial fixtures. The maximum desirable level of manganese is 0.05 mg/L, and the point at which it creates an undesirable taste is about 5 mg/L.

Operational Problems

The presence of iron in a water distribution system may also be caused by corrosion of metal pipes in the system. In this case, the problem must be corrected by corrosion control, as discussed in Chapter 11.

If a disinfectant is added to the water, or if iron and manganese are fully or partially oxidized by any means before entering the distribution system (Figure 8-2), the oxidized iron and manganese will precipitate in the distribution system. The following problems may occur:

- Much of the precipitate could settle out in the mains. The worst problems will usually be in dead-end mains, where velocity is the lowest. If the iron and manganese problem in the system is not very serious, sometimes only the customers on dead ends will continually have a problem with rusty water.

- Sudden demands for extra water, such as the opening of a fire hydrant, may disrupt the normal flow in the system. The sediment that has accumulated on the bottom of mains will then be put back into suspension. Parts of the system, or even the whole system, will then have rusty water for a few hours or even a day or two. Customers will register complaints during and after the event—particularly those who were doing laundry at the time.

The presence of iron and manganese in the distribution system, in either the dissolved or oxidized state, can also provide a food source for bacterial growth in the system. The bacterial slimes that form can have the following detrimental effects:

- Reduction in pipeline flow capacity
- Clogging of meters and valves
- Further discoloration of the water as a result of the bacterial growth
- Increased chlorine demand

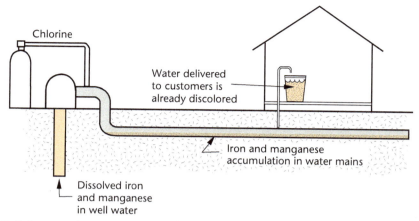

Figure 8-2 Iron and manganese oxidized by addition of chlorine

Control Processes

Iron and manganese are controlled or removed from water using sequestration for control and precipitation for removal. Precipitation is discussed first.

Precipitation

In the precipitation process, the soluble forms of iron and manganese are oxidized to insoluble ferric and manganic compounds, similar to the mechanism that rusts iron. Oxidized iron and manganese precipitate as ferric oxide or oxyhydroxides and manganese hydroxide. Following oxidation, the precipitates are removed from the water by a combination of settling and filtration, direct filtration, or membrane filtration.

Both iron and manganese are effectively removed by lime softening, but costs associated with chemical usage and sludge disposal usually make the process too costly to use unless water softening is also desired. This process is discussed in detail in Chapter 12.

Oxidation

In the precipitation of iron and manganese from water, the first step is oxidation to the insoluble state. The Stage 1 Disinfectants/Disinfection By-products (D/DBP) Rule limits trihalomethanes and haloacetic acids. Iron and manganese treatment strategies must be evaluated in light of this rule.

Air is often used to oxidize iron. If manganese is present, chlorine, either as sodium or calcium hypochlorite, ozone, chlorine dioxide, or potassium permanganate is required.

Unless there is past experience with a specific water supply, each oxidant must be evaluated using bench-scale jar testing to select the best process conditions.

The pH of the water must be adjusted to an optimum value, determined not only by the solubility of the precipitate but also by its charge. It may be necessary to determine the zeta potential in selecting the optimum pH adjustment. For example, lime is usually more effective than caustic soda at the same pH, and this may very well be attributed to the charge on the particles and charge neutralization.

The optimum pH value will establish detention time—that is, the time required for oxidation to occur at a given concentration (percentage of stoichiometric). A number of relationships involving the concentrations of iron or manganese and the pH, temperature, and oxidant used affect the detention time.

In cases where only iron is present at low concentration, the reaction is almost instantaneous. In other cases, particularly where manganese is present, detention times up to an hour may be required for complete oxidation. If the reaction has not been completed when the water is filtered, the soluble forms will pass through the filter and will later precipitate in the distribution system.

Once the iron has been oxidized and precipitated, the volume of sludge produced must be examined to determine whether the treated water can then be clarified by direct filtration or will require treatment through a sedimentation tank prior to filtration. In general, where the iron is less than 5 mg/L, the oxidized water can be fed directly to a filter. The filter used is either a mixed-media filter provided with air scour devices so that the bed can be kept clean, a pleated membrane filter, or a hollow-fiber membrane operated in a dead-end mode.

Very effective oxidation of iron and manganese can be achieved by using ozone. However, ozone alone is rarely used for this purpose because of the high cost of equipment, operation, and maintenance. Ozone can also be used to control trace organics, if this happens to be a problem in the water. Another factor that must be considered is that excessive amounts of ozone can oxidize manganese to permanganate, which will cause pink water.

Chlorine dioxide is a powerful oxidant, second only to ozone in biocidal efficacy, but without ozone's high costs. Chlorine dioxide does not generate ozonation by-products or biodegradable organic by-products, such as aldehydes and carboxylic acids. A by-product of concern in using chlorine dioxide as an oxidant is the formation of chlorite, a reduction by-product regulated under the Stage 1 D/DBP Rule at 1.0 mg/L maximum contaminant level. Chlorine dioxide instantaneously reacts with soluble iron and manganese to form insoluble precipitates easily removed by filtration.

Potassium permanganate is very effective in oxidizing both iron and manganese, and the reaction is rapid. Another benefit of using permanganate is that it reacts with hydrogen sulfide, cyanides, phenols, and other taste-and-odor compounds if present. Again, no trihalomethanes are known to be formed. Care, however, must be taken not to overfeed permanganate, as purple water will be discharged to the distribution system.

Removal

After the precipitates of iron and manganese are formed by the oxidation process, they are removed by filtration.

Granular Media Filters Granular media filters are generally used for removing iron and manganese precipitates. If the solids concentration is relatively low (under approximately 5 mg/L), then the water can usually be processed directly by filtration, without sedimentation (Figure 8-3). If the solids concentration is higher, the water must be clarified using a sedimentation step to remove as much precipitate as possible before the water is filtered. Lime is often added to provide alkalinity and facilitate iron precipitation. If the loading is not properly reduced through sedimentation, then filter backwashing will be excessive.

Manganese Greensand Filters Manganese greensand filters use a special type of medium that removes iron and manganese by a combination of both adsorption and oxidation. In the process, permanganate is added ahead of the greensand filter to allow the grains of the medium to become coated with oxidation

granular media filter
Materials used to filter raw water, including granular filter media sand, crushed quartz, garnet sand, manganese greensand, and filter coal, typical ranging in sizes from 0.25 mm to 1.20 mm, though some sizes can be larger or smaller than this range. Water treatment uses other granular filter media, but this list cites those most often used.

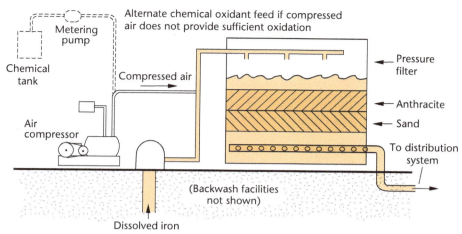

Figure 8-3 Low concentration of iron oxidized with compressed air or oxidant chemical and removed with a pressure filter

products. The oxidized greensand then adsorbs the dissolved iron and manganese from the water, after which the substances are oxidized with permanganate and removed by the filtering action of the filter bed. A potassium permanganate backwash is used to regenerate the bed, or permanganate is fed continuously in a small dose.

Greensand grains are somewhat smaller than silica sand, so the head loss can quickly become excessive under a heavy loading. The length of filter runs can be increased by adding a layer of anthracite above the greensand.

When the lowest concentrations of iron and manganese are required, and when footprint and chemical disposal issues are critical, membrane filtration should be considered. Membrane systems are able to reduce pumping costs by operating without breaking head; that is, the water is pumped directly from the well to the membrane and subsequently to the distribution system without the need to collect filtered water in a clearwell or other storage vessel.

Pleated Membrane Microfiltration   An emerging microfiltration technology offers the economy of pleated media filtration with the removal performance of fine membranes. Although the patented pleated filter configuration was specifically designed for iron removal from groundwater, this technology is currently in operation in France, processing 40 million gallons per day (mgd; 150 ML/d) of river water as pretreatment to final-stage nanofilters.

The membrane module, known as Septra XS, is a high-area, coreless, single, open-ended, pleated cartridge with an outside-to-inside flow pattern. The membrane traps within the filter matrix all debris and particles 0.1 micron and larger. The filter system is operated in the dead-end mode at high pressure, which allows treatment without breaking head. Every 2–4 hours, air is injected at the feed inlet, mixing with the water in the filter to remove any iron or particulate particles trapped in the filter matrix. This infrequent cleaning maintains a stable flux while keeping recovery at or above 99 percent.

Recovery is the ratio of the volume of product water, or filtrate produced, to the volume of raw water treated. Process recovery is a function of the filtrate flow rate, length of the filtration cycle, and volume of water used in a backwash. At 99 percent recovery, less than 1 percent of feedwater must be handled as waste. The high flow capacity and high-pressure properties of the filter allow the unit to operate without excessive head loss.

Membrane Filtration   Hollow-fiber microfiltration and ultrafiltration membrane technology are effective in iron and manganese removal.

Microfiltration systems use uniquely designed filtration modules with a hollow-fiber membrane made of materials such as polyvinylidene fluoride (PVDF). After suitable oxidation with aeration, chlorine, potassium, permanganate, ozone, or chlorine dioxide, microfiltration systems with pore sizes about 0.1 micron remove turbidity, iron, and manganese from groundwaters.

In the case of PVDF hollow fibers, no requirement exists to remove the oxidant prior to contact with the membrane. The microfiltration systems are highly permeable, resulting in high water production rates with a very small footprint and minimal operator attention.

Ultrafiltration has been showcased together with ozone for iron and manganese removal. Typically, cross-flow energy costs make this technology operationally impractical.

In theory, nanofiltration and reverse osmosis remove iron and manganese in a reduced state without oxidation. These technologies are operationally impractical because of membrane fouling concerns, and capital and operating costs.

Sequestration

In the sequestration process, polyphosphates or sodium silicates are added before the water is exposed to air or disinfectants. The total phosphates applied should not exceed the amount specified by the chemical supplier. If effective, sequestration tends to keep iron and manganese soluble in the finished water. Since it does not *remove* the iron and manganese, bacterial slimes may still form in the distribution system as a result of bacterial growth. In addition, a chlorine residual of at least 0.2 mg/L should be maintained in the system at all times.

Sequestration is effective only for groundwater with a relatively low level of dissolved iron and manganese and no dissolved oxygen. It is not usually recommended if the concentration of iron, manganese, or a combination of the two exceeds 1.0 mg/L.

Control Facilities

This section discusses types of equipment used to control iron and manganese.

Aeration Equipment

Aeration equipment that can be used for the oxidation of iron and manganese is detailed in Chapter 15. The medium used in an aerator, such as coke or broken stone in a cascade aerator, becomes coated with iron or manganese hydroxide over a period of time. It has been found that this promotes catalytic precipitation of iron and manganese from the raw water. The process of aeration not only exposes the dissolved iron and manganese to oxygen, but also removes carbon dioxide from the water, which enhances the oxidizing action because it raises the pH of the water. It takes only about 0.14 mg/L oxygen per milligram per liter of iron to oxidize iron so long as the pH is above about 6.5. Oxidation of manganese by aeration is not effective at a pH below about 9.5.

It is not usually advisable to aerate water any more than necessary to achieve oxidation. Although it is not corrosive when it is oxygen-free, low-alkalinity water becomes quite corrosive when it is heavily saturated with oxygen.

Most aeration schemes require that the water be pumped twice. It must first be pumped from the source and through the aeration device, and then pumped again to produce the required distribution system pressure. Line diagrams of typical installations are shown in Figures 8-4 and 8-5.

If only a low level of iron is present and it has oxidized quickly, it may be possible to aerate and filter the iron-bearing water under pressure. As illustrated in Figure 8-3, compressed air can be pumped into the water entering the filter or into the space above the filter. It can produce sufficient oxidation to precipitate the iron in the short time before the water passes through the filter media. The filter in this type of installation uses relatively coarse media.

Chemical Oxidation Equipment

Chlorine gas from a chlorinator or chlorine dioxide can be applied directly to the raw-water pipeline or contact basin. Liquid chlorine can be fed using a metering pump.

Potassium permanganate can also be injected into a pipeline or gravity-fed into a detention basin (Figure 8-6). The permanganate can be fed as needed by a dry-chemical feeder, or batches of concentrated solution may be prepared and fed as a liquid. If ozone is used, transfer of oxygen to the water must take place in a reaction basin.

If an oxidant other than chlorine is used for the iron and manganese removal process, a system could, by choice or by state mandate, be required to add postchlorination before the water leaves the treatment plant to maintain a residual in the distribution system.

Detention Equipment

Unless pilot tests have shown it to be unnecessary, a detention or contact chamber is needed to hold the water temporarily after aeration or the addition of an oxidant. A contact time of 20 minutes is about average, but as long as 1 hour

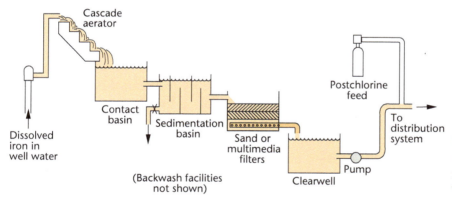

Figure 8-4 Iron removal using aeration, sedimentation, and filtration

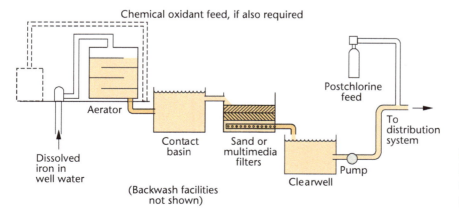

Figure 8-5 Iron removal by oxidation using a contact basin but no sedimentation

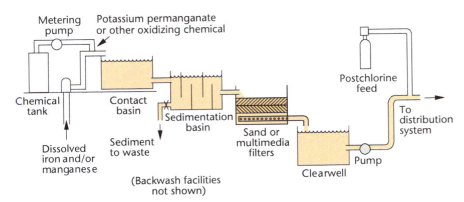

Figure 8-6 Iron and manganese removed by chemical oxidation, sedimentation, and filtration

may be required under some circumstances to allow the oxidation reaction to be completed.

The chamber is usually a concrete, steel, or fiberglass tank with sufficient capacity to hold the water for the required time at the maximum flow rate. If an aerator is used to provide oxidation, the detention tank can be placed directly below the aerator. If the concentration of iron and manganese is relatively low, flow from the contact chamber can be applied directly to the filter. If the level is higher, the waste then flows to a sedimentation basin.

Sedimentation Basins

Unless shown to be unnecessary by a pilot-plant study or past experience, a sedimentation basin is required after the oxidation process to allow as much precipitate as possible to be removed before filtration. Sedimentation basins are generally constructed similarly to basins for conventional filtration systems, as detailed in Chapter 5. Plain sedimentation is sufficient on some systems, but the required settling period may be as long as 12–24 hours.

If iron or manganese removal is particularly difficult, additional treatment may be required to hasten sedimentation. Some common methods are as follows:

- Lime, or lime plus alum, can be added and mixed for 20–30 minutes before sedimentation. This raises the pH and facilitates precipitation. If the pH has been sufficiently increased to cause a chemical balance of the water, recarbonation may be required for stabilization before the water is released to the distribution system.
- Alum can be mixed into the water before sedimentation.
- Alum and bentonite can be mixed into the water before sedimentation.
- Alum and specially treated silica can be mixed into the water before sedimentation.

Sand Filtration Equipment

Filtration using either sand alone or sand capped with anthracite can be used for open or tank-type filters, as discussed in detail in Chapter 6. The media that work most efficiently may not be the same gradation as used in the conventional filtration of surface water.

Manganese Greensand Equipment

Granular filtration using manganese greensand includes a rather standard-looking filter unit, except that the unit uses special filtering media. The greensand is usually capped with about 6 in. (50 mm) of anthracite to act as a roughing filter medium. No preliminary contact or sedimentation chambers are necessary. Potassium permanganate is continuously fed as far in advance of the filter as possible. Other oxidizing agents or processes, such as chlorination or aeration, can be used prior to the permanganate feed in order to reduce the cost of materials.

The filter must be furnished with provisions for air-backwash, and also with sample taps at various points opposite the media bed for monitoring the progression of permanganate through the filter. The filters are backwashed with a permanganate solution to regenerate the bed.

Pleated Membrane Filtration Equipment

Pleated filter microfiltration systems are production-scale, membrane filtration systems that use pleated backwashable membranes made of PVDF. Equipment includes racks that hold 2, 5, and 10 modules with associated ancillary equipment. See Figure 8-7.

The filtration skid is constructed with a painted steel frame and stainless-steel or Schedule 80 polyvinyl chloride (PVC) piping. Valves are manual and air operated. Tanks are polyethylene. Instrumentation includes programmable logic controller (PLC) controls and software. Standard system design pressure is 45 psi (300 kPa). A 150-psi (1,000-kPa), high-pressure unit is available.

Membrane Equipment

Membrane equipment used for iron and manganese removal is described in Chapter 16. A typical microfiltration skid is shown in Figure 8-8.

Sequestering Chemical Feed Equipment

Polyphosphates and sodium silicates used for sequestering iron and manganese are normally purchased as dry chemicals and prepared as a solution. The solution is then fed into the system by a metering pump (Figure 8-9). The feed point should

Figure 8-7 Septra CB rack assembly, 10 modules, 450 gpm nominal capacity

Courtesy of Pall Corp.

Figure 8-8 Typical microfiltration skid

Courtesy of Ionics, Inc., Water Systems Division.

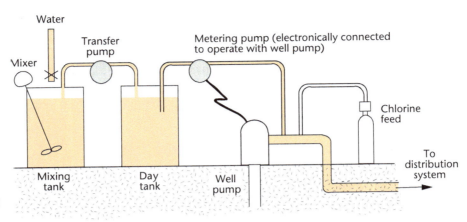

Figure 8-9 Sequestering chemical feed equipment

be located immediately after the water leaves the well to minimize oxidation of the iron and manganese before the sequestering chemical is added.

Regulations

The Safe Drinking Water Act sets no maximum contaminant levels (MCLs) for iron or manganese in drinking water. Iron and manganese do have a secondary maximum contaminant level (SMCL) attached to them, however. SMCLs are established only as guidelines to assist public water systems in managing their drinking water for aesthetic considerations, such as taste, color, and odor. These contaminants are not considered to present a risk to human health at the SMCL. However, the following SMCLs have been established by the US Environmental Protection Agency (USEPA):

- Iron, at 0.03 mg/L
- Manganese, at 0.05 mg/L

USEPA strongly recommends that iron and manganese be limited to below noticeable limits to avoid tastes, odors, color, staining of fixtures, or other objectionable qualities of the water. There is also fear that customers who are bothered by aesthetically displeasing water from a public water system will use water from another source that is aesthetically pleasing but may not be chemically or microbiologically safe.

Many states have established mandatory limits on the amount of iron and manganese in drinking water. In some cases, the limit is the same as the federal secondary limit, and in other cases it is somewhat more stringent. The state public water supply program should be consulted for information on current state requirements.

All chemicals used in iron and manganese control must meet AWWA and NSF International standards.

Manganese Greensand Filter Operation

Manganese greensand has been used for several decades in North America, often specifically for iron and manganese removal. Manganese greensand is a purple–black granular filter medium processed from glauconite sand. The glauconite is synthetically coated with a thin layer of manganese dioxide—$MnO_2(s)$—and some particles have a definite green color, giving the material its common name. The only North American manufacturer of manganese greensand is located in New Jersey, USA.

Physical Characteristics of Manganese Greensand

Glauconite exhibits an ion exchange capacity that allows the surface to be saturated with manganous ions. Following saturation, the glauconite is soaked in a strong oxidizing solution, which transforms the manganous ion to the insoluble $MnO_2(s)$ form. The surface coating makes up about 4.0 mg of a 1,000-mg greensand sample; about 0.4 percent of the weight of a particle of greensand is actually $MnO_2(s)$. Since $MnO_2(s)$ is black, particles not covered or only partially covered with the substance are easily identified by the greenish color of the underlying glauconite sand.

Manganese greensand has an effective size of 0.30–0.35 mm, a uniformity coefficient of less than 1.60, and a specific gravity of approximately 2.4.

Installation of Manganese Greensand

The process of installing a manganese greensand filtration system is described as follows:

1. *Initial backwash.* After placement of the greensand in a filter vessel, and prior to placing any filter coal, the greensand must be thoroughly backwashed and skimmed (i.e., removed from the surface). Commence the backwash cycle and increase the backwash rate to full flow (called "ramping up"), and continue at full rate until the backwash effluent runs clear. The backwash water will start off black and pass through shades of gray, running light gray for some time. Depending on the depth of the manganese greensand, this initial backwashing could last from 15 to 30 minutes. However, this process of removing fines (i.e., manufacturing dust, particles too small to be good filtering material, and other small impurities) is critical to optimized filter runs under actual operating conditions.

2. *Greensand stratification.* After the backwash water has cleared, ramp down the backwash rate slowly to ensure that the grains of greensand stratify. During backwashing, most of the greensand fines migrate to the surface of the filter bed, where they must be physically removed in a step known as skimming.

3. *Skimming greensand.* At least 2.5 cm (1 in.) per 30 cm (12 in.) in the installed greensand's depth should be removed, because extending the initial backwash cycle will not remove the majority of fines from the greensand surface. Example: If 46 cm (18 in.) of greensand has been installed, at least 4 cm (1.5 in.) should be skimmed. (Another term used to describe the physical removal of fines is undercutting.) Skimming is best accomplished with a flat-mouthed shovel.

The skimming step is critical, because failure to remove the fines will result in shortened filter runs. Left in the filter, the fines form a very dense layer at the top of the greensand. In a relatively short time, this dense layer becomes even denser with accumulation of turbidity from the raw water, oxidized iron, agglomerated iron and manganese, and other fine particulates. As its density increases, this layer restricts the flow of water and builds up head pressure. Backwashing becomes necessary long before the rest of the bed's capacity is reached, or it leads to a pressure rise that can fracture the glauconite and the $MnO_2(s)$ coating.

Skimming granular filter media (including manganese greensand) requires the use of the following equipment:

- *Plywood.* Small pieces of plywood (about 60 cm × 60 cm [24 in. × 24 in.]) should be placed on the top of the media to avoid walking directly on the bed's surface and sinking down into it.
- *Flat-mouthed shovel.* A scraping motion can be used to draw the surface material to a central location, from which it is then removed. As an alternative, the shovel can be used to undercut the surface, removing material from about 180 cm² (29 in.²) of area per scoop.
- *Bucket.* The easiest way to dispose of the fines is to place each scoop into a plastic bucket, then periodically dump the bucket into another container on the floor of the water treatment plant for final disposal.

Conditioning Manganese Greensand

Virgin greensand (i.e., new and unused material) is not shipped in a regenerated form. Therefore, it must be conditioned prior to being put into service. The manufacturer recommends soaking the bed for not less than 1 hour in a solution containing about 60 g (2 oz) of potassium permanganate ($KMnO_4$) per 28,320 cm³ (1 ft³) of media.

Greensand can also be regenerated using chlorine. Typically, a new bed is soaked in a solution containing about 100 mg/L Cl_2 for several hours.

Manganese Greensand Operating Modes

The typical operating mode for manganese greensand treatment involves preoxidation followed by filtration. In this sequence, called continuous regeneration, a strong oxidant such as $KMnO_4$ is added ahead of the greensand filter. In theory, by continuous preoxidation of both the iron and the manganese to an insoluble particulate form, both the insoluble metals are removed by physical filtration/straining action through the greensand media. Any iron or manganese that is not oxidized (perhaps due to underfeeding of the oxidant) is adsorbed onto the $MnO_2(s)$ surface on the greensand particles.

Operating in this mode, a filter bed accumulates solids, which are then removed during the backwash cycle. The inherent stickiness of most oxidized iron and manganese compounds makes removal with backwash water alone a difficult task. Backwash assisted by air scour is recommended by the manufacturer, so that the bed is kept as clean as possible. Concerns about removing the medium's $MnO_2(s)$ coating by excessive air scouring could be alleviated by limiting air scour to low rates, short periods, infrequent intervals, or a combination of the three. Keeping the media clean is the key to a sustainable filtration process and extended media life.

A second operating mode for manganese greensand is oxidized iron filtration and manganese removal by adsorption. Typically, raw water is aerated to oxidize the iron, the major portion of which is then removed in the coal layer that tops the greensand. The manganese in solution remains until it is adsorbed onto the fully regenerated $MnO_2(s)$ coating. This sequence is called intermittent regeneration. Formulas have been developed to calculate appropriate filter run lengths, but a practical guideline calls for backwashing and regeneration when the manganese level in the filter effluent reaches 0.05 mg/L. Operators interested in optimum water quality end filter runs at some point before 0.05 mg/L of manganese is reached in the filter effluent.

Intermittent regeneration is preferred where raw water constituents interfere with the preoxidation/filtration process. The specific nature of these interferences is not fully understood. Difficulties that appear in one region do not appear to cause problems in other areas.

In some regions, levels of organic carbons above about 2 mg/L in groundwater are almost always accompanied by both ammonia and hydrogen sulfide. This combination frequently presents difficulties for efforts to remove iron and/or manganese to satisfactory levels. Are the iron and manganese organically complexed in such situations? If they behave as though they are, the question is academic. Pilot testing should guide the choice of a removal process that works, rather than devoting resources to nailing down the precise chemical reason for the removal difficulty.

A third mode of operation for manganese greensand filtration involves oxidation of iron by aeration, followed by a chlorine feed at least sufficient to continuously regenerate the $MnO_2(s)$ surfaces, on which the manganese is then removed by adsorption. In this mode, sufficient chlorine can be fed to take care of both regeneration and other chemical demands, while providing the necessary free chlorine residual in the filter effluent. Typically, dosing that gives a free chlorine residual in the filter effluent of 0.5 mg/L results in continuous regeneration of the $MnO_2(s)$ surfaces.

Regeneration of Manganese Greensand

In any operating mode, to continue removing iron and manganese, manganese greensand requires regeneration. The manufacturer recommends continuous regeneration for well waters where iron removal is the main objective with or without the presence of manganese. This method involves feeding sufficient $KMnO_4$ and/or Cl_2 to satisfy all chemical demands, including regeneration of sites on the $MnO_2(s)$ coating occupied by adsorbed iron and/or manganese.

Actual filter audits have demonstrated a need for careful dosing. Effluents from some filters receiving slightly pink water have contained higher levels of manganese than were found in the raw water. This effect resulted from $MnO_2(s)$ surfaces coated with oxidized iron, scale formation on the greensand particles, or a high percentage of manganese greensand particles backwashed away as the grains were replaced by heavier mudballs double the size of greensand (which have the appearance of manganese greensand when wet). All three conditions can drastically reduce $KMnO_4$ demand. In other words, a slight pink feed provided too much regenerant for a filter bed with a very low regenerant demand. Because $KMnO_4$ is about 35 percent manganese, any overfeed of it may elevate manganese residuals in the finished water.

For Cl_2 dosing, the rule of thumb is to prefeed enough to satisfy all chemical demands and give a free chlorine residual in the water leaving the treatment plant adequate to comply with regulatory standards.

The greensand manufacturer recommends intermittent regeneration for well waters needing treatment for manganese alone or for manganese with small amounts of iron. Briefly, intermittent regeneration involves addition of a predetermined amount of $KMnO_4$ or Cl_2 in solution applied to the manganese greensand bed after a specified quantity of water has been treated. The company's recommended dosage is about 60 g (2 oz) of $KMnO_4$ per 28,320 cm³ (1 ft³) of manganese greensand. A Cl_2 solution of about 100 mg/L free chlorine can also be used.

These rules of thumb do not reflect a relationship between the regenerant used and the amount of iron and manganese accumulated in the bed. Some formulas take account of the molecular weights of the metals to be removed and the oxidants to be applied, allowing calculation of amounts of oxidant for specific quantities of water treated and metals removed, assuming a fully regenerated bed at the start of the filter run and the absence of other demands on the oxidant of choice. These formulas are best applied according to site-specific considerations derived from actual examples, since the results are then often weighted by other raw-water considerations or oxidant use upstream of the filter. Consult your treatment specialist or process consultant if these calculations are required.

A 1991 research project by led William R. Knocke reached the following conclusion:

(1) The sorption of Mn(II) by $MnO_x(s)$-coated filter media is very rapid. Both the sorption kinetics and sorption capacity increase with increasing solution pH or surface $MnO_x(s)$ concentration or both.

(2) In the absence of a filter-applied oxidant, Mn(II) removal is by adsorption alone. There was no evidence to substantiate any auto-oxidative reaction between the sorbed Mn(II) and the $MnOx(s)$ surface over the pH range examined. . . .

(3) When free chlorine is present, the oxide surface is continually regenerated, promoting efficient Mn(II) removal over extended periods of time. This means that treatment facilities that practice prefilter chlorination are maintaining these oxide surfaces in a viable state for continuous Mn(II) removal. Conversely, treatment facilities that decrease or totally eliminate prefilter chlorination may subsequently be faced with elevated Mn(II) concentrations in the effluent waters (Knocke, Occiano, and Hungate 1991).

Topping Coal

The use of a filter coal (anthracite) layer on top of a manganese greensand bed is often a critical step in reducing the solids load (i.e., filtering out particulate matter from pretreated water), promoting optimum filter run lengths. The effective size, uniformity coefficient, and specific gravity of the topping coal are important choices. Poor combinations result in intermixing of the two media. A higher degree of intermixing reduces the capability of each medium to perform its intended function.

A filter coal layer over manganese greensand gives two benefits. The first results because on average the coal particles are about three times larger than the greensand particles (0.9 mm compared to 0.3 mm). Because of this difference, the spaces between the coal particles are also significantly greater, so they can hold a greater volume of filtered material than the manganese greensand can hold.

The space between particles of granular filter media is referred to as its void area (although some use the term bed porosity). The amount of filtered material needed to fill those void areas is known as the filter medium's solids holding capacity. Coal can hold much more filtered material than manganese greensand before the flow of water becomes restricted. Typically, filter coal with an ES of 0.7–0.8 mm has a void volume of approximately 60 percent. A volume of manganese greensand can hold much less than the same amount of a compatible filter coal before the flow is restricted. Typically, the void volume of manganese greensand is 35–40 percent, according to the manufacturer. Without a coal layer, the restricted flow through the manganese greensand bed can build up pressures high enough to fracture the manganese greensand particles and their $MnO_2(s)$ coating.

The second benefit of a coal layer results because the void area in the coal provides a place for oxidized iron and manganese to flocculate (i.e., bunch up). Pretreatment of the raw water aids filtration. For example, Cl_2 or $KMnO_4$ is added to oxidize the iron. As the iron goes through a chemical change, it usually forms solid particles that join together into bunches big enough to get caught between the particles of filter media. As the oxidized iron particles jostle through the coal layer, they collide with each other, stick together and to the coal particles, and become trapped. This description simplifies removal of particulate.

If the oxidized iron is permitted to flow down into the manganese greensand layer, it tends to coat the particles with iron oxides over time. No further manganese can be adsorbed by the $MnO_2(s)$ at these locations, and the manganese greensand gradually loses its ability to remove manganese.

Some exceptions complicate filtration. Some iron species under certain oxidation conditions remain fine enough to sift right through the filter bed. Some of the possible solutions are increasing detention time (i.e., the length of time the oxidizing chemical is in contact with the iron before the water reaches the filter), adding a flocculant (for example, alum) following oxidation, and/or adding a polymer (sometimes referred to as a filter aid) before filtration.

Despite variations, a properly designed coal layer is an important tool for keeping the manganese greensand clean and extending its useful life. Proper design of the filter coal layer must ensure the optimum hydraulic backwash rate, since the rate for the filter coal is slightly less than the optimum rate for manganese greensand. Since greensand is produced by only one manufacturer, operators can rely on a predictable consistency of particle size and specific gravity, which aids in calculations of appropriate backwash rates. The same is not true for topping coal. Specific gravities of coal samples can vary from 1.3 to 1.8, and uniformity coefficients can vary from 1.3 to 2.0, depending on the source of supply. An optimum design includes topping coal as coarse as possible while retaining the ability to fluidize at a backwash rate just below the fluidization rate for the manganese greensand. To achieve this balance, coal with the right effective size, uniformity coefficient, and specific gravity must be chosen.

Experience suggests the following general guidelines:

1. Both the glauconite mineral sand grains and the $MnO_2(s)$ can be fractured at pressure differentials over 8 psi (55 kPa).

2. The fine particle size distribution of manganese greensand and the typical depth installed in many filters result in a high clean bed head loss. This means that water is much more reluctant to flow, even through clean manganese greensand, than through typical sand and/or coal, which have larger particle sizes. To achieve acceptable filter run lengths, a clean bed is important, which means only minimum backwash flow maldistribution can be tolerated and the backwash volume must meet specifications. Failure on either count results in reduced life of the manganese greensand.

3. If pressure differentials result in particle fracturing, fines are created and accumulate on the surface of the bed. This accumulation in turn reduces filter run lengths by raising clean bed head loss. If the production of new fines resulting from excessive pressure differentials is not identified and corrected, the cycle will repeat itself until replacement of the manganese greensand is necessary.

4. The $MnO_2(s)$ surface coating is also susceptible to attrition and surface iron fouling. Careful precautions are needed to maintain this surface coating in a clean condition, or its adsorptive capacities will be reduced or lost.

5. Some worry about attrition resulting from repeated air scouring of manganese greensand, although no available documentation substantiates the claim. The manufacturer's literature (Inversand, n.d.) recommends air/water scour as an option, using 4.1–10.2 (L/s)/m² (0.8–2.0 SCFM/ft²) of airflow with a simultaneous treated-water backwash at 9.8–12.2 m/h

(4–5 gpm/ft^2). In practical applications, an air scour rate of 10.2 (L/s)/m^2 (2 SCFM/ft^2) or more is required to generate the kind of vigorous action needed to loosen filtered particulate.

6. In order to maintain adsorptive capacity in manganese greensand, the bed must be regenerated, either continuously or intermittently following a backwash cycle. Once the MnO_2(s) coating on each individual particle of glauconite has adsorbed all the soluble iron and manganese it can, the bed acts only as a strainer. Regeneration results from bringing an oxidant into contact with the manganese adsorbed, which changes its chemical character to a form that can be removed by backwashing, perhaps with air scour. The same removal mechanism applies to adsorbed iron. Cl_2 and $KMnO_4$ are commonly used regenerants.

Process Monitoring

Equipment should always be available to test for the iron content of water to a minimum of 0.03 mg/L and the manganese content to a minimum of 0.005 mg/L. If the iron and manganese concentration in the raw water is variable, the water may have to be tested rather frequently to adjust the process for thorough, yet economical treatment.

When polyphosphate or sodium silicate is used in the appropriate analysis, equipment must be available for periodic testing of process control and to ensure that the maximum limit is not exceeded.

Operating Problems

Although the iron and manganese levels in groundwater are generally stable, in a few cases, they have drastically increased over a period of years. The design of a control method should therefore be flexible enough that the treatment can be changed to meet the demands of worsening water quality.

Record Keeping

If iron or manganese are being removed, the following important records should be maintained:

- Results of periodic analysis of raw water to monitor for changes in the iron and manganese concentration
- Concentration of the oxidizing chemical being fed
- Detention time being provided between oxidation of the water and sedimentation
- Concentration of any supplemental chemicals added to improve sedimentation
- Sedimentation time being provided
- Length of filter runs
- Quantity of water treated
- Results of periodic analysis of distribution system samples for iron and manganese concentration

- Details of all distribution system flushing done to control discolored water
- All customer complaints of discolored water

If iron and manganese are being controlled by the addition of a sequestering chemical, records should include the following:

- Results of periodic analysis of raw water to monitor for changes in the iron or manganese concentration
- Brand of chemical being used and the concentration being fed (noted daily)
- Quantity of water treated
- Details of all distribution system flushing done to control discolored water
- All customer complaints of discolored water

Study Questions

1. Which of the following would immediately occur if newly installed manganese greensand was not skimmed of the fines after backwashing and stratification steps were completed?
   a. Uneven flow through the bed
   b. Cracks in the bed
   c. Mudball formation
   d. Shorter filter runs

2. Manganese greensand filters can be regenerated by using
   a. a surface wash and an air-water backwash.
   b. brine water during backwashing.
   c. potassium permanganate solution during backwashing.
   d. first a brine solution during the first backwashing cycle followed by potassium permanganate solution for the second backwash cycle.

3. Which material is manganese greensand and which is the coating?
   a. Quartz sand coated with manganese hydroxide [$Mn(OH)_2$]
   b. Garnet sand coated with manganese dioxide [MnO_2]
   c. Ilmenite sand coated with manganese hydroxide
   d. Glauconite sand coated with manganese dioxide

4. Depending on water temperature, what is the typical backwashing flow rate for a manganese greensand filter bed?
   a. 7–8 gpm/ft^2
   b. 8–10 gpm/ft^2
   c. 10–12 gpm/ft^2
   d. 12–14 gpm/ft^2

5. The length of filter runs for manganese greensand filters can be increased by
   a. adding a high-molecular-weight polymer filter aid.
   b. keeping the pH above 9.0 and lowering it after filtration.
   c. keeping the pH below 7.3 and raising it after filtration.
   d. adding a layer of anthracite above the greensand.

6. Sequestration is effective only for groundwater with a relatively
   a. low level of dissolved iron and manganese and high level of dissolved oxygen.
   b. high level of dissolved iron and manganese and no dissolved oxygen.
   c. low level of dissolved iron and manganese and no dissolved oxygen.
   d. high level of dissolved iron and manganese and high level of dissolved oxygen.

7. Iron and manganese treatment strategies must be evaluated in light of which regulation?

8. Which type of filter uses a special type of medium that removes iron and manganese by a combination of both adsorption and oxidation?

9. Unless pilot tests have shown it to be unnecessary, what device is needed to hold the water temporarily after aeration or the addition of an oxidant?

10. What type of regeneration is preferred where raw water constituents interfere with the preoxidation/filtration process?

Chapter 9

Fluoridation Process Operation

Operation of the Fluoridation Process

Proper operation and maintenance of equipment are required to ensure uninterrupted and unvarying feed of fluoride chemicals. The following paragraphs give general guidelines for operation and maintenance.

Dry Feeders

Dry-chemical feeders should be inspected and cleaned routinely to prevent breakdowns. The belts, rolls, and disks or screws must be regularly inspected for signs of wear. Worn parts should be replaced before failure occurs. A lubrication schedule should be established based on the manufacturer's recommendations. The dissolving tank should periodically be inspected for precipitate buildup and cleaned if required. If the deposits are a result of hardness, the makeup water may have to be softened.

The calibration of dry feeders should be checked occasionally, particularly if the plant is having a problem maintaining a constant fluoride concentration. To do this, a small set of scales and a watch that indicates seconds are needed. A shallow pan or sheet of cardboard is inserted between the feeder's measuring mechanism and dissolving chamber to collect a sample over a short period of time. Several samples are collected in intervals, such as 5 minutes each; each sample is weighed and all of them are totaled. The weights of the individual samples will indicate the uniformity of feed, which should show less than 10 percent variation. The total should check against the rate setting on the feeder.

Saturators

The operator must periodically check to ensure that at least 6 in. (150 mm) of chemical is maintained in the saturator at all times. Saturators that treat over 100 gpm (6.3 L/sec) require a chemical depth of at least 10 in. (250 m). Lines drawn on the outside of the translucent containers will help in determining when to add the chemical.

Only crystalline sodium fluoride should be used in downflow saturators. Powdered sodium fluoride will quickly clog the gravel bed. Any form of sodium fluoride can be used with upflow saturators, but crystalline grade is most often used because less dust is produced. Sodium fluorosilicate should never be used with a saturator because it will not dissolve to produce a 4 percent solution. The saturator tank and gravel bed in a downflow saturator should be cleaned

saturator

A piece of equipment that feeds a sodium fluoride solution into water for fluoridation. A layer of sodium fluoride is placed in a plastic tank and water is allowed to trickle through the layer, forming a solution of constant concentration that is fed to the water system.

from one to three times a year depending on how rapidly the hardness scale builds up.

Metering pumps are quite reliable, but they should receive routine maintenance according to the manufacturer's recommendations. The pump head should periodically be dismantled for cleaning, and the check valves and diaphragm should be inspected and replaced if worn or cracked.

Fluoride Injection Point

The fluoride injection point should normally be located so that the chemical is applied after water has received complete treatment. In particular, it should not be fed before alum coagulation, lime softening, or ion exchange processes. If water treatment chemicals containing calcium (lime or calcium hypochlorite) are used in the treatment process, the fluoride injection point should be as far away as possible to prevent precipitation.

The feed point used in most treatment plants is just ahead of the clearwell. If fluorosilicic acid is applied into a horizontal pipeline, the point of application should be located in the lower half of the pipe. Injectors should be cleaned one to three times a year to prevent blockage caused by scale formation.

Chemical Storage

Chemical storage areas must be kept clean and orderly. Poor storage conditions, as shown in Figure 9-1, can cause safety hazards and loss of chemicals. The storage area for fluoride chemicals should be isolated from areas used to store other chemicals to avoid any possible mix-up when feeders are being charged. Bags of dry chemicals should be piled neatly on pallets as close to the feeding equipment as possible (Figure 9-2). Whenever possible, whole bags should be emptied into hoppers because partially emptied bags are difficult to store without spilling and generating dust.

Figure 9-1 Poor storage of chemicals

Figure 9-2 Proper chemical storage

Fluoridation Operating Problems

Two problems commonly encountered in fluoridation are (1) varying fluoride concentration and (2) measured concentrations that differ from those computed based on the feed rate.

Variable Fluoride Concentration

Fluoride levels detected in the distribution system will vary considerably for a period of time after fluoride is first fed at the treatment plant. This is partly because the fluoridated water is diluted by being mixed with the unfluoridated water held in storage.

Varying concentrations detected in the system may also indicate that a dry feeder needs recalibration. A variation of 0.2–0.3 mg/L over a 2- to 3-day period is not a serious problem; however, the operator should investigate the variation because it may indicate a more serious problem.

Low Fluoride Concentration

When the fluoride concentrations measured by laboratory tests are consistently lower than those calculated based on the feed rate, a number of problems may be present.

Assuming that the calculations, weight, and flow data are correct, the problem may be caused by something interfering with the laboratory test procedure. For example, aluminum introduced by alum in the coagulation process sometimes interferes noticeably with the test method and causes erroneously low readings. The fluoride concentration in samples should be rechecked by a comparison of results with those of the state or local health department laboratory to determine if this is the cause of the problem.

One of the most common causes of low readings is a fluoride underdose caused by inadequate chemical depth in a saturator or incomplete mixing in a dissolving tank. Deposits of undissolved chemical are an indication of incomplete mixing in a dissolving tank.

If the fluoride level is low in a sample collected from the distribution system, unfluoridated water could be mixing with the treated water. Groundwater sources are used periodically in some systems to supplant normal flows from the treatment plant. If the groundwater is low in natural fluoride, it will lower the concentration detected in parts of the system.

If fluoride is added before filtration, a significant fluoride loss will occur, resulting in low readings. Either the injection point will have to be moved downstream of the filters or a higher concentration of fluoride will have to be used to compensate for the loss.

High Fluoride Concentration

Several problems could be indicated if testing indicates a fluoride concentration consistently higher than the calculated concentration.

Polyphosphates used for water stabilization can cause high readings when the SPADNS method is used. This can be checked by the electrode method or by a comparison of results with those of the state or local health department.

Sometimes high readings occur because the natural fluoride in the water has not been measured or considered in the dosing calculations. The natural fluoride level in some surface water sources can vary greatly. If this is the case,

SPADNS method
A procedure used to determine the concentration of fluoride ion in water; a color change takes place following addition of a chemical reagent. SPADNS is the chemical reagent used in the test.

the level may have to be measured daily so the correct dosage can be calculated. The natural fluoride level in most groundwater varies only slightly from month to month.

Danger of Fluoride Poisoning

There have been several reported episodes of acute fluoride poisoning in the United States, caused by fluoride equipment that was improperly designed or operated or that malfunctioned. These cases were all caused by a fluoride feeder accidentally or purposely not being turned off when the supply pump stopped. As a result, high concentrations of fluoride built up in the supply line and later produced concentrations as high as 375 mg/L in water delivered to the system when the pump was turned back on. Although there were no fatalities, several hundred people became ill during these episodes.

In addition, at least one death has been attributed to excessive levels of fluoride in a public water system. In this case, several patients became ill while they were connected to kidney dialysis machines, and one later died. The official cause of death was ruled "acute fluoride intoxication," caused by a water system fluoride level about 15 times above normal as a result of a malfunction of fluoridation equipment. Although there were no other reports of sickness in this case, the occurrence does point out the extreme consequences that can result from not carefully controlling and monitoring fluoride feed.

Control Tests

The types of testing for fluoride concentration that are necessary in a water system include (1) measurement of the fluoride level before fluoride is added (to determine the natural fluoride level) and (2) measurement of the level after fluoridation to confirm that the correct amount of chemical is being added.

Daily Control Testing

The fluoride concentration of treated water should be measured daily to ensure that the level being delivered to customers conforms to the limits specified by the state.

If experience has shown that a surface water source has a variable natural fluoride content, daily testing of the raw water may be necessary to calculate the dosage needed for the fluoridation equipment. If experience has shown this dosage to be relatively stable, the raw water will need to be checked only periodically to determine whether there has been any change.

If a water system is drawing from several wells, the natural fluoride may vary from well to well, so the utility must compensate based on which wells are being used.

Continuous Control Testing

Continuous testing can be performed by automatic monitors. An automatic monitor is usually connected to a recording chart, and it also has connections for an alarm that will alert the operator if the level varies greatly from preset limits.

Plant operators using a water source that has variable natural fluoride levels can improve their operation by using a continuous monitor on the raw water.

In this way, they can anticipate changes and make adjustments to their fluoride solution feed rate.

Monitors equipped with control equipment can also be used to automatically control the fluoride feeder. Even though a plant has an automatic monitor, manual sampling should still be performed periodically as a check on the accuracy of the monitor.

Safety Precautions

Fluoridation chemicals present a potential health hazard to water plant operators through overexposure from ingestion, inhalation, or bodily contact.

Ingestion

Dry fluoride chemicals can be ingested through contaminated food or drink. Fluoride chemicals could be mistaken for sugar or salt if meals are eaten in areas where fluorides are stored or applied. No one should be allowed to eat, drink, or smoke in areas where fluoride chemicals are stored, handled, or applied.

Personnel who handle fluoride should be instructed not to touch their faces until they have washed thoroughly.

Inhalation

Accidental inhalation of dust is very possible in a water plant using dry chemicals. The causes of dust should be minimized, and operators should wear dust masks when filling feeder hoppers and disposing of bags.

Bags should not be dropped. An even slit should be made at the top of the bag to avoid its tearing down the side, and the contents should be poured gently into the hopper. Crystalline chemicals should be used when possible, because they produce much less dust than the powdered forms. Even if masks are worn and there is no visible dust, the area should be well ventilated.

Bodily Contact

To avoid bodily contact, personnel should wear the following when handling fluoride chemicals:

- Chemical goggles
- Respirator or mask approved by the National Institute of Occupational Safety and Health (NIOSH)
- Rubber gloves with long gauntlets, a rubber apron, and rubber boots
- Clothing that covers the skin as completely as possible
- Tight covers over open cuts or sores

Acid Handling

Fluorosilicic acid requires very special handling. Acid spilled on the skin or splashed into the eyes is a serious hazard, as is inhaling the vapors.

Fluoride is not absorbed through the skin, but the acid is very corrosive and can burn the skin. If acid is splashed on the body, it should immediately be rinsed with water. An emergency shower and eye wash should be provided for this purpose in areas where acid must be handled.

Record Keeping

Typical records that should be maintained to monitor the fluoridation process include the following:

- Daily analyses of raw-water fluoride concentration, unless it is known to be stable (If there is more than one source and the levels are known to differ, the concentration must be measured for each source.)
- Daily analysis of finished water fluoride concentration
- Daily records of the amount of chemical fed (in pounds or kilograms)
- Records by the operating shift of the chemical feed rate setting
- Daily computation of the theoretical concentration, based on the weight of chemical fed and the volume of water produced, as well as a comparison with analyzed values

Study Questions

1. If the natural fluoride content of the raw water is variable, the concentration of the raw water should be measured
   a. every 8 hours.
   b. every 12 hours.
   c. every day.
   d. continuously.

2. People's teeth are likely to become pitted when the fluoride concentration in drinking water goes above which amount?
   a. 3 mg/L
   b. 4 mg/L
   c. 5 mg/L
   d. 6 mg/L

3. How often should the fluoride concentration of treated water be measured?
   a. Every 8 hours
   b. Every 12 hours
   c. Every day
   d. Continuously

4. Where should the fluoride injection point be located?
   a. Right after flocculation
   b. After sedimentation, but before lime softening
   c. Before filtration
   d. After water has received complete treatment

5. The operator must periodically check to ensure that at least _____ of chemical is maintained in the saturator at all times.
   a. 1 in. (25 mm)
   b. 3 in. (75 mm)
   c. 4 in. (100 mm)
   d. 6 in. (150 mm)

6. When adding fluoride, when should the fluoride level be measured to test for fluoride concentration?

7. Metering pumps are quite reliable, but they should receive routine maintenance according to what authority?

8. If fluoride is added before filtration, a significant fluoride loss will occur, resulting in what?

9. List three recordings that should be maintained to monitor the fluoridation process.

Chapter 10
Water Quality Testing

Testing and Laboratory Procedures

Reasons for Testing

Water plant operators take samples every day. Water in the plant and the distribution system is sampled frequently for two reasons: to ensure the water is safe for human consumption through compliance testing and and to measure the efficiency of the treatment process. Some samples, such as those for disinfectant residuals and turbidity, are tested on-site and continuously; others, such as those for disinfection by-products or metals, may be sent periodically to a laboratory for analysis. The size and resources of the individual utility will affect the choices that the operator makes regarding sampling.

Sampling

Reliable test results depend on proper sampling procedures and techniques. Samples are supposed to accurately represent the quality of the water being sampled. Consequently, consideration must be given to the way the sample is collected, the volume taken, any storage requirements for the sample, and sampling locations and frequencies.

Sample Collection

Improper sampling is a common cause of error in water quality analysis. Because the results of an analysis can show only what is actually in the sample, the sample must have the same content as the water from which it was taken. Water laboratory analysts refer to this as a *representative sample*, and it is defined as being one of two types: grab sample or composite sample.

Grab Samples   Grab samples are single-volume samples collected at one time from one place. An operator taking a 100-mL sample from a distribution system for bacteriological analysis is taking a grab sample. Similarly, an operator taking a few milliliters of settled water for turbidity analysis is taking a grab sample. As such, these samples represent the quality of water in those locations only at the time that the sample was obtained. For this reason, grab samples are best used when the quality of the water is not expected to change significantly or where changes take place slowly. A grab sample might not give an accurate representation of the quality (over time) of a flowing river that is subject to intermittent pollution from upstream wastewater plants or chance rainfall events. The frequency of grab sampling is determined by studying the history of changes in water quality so that adequate representation of the quality can be made.

grab sample
A single water sample collected at one time from a single point.

Composite Samples A composite sample consists of multiple grab samples taken from the same location at different times and mixed together to make one sample. Time-composite samples are made up of equal-volume grab samples taken at different times. An example of a time-composite sample is 100-mL samples of backwash water taken at 1-minute intervals for the length of the wash and mixed together to measure total solids. Flow-proportional composite samples consist of different volumes of sample taken from the same location at different times; the volume collected each time is dependent on the flow rate of the water at the sample location. When composite samples of raw water in a water treatment plant are needed and the raw water flow rate varies over the day, a proportional composite sample is sometimes taken because it more accurately represents the water quality over time.

Composite samples allow the laboratory analyst to determine the average concentration of a constituent over time, without having to perform the analysis every time a sample is taken. Automatic samplers can be used to take samples at night, which enables the chemist who comes to work in the morning to perform the analysis.

There are drawbacks to composite samples. The high and low points, or ranges, of a contaminant cannot be known because they are averaged. Also, some constituents, such as pH, dissolved gases, chlorine residual, and temperature, begin to change immediately after the sample is taken and therefore should not be composited. Bacteriological samples, such as those taken for coliform analysis, must always be taken as grab samples because the number of bacteria in the samples begins to change immediately after sampling.

Sample Volume

The volume of sample will vary according to the testing procedures used and often is set by regulations. For example, coliform compliance sampling requires a 100-mL grab sample. The laboratory analyst will provide instructions for sample volume.

Sample Storage

Glass and carbonate plastic sample containers are most often used in water utilities, but glass is avoided where there is a greater chance of breakage, such as in shipping. Also, glass is not used when sampling for fluoride because fluoride will adhere to the glass and not provide accurate results. Plastic can also be problematic because certain organic chemicals can permeate the plastic, and therefore the organic chemicals will not be detected in the subsequent analysis. Some types of plastics may actually release organic chemicals into the sample and so invalidate the results.

Samples for total trihalomethane and haloacetic acid analyses are taken in glass vials, usually colored brown to filter out sunlight, and shipped in coolers. Often, sample containers have preservatives, quenching agents, or pH-adjustment chemicals in them to facilitate sample preservation and/or storage, which extend the shelf life of the sample, allowing more time for analysis. This in turn allows the analyst to collect all the necessary samples and then perform a single analysis, greatly reducing errors that may otherwise occur during calibration. It also is more efficient and less costly to perform one calibration for multiple samples than it is to calibrate for each individual sample.

Regulated parameters have specific directions for sampling, preservation, and storage before testing. These requirements are listed in the regulatory agency

composite sample
A series of individual or grab samples taken at different times from the same sampling point and mixed together.

compliance regulations. Some agencies have adopted the USEPA requirements and some have separate, additional requirements. Many, including USEPA, have referred to *Standard Methods for the Examination of Water and Wastewater*. Consult the regulatory agency for requirements in the utility's location.

Sample Location and Frequency

Selecting representative sampling points and determining sampling frequency are important steps in attaining meaningful water quality data. Samples are usually collected from three areas within the treatment system: the raw or source water, the plant, and the distribution system. Because sources and treatment schemes differ from system to system, no guidelines are given in this handbook. Local regulatory agencies usually provide a minimum frequency for sampling, and most operators take samples with a greater frequency than is required.

In-plant sample locations should be chosen to allow for the measurement of the performance of each unit process. Operators should take care not to sample immediately after chemical addition but rather at a point where proper mixing has been achieved. Sampling of combined filter effluent turbidity should take place before post-treatment chemicals are added to avoid measuring added turbidity, such as from lime or phosphates.

Distribution system samples are intended to measure the quality of the water that is delivered to customers at various points in the system. In general, samples should not be taken from customers' taps. The exception is for lead and copper testing and for water quality complaint sampling. The inlet and outlet of storage tanks are good sampling sites because they give information about the effects of storage. The direction of flow should be noted when sampling, i.e., is the tank filling or emptying at the time of sampling. Bacteriological sampling and the associated chlorine residual testing are regulated. Frequency of sampling is based on population served.

Quality Assurance and Quality Control

Standard Methods for the Examination of Water and Wastewater defines quality assurance (QA) as "a definitive plan for laboratory operation that specifies the measures used to produce data of known precision and bias." It defines quality control (QC) as a "set of measures within a sample analysis methodology to assure that the process is in control."

A laboratory's QA program consists of a QA manual, written procedures, work instructions, and records. For example, an organizational chart that lists the training, capabilities, and responsibilities of each lab analyst would be part of a QA program. Quality control is also part of the overall QA program.

In a QC program, each analyst demonstrates his or her capability to obtain acceptable results for an analysis. This is demonstrated by analyzing a set of "blind" standards. Analysts are given samples that contain a specific amount of a chemical. The analyst runs the samples, and the results are reported to an agency that knows the correct levels of chemicals in the samples. The agency then rates the laboratory and the analyst on their performance. This type of testing is commonly called a *performance evaluation* or PE. PEs are typically run twice a year and determine if a laboratory is "certified" to run the analysis.

Another type of continuous laboratory evaluation is performed through the use of "known" standards. Known standards are samples that contain a predetermined amount of a chemical. These standards along with each set of samples are run by an analyst. This comparison allows an analyst to determine how well

quality assurance (QA)
A plan for laboratory operation that specifies the measures used to produce data of known precision and bias.

quality control (QC)
A laboratory program of continually checking techniques and calibrating instruments to ensure consistency in analytical results.

a particular set of analyses has been performed in each analytical trial. Power fluctuations or other daily hazards may affect laboratory equipment, and the evaluation of known samples will let an analyst determine if a particular run is satisfactory. Sometimes these known samples are referred to as *standards*. For example, known turbidity standards are run before each set of filter turbidity grab samples is run, thus allowing an operator to determine if the instrument is properly measuring turbidity.

A final set of samples is used to determine data quality. These samples are referred to as *spiked samples, laboratory control samples,* or *laboratory fortified matrix (LFM) samples*. Spiked samples are made in the laboratory by adding a known quantity of the chemical being measured to the collected samples. The analyst splits one sample into two aliquots and then adds a precise amount of the chemical being analyzed to one of the aliquots. The two aliquots are run and the results compared. The sample to which the chemical was spiked should have a level of chemical equal to the known addition plus the results of the unaltered sample. This type of sample is important because it allows the analyst to determine if some other chemistry in the sample is interfering with how the instrument measures the contaminant. Regulations and reference to *Standard Methods for the Examination of Water and Wastewater* determine the frequency of this type of QA sampling. Different analyses require different numbers of QA samples; this will affect the number of samples an operator may be required to take for any given parameter.

Another way to evaluate laboratory performance is by reviewing historical analytical performance. In differentiating quality *assurance* and quality *control*, this historical review is generally considered the latter, quality *control*. The results of known samples and spiked samples are graphed and reviewed for trends over time. This is important because the efficiency of analytical equipment can decrease over time, or an analyst can improve his or her performance with experience. For example, a decrease in the amount of a known sample result over time can indicate that a probe or detector needs to be replaced.

The importance of laboratory QA/QC to overall plant operations cannot be overstressed. It is critical that test results be reliable because they are a measure of how a plant is performing and a measure of the safety of the water the plant is producing. Performance must be measured accurately, and the accuracy of analysis depends on sample quality and analytical quality. All laboratories should strictly adhere to the laboratory practices described in this section and be willing to produce QA/QC information as requested.

Common Water Utility Tests

Following is a brief description of some of the more common parameters that water plant operators frequently measure, including their significance.

Alkalinity   The buffering capacity of the water is measured by its alkalinity. This test can determine the concentration of carbonate (CO_3^{-2}), bicarbonate (HCO_3^-), and hydroxide (OH^-) alkalinity. These measurements are useful to determine the corrosive nature of the water in combination with other factors and to optimize the lime–soda ash softening process.

The test involves careful titration of a measured quantity of water with a standard 0.02 N sulfuric acid solution to pH end points of 8.3 and 4.5 (these are indicated by color change indicators or by using a pH meter). The carbonate, bicarbonate, and hydroxide alkalinity levels are then calculated from the results.

Carbon Dioxide Water that contains high concentrations of free carbon dioxide (CO_2) can cause the consumption of lime when using this method of softening. Also, carbon dioxide is a factor affecting corrosion.

The test usually used is to titrate a measured water sample with a standard solution of sodium hydroxide. The end point is signified by a change in color for phenolphthalein or a pH of 8.3.

Chlorine (Free or Total) For water plants that use chlorine for disinfection or oxidation, this is one of the most important tests performed by operators. After any addition of chlorine, a measurement should be taken routinely. This test will verify the correct dosage and reveal changes that may affect plant performance and the safety of the water supply. Many plants use both free and total (or combined) forms of chlorine in their processes.

The test most often used is the N,N diethyl p-phenylenediamine sulfate (DPD) color test. The DPD (either for free or total chlorine testing) is added to a water sample, and the intensity of the color indicates the amount of chlorine present in the sample. Most plants use a digital read-out colorimeter to give an accurate result. Some color comparison portable test devices are used as well; however, they are not always accurate. Another test method is amperometric titration. This test is usually performed in the laboratory. A special meter is used to determine the end points when titrating a measured sample with a standard phenylarsine oxide (PAO) solution. This method is very accurate and is capable of determining many chlorine species that may be present. There are several other chlorine residual test methods that can be considered (see *Standard Methods for the Examination of Water and Wastewater*).

Both the DPD color test and amperometric titration are used in online instruments for continuous chlorine monitoring. As with any instrument, these devices must be calibrated and checked frequently to ensure accuracy.

Chlorine demand can be determined using the residual test method. This measurement can be used to predict residual chlorine over a specified time. Jar testing apparatus may be used for this test. A sample is taken and the chlorine residual is measured immediately. After a specified time, the residual is measured again and the difference is the demand. Care is needed to duplicate the conditions (light, temperature, holding time) that are of interest.

Chlorine Dioxide Chlorine dioxide residual is limited to 0.8 mg/L (USEPA MRDL), and a by-product, chlorite, is also regulated with an maximum contaminant level (MCL) of 1.0 mg/L.

The test methods for chlorine dioxide are similar to those for chlorine. There are DPD and amperometric test methods, but some of the test conditions have been modified to yield chlorine dioxide–specific results. Also, there is an ion chromatography method that requires trained technicians in a certified laboratory.

Coliform Coliforms are a group of bacteria that produces gas bubbles in lactose or lauryl tryptose broth at 35.5°C (96°F) within 24 to 48 hours. They are considered indicator organisms, meaning that their presence may indicate the presence of other more harmful bacteria and organisms. Because total coliforms are easier to analyze in the average water utility laboratory than the actual disease-causing microorganisms, total coliform testing is used in place of more tedious, more expensive testing for these other organisms.

Most laboratories use one of two methods to test for coliforms. One method is the membrane filter technique, in which water samples are passed through a 0.45-μm filter. The filter paper is fine enough to trap bacterial particles as the water passes through. The filter paper is then placed onto a growth medium (such as M-endo) and incubated. Any bacterial colonies that are present will grow in size, will be visible to the naked eye, and can be counted after a period of time (usually 24 hours).

Another method is the MMO-MUG (minimal medium) technique. This method allows for inoculation of water sample bottles with powder. The substance in the bottle will feed any total coliform that may be present and produce a color change during incubation.

Samples for coliform testing are always collected in sterile bottles and in quantities sufficient for testing (Figure 10-1). Most bacteriological samples require a minimum of 100 mL (approximately 4 oz) for analysis. Samples should be analyzed the same day as they are collected but can be refrigerated for 8 hours prior to analysis. Coliform samples are taken in the plant at various stages to test for process efficiency; they are also taken in the distribution system for regulatory compliance. The number of samples that must be taken in the distribution system is a function of the population served. Many water treatment plants also take coliform samples of the raw or source water.

WATCH THE VIDEO
Coliform Sampling (www.awwa.org/wsovideoclips)

Conductivity   This test (more correctly named *specific conductance*) measures the ability of water to conduct electricity. This is an indirect measure of the ions or minerals in the water. It is sometimes used to estimate the total dissolved solids (TDS) content due to its ease of measurement and the availability of inexpensive mobile field instruments.

A conductivity meter is connected to an electrode, and this is immersed in a water sample. An instrument setting to match the sample temperature is often used to adjust for extremes. The conductivity reading is in microsiemens per centimeter (µs/cm).

Cryptosporidium   Cryptosporidium is a parasite regulated under the Long-Term 2 Enhanced Surface Water Treatment Rule (LT2ESWTR). It is a regulated pathogen, and its measurement is necessary to ensure adequate treatment depending on the occurrence in the untreated water supply.

Figure 10-1 Autoclave used for sterilization. Bacteriological equipment must be sterilized in steam under pressure.
Source: Conneaut, Ohio, Water Department.

The method involves filtration, specialized separation, and identification methods. This test can be performed only by laboratories approved for this method. Plant operations personnel will use the test results to determine compliance with the LT2ESWTR.

Disinfectant By-products (DBPs)   Several organic compounds are created when chlorine is used in water treatment. Two groups of these by-products are regulated: trihalomethanes and haloacetic acids. Compounds in both groups have been classified as probable carcinogens.

The test methods for both groups involve procedures that must be performed by trained analysts in approved laboratories. Compliance testing is, therefore, not usually performed by plant operating personnel. Online instruments are available for continuous monitoring, but the instruments should be checked periodically by the same approved laboratory procedures. Operator-performed screening tests can be used to indicate levels for plant monitoring, but these methods are not approved for compliance testing.

Dissolved Oxygen   The amount of dissolved oxygen (DO) in the source water may be an indicator of the condition of the lake or reservoir being used. Lack of oxygen may be a predictor of water quality problems due to anoxic conditions. The DO concentration in the treated water may be a factor contributing to the rate of corrosion.

The DO test involves the use of a dedicated meter and specific sensor probe. Also, there are color methods where the intensity of the color from the addition of the indicator is a measure of the concentration present in the water.

Fluoride   Fluoride reduces tooth decay. Fluoride may be naturally occurring or added during water treatment as directed by local health authorities. The optimum amount of fluoride needed to provide health benefits depends on several factors. The MCL for fluoride in drinking water is 4.0 mg/L.

There are both instrumental and colorimetric tests for fluoride in water. Operations personnel usually use the color methods. The color indicator is added to the water and the color intensity measures the amount of flouride present.

Giardia lamblia   *Giardia lamblia* is a parasite found in untreated water supplies, and it is pathogenic. It is regulated by a treatment technique in which disinfectant concentration and contact time are prescribed to ensure adequate inactivation. Separate requirements are specified for chlorine, chloramine, ozone, and ultraviolet radiation.

The method involves filtration, specialized separation, and identification methods. This test can be performed only by laboratories approved for this method. Plant operations personnel will use the test results to determine compliance with the Surface Water Treatment Rule (SWTR).

Hardness   Hardness is attributed primarily to the amount of calcium and magnesium in the water. Softening treatment plants must carefully monitor hardness as a process control. Most other water systems regularly monitor for this parameter as a basic measure of water quality.

There are several methods for hardness testing. Although hardness can be determined by separately analyzing the calcium and magnesium levels, most water systems use a color titration method for total hardness results. A measured sample is titrated with a standard solution (EDTA) to a color change end point.

Inorganics (Heavy Metals)   Some metals are toxic in high amounts (such as cadmium or chromium) and others are not (such as iron or manganese). There are MCL regulations for many heavy metals and secondary limits for those that can cause aesthetic concerns. Most regulated metals are included in the primary drinking water regulations, and some, such as lead and copper, have special rules.

The test methods for metals are tremendously varied. Instrumental and color methods predominate. Only approved methods can be used for compliance testing. Operations personnel may elect to use unapproved screening methods for process monitoring. The more precise methods require trained analysts and must be performed by approved laboratories.

Iron and Manganese   These metals are often grouped together because they are similar in their role in water quality and often occur together. Both cause staining of fixtures and are not generally toxic at normally encountered concentrations. Treatment processes have been developed to remove one or both of these substances. Secondary standards (nonenforceable) are established for both.

There are several test methods for both of these metals. Many treatment plants use a color method for process control. These methods are typical where an indicator is used, and the intensity of the color is measured to determine the amount present.

Lead and Copper   These metals are grouped together because of the US Environmental Protection Agency (USEPA) regulatory rule requirements. Both substances are toxic at elevated levels; thus, they are regulated. Generally, these contaminants are somewhat unique in that they are not often found in source water but are the result of dissolution from piping components (see the Lead and Copper Rule description in Chapter 3). Therefore, testing is performed primarily in the distribution system rather than the treatment plant.

The test methods vary from instrumental laboratory procedures to simple field-test color methods. The laboratory methods require trained analysts and are not usually conducted by operations personnel. Screening tests using colorimetric methods are often employed by plant operators for process control.

Microbiological Organisms

Many microbiological organisms are found in untreated water. Bacteria, viruses, protozoan, algae, and plankton are examples of the multitude of possibilities. Some of these are pathogenic; others can produce toxic substances. The array of consequences is very broad. Some are regulated in drinking water. The coliform group of bacteria is used to indicate the possibility of contamination. Another useful bacteria indicator is the heterotrophic plate count. This is a general bacteria population measurement that may indicate the presence of other pathogens.

Test methods are as varied as the organisms. Many involve growing colonies on selective nutrients and counting the number from a measured sample. Other samples may be examined by microscope and the organisms identified by experts. Some chemical by-products are analyzed by instrumental methods capable of detecting minute amounts. Generally, trained analysts must perform most of these procedures. The exceptions are the coliform tests that have been developed using color changes to both detect and enumerate these bacteria.

Nitrate, Nitrite, and Ammonia   These three inorganic nitrogen compounds are often encountered in water supplies. Contamination from agricultural activities is often the source in surface water, but another significant source is wastewater discharges. Nitrate and nitrite are regulated contaminants with enforceable MCLs. Ammonia is not regulated in drinking water but is a compound that may encourage the growth of nitrifying bacteria. Ammonia is often associated with the chloramination process.

Several test methods are available for these substances. Operational control testing is usually conducted using colorimetric methods. Special precautions are needed when testing for free ammonia in the presence of chloramines in order to ensure accurate results.

Orthophosphate and Polyphosphate Phosphate can be naturally occurring in surface water, where it may be the result of urban runoff or waste discharges. Orthophosphate and polyphosphate are often used as corrosion inhibitors in water supplies. These substances are not considered toxic, but the amount is carefully monitored to ensure optimum effectiveness.

There are instrumental methods for phosphate that require approved laboratories and trained analysts. Most operational control testing uses colorimetric tests. Selecting the correct test may require knowledge of the type of phosphate used for corrosion control.

pH This is one of the most common tests performed in water treatment and drinking water monitoring. The pH value (Chapters 4 and 11) is an indicator of the acidity or alkalinity of the water. There is not an MCL for pH in drinking water. However, there is a secondary standard range of 6.5–8.5.

ph testing uses a scale from 0 to 14, with the midpoint of 7 being neutral (i.e., the acidity and alkalinity are balanced). Below 7, the acidity of the water predominates; above 7, the alkalinity of the water predominates. With each unit increase or decrease, the concentration or intensity changes 10-fold.

For example, for the pH to change from 5 to 6, the acidity must decrease by a factor of 10. The pH of water is significant because it affects the efficiency of chlorination, coagulation, softening, and corrosion control. Also, pH testing can provide early warning of unit process failure. For example, the addition of alum to the rapid-mix stage should produce a predictable drop in pH. If it does not, a malfunction of coagulant feed could be indicated.

Samples for pH should be collected in glass or plastic containers and analyzed as quickly as possible. Samples should not be agitated because dissolved carbon dioxide could be liberated, which will change the pH.

A pH meter is used for the test in combination with a suitable probe. The meter must be periodically calibrated using a known standard. The pH value for a water sample may change while standing due to a change in temperature or exposure to air. Therefore, measurements are usually taken immediately upon sampling.

Radiologic Substances Contamination from radiological substances is more common in groundwater than surface water. Several of these substances have MCL regulations. There are numerous possible radioactive isotopes that may be encountered; therefore, screening water supplies for general radioactivity (gross alpha and gross beta activity) may be prudent before employing specific substance testing.

Testing for radioactive substances often requires specialized equipment and procedures. Approved laboratories are required for compliance testing, and analysts with specific training are needed. Test results for radioactivity are sometimes expressed as pCi/L (picoCuries/L), but, depending on the substance, μg/L or mrem/yr may be used.

Solids—Settleable, Dissolved, Suspended   The solids contained in a water supply are all of the substances that are not H_2O. These can be inorganic, organic, volatile, nonvolatile, suspended, settleable, or dissolved. There are not MCLs for solids content, but there is a secondary standard for TDS levels of 500 mg/L.

The test method involves taking a measured sample and heating it to remove the water and then weighing the remaining residue (solids). The temperature used for heating can define the volatility of the residue. Also, filtration before heating can be used to give a dissolved result. TDS is the most used solids measurement in drinking water. TDS is a gross measure of the inorganic content of the water (because organic substances usually are a minor part of the total).

Synthetic Organic Chemicals   This is a large group of organic compounds that is regulated and has MCL standards. Many of these compounds are pesticides and herbicides that for the most part are not currently manufactured in the United States. Most of these compounds are probable carcinogens and, therefore, have MCLG limits of 0. These must be tested according to the requirements of the Primary Drinking Water Regulations.

The test methods for this group of compounds involve complex chemistry instrumental procedures. Only approved laboratories with trained analysts can perform these tests. A few of these substances have intricate rapid test methods that are not approved for regulatory compliance. These methods may be suitable for screening surveys or other occurrence evaluations.

Taste and Odor   Consumers often react to taste and odor as their only way to evaluate the quality of the drinking water. A disagreeable response may result in a poor perception, so it is imperative that the water utility seek to provide water that is pleasing to its customers. There is no regulatory MCL for taste or odor, but there is a secondary standard MCL for threshold odor number (TON) of 3.

There are several tests for taste and odor. One is the flavor profile analysis (FPA). This test involves a trained panel to routinely evaluate the water. Several utilities use this method with good results. However, most utilities find this method to be labor intensive and instead use the older TON method. This method enlists a panel to smell water samples of various dilutions. The consensus of the dilution where an odor was detected is the TON for the sample. Although this test can be a bit subjective, it can also be useful to detect odor problems and to assess the effectiveness of treatment.

 WATCH THE VIDEO
Taste and Odor (www.awwa.org/wsovideoclips)

Temperature   This is probably the test most frequently performed on drinking water. There is no MCL for temperature. Differences in temperature can indicate probable water quality problems. Also, water supply changes are often linked to temperature.

A thermometer is used and the water must be tested at the sample location. The result is usually expressed in °C because this scale is used for many other test procedures.

Total Organic Carbon   This parameter is a nonspecific measure of the organic content of the water. There is no MCL for total organic carbon (TOC), but the Disinfectants and Disinfection By-products Rule (DBPR) uses TOC measurements to determine if precursor removal is needed to comply with the rule. Enhanced coagulation or other means may be required to reduce the TOC.

The test uses a TOC instrument either in the laboratory or online. Calibration is required by analyzing known standards and adjusting the instrument to provide an accurate result. Specific training is necessary to calibrate and use the instrument to provide the best results.

Volatile Organic Chemicals Several volatile organic chemicals are regulated and have MCLs. Although these compounds are volatile, it is possible that water under pressure may contain these substances, and when the pressure is released, the compounds could be inhaled.

Test methods for this class of compounds involve careful sample collection (zero headspace). The samples are purged with an inert gas, and the vapor is analyzed by gas chromatograph (instrument). This test must be performed in an approved laboratory by trained analysts. Operational personnel may assess the results and compare them to regulatory standards.

Physical and Aggregate Properties of Water

Calcium Carbonate Stability

The principal scale-forming substance in water is calcium carbonate ($CaCO_3$). Water is considered stable when it will neither dissolve nor deposit calcium carbonate. This point is referred to as *calcium carbonate stability* or the *equilibrium point*. The reactions and behavior of calcium carbonate and calcium bicarbonate are therefore important in water supplies. The actual amount of calcium carbonate that will remain in solution in water depends on several characteristics of the water: calcium content, alkalinity, pH, temperature, and TDS.

Significance

Scale formation can cause serious problems in water distribution mains and household plumbing systems by restricting flow, plugging valves, and fouling water heaters and boilers. Corrosion can cause premature pipe or equipment failure. Public health and aesthetic problems can also result if water is corrosive, because pipe materials (e.g., lead, cadmium, and iron) will dissolve into the water.

Several methods can be used to determine the calcium carbonate stability of water. A popular method is the Langelier saturation index (LSI). The LSI is equal to the measured pH (of the water) minus the pH_s (saturation). The pH_s is the theoretical pH at which calcium carbonate will neither dissolve into nor precipitate from water. Water at the pH_s is considered stable. Therefore, if $pH - pH_s = 0$, the water is in equilibrium and will neither dissolve calcium carbonate nor deposit it on the pipes.

If $pH - pH_s > 0$ (positive value), the water is not in equilibrium and will tend to deposit calcium carbonate on mains and other piping surfaces. If $pH - pH_s < 0$ (negative value), the water is also not in equilibrium and will tend to dissolve the calcium carbonate it contacts; no coating will be deposited on the distribution pipes, and if the pipes are not protected, they may corrode.

The calcium carbonate stability of water is maintained in the distribution system by adjusting the LSI of the water to a slightly positive value so that a slight deposit of calcium carbonate will be maintained on pipe walls. Adjustment is usually made by adding lime, soda ash, or caustic soda.

Indices other than the LSI use alkalinity as part of the equation or method to determine the stability of the water, especially its corrosiveness. One is the *marble test*, in which calcium carbonate (limestone) is dissolved in the water sample and the initial alkalinity is compared with the final alkalinity. The Ryzner index is used to perform a similar calculation to the LSI; it indicates the corrosiveness of the water compared to the pH_s.

Alkalinity is also important in determining how effectively the water reacts with coagulants for plant treatment. The alkalinity ions act as a "reservoir" of molecules that are available to react with coagulants or other chemicals such as disinfectant to reduce the pH change to the water that these chemical may affect; this effect is known as the "buffering capacity." Low-alkalinity waters have less buffering capacity and can produce less floc or weak floc without the addition of chemicals to increase the buffering capacity of the water. This is because many coagulants are acidic compounds.

Sampling

If calcium carbonate stability maintenance is used for corrosion control, finished water at the treatment plant and in the distribution system should be evaluated routinely for calcium carbonate stability. Evaluation is particularly important when treatment plant unit processes or chemical doses are changed. If the LSI indicates unfavorable conditions, process adjustments should be made. It is very important to remember that the LSI is only an indicator of stability; it is not an exact measure of corrosivity or of calcium carbonate deposition. The LSI is developed from results of alkalinity, pH, temperature, calcium content, and TDS (dissolved residue) monitoring.

Methods of Determination

If the temperature, TDS, calcium content, and alkalinity of the water are known, the pH_s can be calculated. The following equation may be used:

$$pH_s = A + B - \log (Ca^{+2}) - \log (alkalinity)$$

In the equation, A and B are constants, and calcium and alkalinity values are expressed in terms of milligrams per liter as calcium carbonate equivalents. Tables 10-1 and 10-2 are used to determine the values of the constants and logarithms.

The actual pH of the water is measured directly with a pH meter, and the LSI is calculated using the formula $LSI = pH - pH_s$.

Table 10-1 Constant *A* as a function of water temperature

Water Temperature, °C	A
0	2.60
4	2.50
8	2.40
12	2.30
16	2.20
20	2.10

Table 10-2 Constant *B* as function of total dissolved solids

Total Dissolved Solids, mg/L	B
0	9.70
100	9.77
200	9.83
400	9.86
800	9.89
1,000	9.90

Example

A sample of water has the following characteristics:

$$Ca^{+2} = 300 \text{ mg/L as } CaCO_3$$
$$\text{Alkalinity} = 200 \text{ mg/L as } CaCO_3$$
$$\text{Temperature} = 16°C$$
$$\text{Dissolved residue} = 600 \text{ mg/L}$$
$$pH = 8.7$$

Determine the saturation index (LSI):

$$pH_s = A + B - \log(Ca^{+2}) - \log(\text{alkalinity})$$
$$pH_s = 2.20 + 9.88 - 2.48 - 2.30$$
$$pH_s = 7.3$$
$$LSI = 8.7 - 7.3 = +1.4$$

An LSI of +1.4 indicates that this water is scale forming.

Coagulant Effectiveness

The removal of suspended solids from surface water is necessary both to make the water aesthetically pleasing to customers and to assist in the elimination of pathogenic organisms. The SWTR, the Total Coliform Rule (TCR), and the Disinfectants and Disinfection By-products Rule (D/DBPR) require more complete removal of turbidity and dissolved organics than was previously practiced by most water systems. This requirement in turn demands more efficient coagulation, flocculation, and sedimentation. Effective coagulation is also a tool in removing organic chemical precursors from the raw water.

Significance

Coagulation and flocculation involve the addition of chemical coagulants such as aluminum sulfate, ferric chloride, or polyelectrolytes to raw water to hasten the settling of suspended matter. Plant operation is most efficient when the lowest turbidity is obtained in finished water with the lowest cost for coagulant chemicals. Several laboratory tests can provide the information necessary to accomplish this goal. These tests allow operators to select optimal chemical dosages in the laboratory rather than using trial and error in the plant.

The tests can also be used to check the adequacy of flash mixing or flocculation mixing in the plant. Test results may indicate when to improve or modify flash mixers and flocculation basins to obtain more efficient operation. (See Chapter 4 for further discussion of these treatment processes.)

When coagulants are evaluated, the goal is to identify the one coagulant (or combination of coagulant and coagulant aids) that will produce low turbidity with the least expensive dose of chemicals. Chemical prices must also be evaluated, because a low dose of an expensive chemical may be more cost effective than a high dose of an inexpensive chemical. One must also consider the cost and ease of disposal of the coagulant and the contaminants removed in the process.

Sampling

The location from which the samples are collected for analysis depends on the procedure being used and the information desired from the test. The tests should be conducted whenever there is a significant change in water quality or when other conditions may require a change in coagulant dose.

Methods of Determination

The following methods are commonly used to determine optimum coagulant effectiveness:

- Jar test
- Zeta potential detector
- Streaming current detector
- Particle counting

Because of its relative complexity, we will save the discussion of jar testing for last.

Zeta Potential   Coagulation and flocculation are an electrochemical process in which the electrical resistance between the suspended particles (colloids) in the water is lowered to the point that they will adhere to each other and settle out as a heavy floc.

In many water plants, technicians use a *zeta meter* to assist in evaluating the effectiveness of coagulant doses. Zeta potential may be viewed as the electrical charge on a suspended particle that allows it to repulse other particles and thus stay suspended. The type and amount of coagulant added reduces the zeta potential, and a zeta meter measures this potential. The closer the reading is to zero, the more the particles tend to settle, thus indicating more effective coagulation.

Normally, the zeta meter is not an online instrument. Samples may be collected at various points in the treatment process to determine coagulant effectiveness, or various coagulant doses may be tested on the raw water in bench-scale experiments. Trained personnel are necessary to operate the zeta meter and interpret the data it produces.

Streaming Current Detector   A streaming current detector (SCD; Figure 10-2) is an online continuous-monitoring device based on the same electromotive principle as is the zeta meter. The detector measures the effectiveness of the coagulant chemical by determining the level of electrical resistance in the treated water after chemical application.

The advantage of having a continuous-monitoring device is that it allows the operator to evaluate changes in the chemical doses as changes in raw-water quality occur. A major concern in installing an SCD is the maintenance (cleaning) of the electrodes and the calibration of the meter to ensure accurate readings. Another advantage of the SCD is that it can be used to automatically control coagulant dose by connecting the output signal from the SCD to a coagulant feed pump. Location of the sample point for online control is critical in that, for best results, the sample must be thoroughly mixed and representative of the treatment process before it enters the instrument.

Particle Counting   Instruments are available that enumerate the concentration of particles in a water supply by size (Figure 10-3). These instruments, known as particle counters, combine particle detection technology with electronic counting technology. A sensor detects the particle and converts the information to an electronic signal that is used by the electronic counter.

Particle counters and turbidimeters are similar in that both use a fixed light source to interact with the particles in water. Turbidimeters use light scattering. Particle counters use the principle of light blocking. Particles that do not reflect light and are not detected by a turbidimeter can be counted by a particle counter.

streaming current detector

An instrument that passes a continuous sample of coagulated water. The measurement is similar in theory to zeta potential determination and provides a reading that can be used to optimize chemical application.

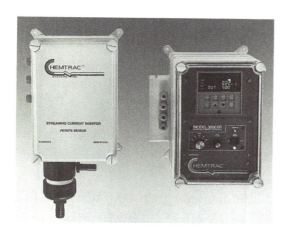

Figure 10-2 Streaming current monitor with remote sensor

Courtesy of Chemtrac Systems, Inc.

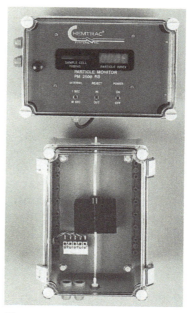

Figure 10-3 Liquid particle monitor with remote sensor

Courtesy of Chemtrac Systems, Inc.

Particle counters and turbidimeters are different in that particle counters provide a quantitative measurement, and turbidimeters provide a qualitative measurement. Particle sensors count individual particles according to their size; turbidimeters do not. Particle counters cannot count particles below a given size. Turbidimeters have the ability to detect smaller particles. Even though the particle counter can tell you the size and number of the particles, it cannot yet tell you what that object is.

The technique of particle counting has been in use for the past decade or longer for monitoring filter performance in regard to filter breakthrough and coagulation efficiency, and it is useful in monitoring pathogen removal. It is used for membrane filtrations equipment in combination with the pressure hold test to monitor membrane integrity. The sizes of cysts and oocysts of the pathogens *Giardia* and *Cryptosporidium* are in the micrometer range easily detected by the particle counter. Reducing the number of particles to above the range of 4–5 µm in the finished water should greatly reduce any occurrence of these organisms.

As water quality regulations prescribe lower and lower contaminant levels, water treatment plant operators will increasingly have to depend on sophisticated control techniques such as particle counting for process control.

 WATCH THE VIDEO
Turbidity and Particle Counting: Particle Counters (www.awwa.org/wsovideoclips)

Jar Testing   Jar testing is perhaps the least understood test but the most useful process tool available to operators. A few simple ideas and techniques need to be mastered. When the plant staff becomes efficient in this process control strategy, they can rapidly respond to treatment upsets. This is especially important with the more stringent turbidity requirements of the IESWTR.

Jar testing can be used for many tasks, such as screening new coagulants and polymers, coupon testing for corrosion control, and biological spiking experiments. The jar testing described here focuses on coagulant dosage control and

jar test
A laboratory procedure for evaluating coagulation, flocculation, and sedimentation processes. Used to estimate the proper coagulant dosage.

oxidant demand. In cases where turbidity breakthrough in a water treatment plant has occurred, an examination of the records often indicates that the staff did not respond adequately to a sudden change in source water turbidity or a sudden source water demand for oxidant. In these cases, the staff simply did not know what dosage of chemicals to use to counteract the changes because they could not perform jar testing.

Jar testing requires a working knowledge of stock solutions. The following discussion explains how to prepare stocks for alum, alum–polymer blends, ferric chloride, chlorine, and potassium permanganate. It also shows how to perform jar tests with these stocks to determine coagulant dosage and chlorine or permanganate demand. The jar test apparatus, shown in Figure 10-4, is the device most commonly used. It consists of a series of six 2-L square jars with stirrers. A common shaft operates the stirring mechanism, ensuring that each jar is mixed identically. Operators should follow the procedures outlined in the manufacturer's recommendations for operation.

Alum   **Alum** dose is usually expressed in milligrams per liter as dry basis. If liquid alum is used, the product is shipped to the plant as approximately 50 percent alum (8.3 percent Al_2O_3). Commercial alum is approximately 642 mg/mL dry basis. To make a working stock solution for jar testing, pipette 15.6 mL into a 1-L volumetric flask and dilute to the 1-L mark with distilled water. 642 mg/mL × 15.6 mL is approximately 10,000 mg, so this stock is 10,000 mg/L, or 10 mg/mL dry-basis alum. Every milliliter of this stock placed into a 2-L beaker for jar testing would produce a dosage of 5 mg/L alum dry basis.

Alum–Polymer Blend   Normally, the dosage for this product is expressed as gallons per million gallons, or parts per million. This is based on the simple idea that 1 gal of coagulant added to each million gallons of raw water is 1 ppm. This

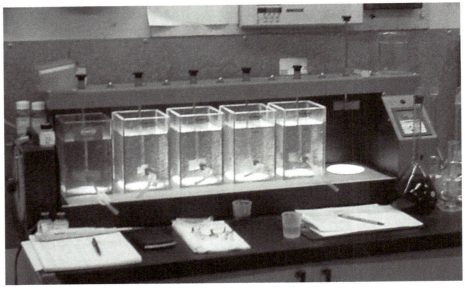

alum

The most common chemical used for coagulation. Also called *aluminum sulfate.*

Figure 10-4 Jar testing equipment is used to determine proper coagulant dosage and for process trial testing. Square jars provide better mixing characteristics than round jars.

Source: Cleveland, Ohio, Division of Water.

expression negates the requirement to know the exact proportions of alum and polymer in the coagulant.

Note that parts per million and milligrams per liter are not the same. Parts per million is a volume-per-volume relationship, while milligrams per liter is a weight-per-volume relationship. As a rule of thumb, a dosage of 18 ppm would deliver approximately 10.6 mg/L dry-basis alum. The calculation is as follows: If 18 ppm of alum–polymer blend is fed into the system, that is the same as 18 gal of product per 1,000,000 gal of water. Alum–polymer blend weighs about 10.9 lb/gal (it may vary), so 18 gal × 10.9 lb/gal = 196.2 lb/mil gal. 196.2 lb/mil gal/8.34 = 23.5 mg/L of liquid alum–polymer blend. Because roughly 45 percent of this is dry alum, 45 percent × 23.5 = 10.6 mg/L dry-basis alum. Because more dry-basis alum is usually required for TOC removal than is needed for turbidity removal, operators should resist lowering blended dosages.

To make a working stock, place 10 mL of the commercial product into a 1-L volumetric flask and fill to the mark with distilled water. This is a 10-parts-per-thousand stock. Each milliliter of this stock placed into a 2-L jar is a dosage of 5 ppm.

Ferric Chloride   Ferric chloride dosage is usually expressed as a milligrams-per-liter product. The material generally comes to the plant as a liquid with a stated percentage that can vary according to manufacturer. Most NSF-certified ferric chloride products are limited to a dosage of 250 mg/L; however, most plants use far less.

To make a working stock solution of ferric chloride from the dry powder, dissolve 2.93 g into 1,000 mL of water and mix thoroughly. The resulting stock is 1,000 mg/L or 1 mg/mL. One milliliter of this stock added to a 2-L jar is a dosage of 0.5 mg/L dry ferric chloride. To make stock solutions from the liquid product, it is best to consult the supplier.

When it is desirable to compare alum and ferric salt coagulants for enhanced coagulation comparisons, make a working stock solution from the liquid product (assuming 38 percent ferric chloride, $FeCl_3$, specific gravity 1.4), transfer 10.27 mL to a 1-L flask, and fill to the mark with laboratory-grade water. This solution is 10 mg/mL ferric chloride. One milliliter of this solution put into a 2-L jar test container will be a dosage of 5 mg/L. The calculation is as follows: 38 percent is 38 g/100 g, and when multiplied by the specific gravity, it yields a concentration of 0.532 g/mL. At 0.532 g/mL × 10.27 mL, a solution is produced that is 5.5 mg/L ferric chloride, which is equivalent to a dosage of 10 mg/L alum. Because the enhanced coagulation requirements are written in 10-mg/L increments of alum, the operator can make easy comparisons.

Chlorine   Gaseous chlorine (stored as compressed liquid) is used for disinfection and for chemical oxidation. The gas is fed through chlorinators equipped with site tubes graduated in pounds per day or in pounds per million gallons. Dosage for chlorine is calculated using the following formula:

$$\text{pounds fed/million gallons/8.34} = \text{milligrams per liter}$$

The impurities in the water will exert a demand for chlorine, so that the residual amount in the water will be less than the amount fed. Other factors, such as temperature and detention time, will also affect the demand for chlorine. For this reason, operators should know how to test for chlorine demand.

To make a working chlorine stock using commercial bleach, make an approximate 2,000-mg/L chlorine stock by adding 40 mL of bleach to a 1-L volumetric flask and fill to the mark with deionized (DI) water. Verify the strength of this stock by adding 1 mL of it to 1 L of DI water. This should test as 2 mg/L free chlorine residual because DI water has no chlorine demand to it. Whatever it tests at, record and use that number when spiking jars for chlorine.

To perform a chlorine demand test, 2 L of raw water should be placed into a square jar and stirred at a slow speed. While stirring, add 2 mL of the stock to the square jar and time the process. After 15 minutes, obtain a sample from the jar for chlorine analysis. The dosage minus the residual is the demand. For example, if the stock solution prepared as above tested as 1.8 mg/L, then 4 mL of that stock placed into 2 L of raw water would produce a dosage of 3.6 mg/L. If, after 15 minutes, the residual chlorine of the jar is 2.0 mg/L, then the demand is 3.6 to 2.0, or 1.6 mg/L.

Note that when performing a simple chlorine demand test, no other oxidants should be added to the jar. Also, commercial bleach has a pH greater than 10 and may increase the pH of the water in the jar.

Potassium Permanganate Potassium permanganate is used as an oxidant for iron and manganese control, some organic precursor control, and taste-and-odor control. Permanganate will enhance the coagulation process as it begins to oxidize some of the constituents in the source water. It is thought that the continuous use of this chemical will further reduce the need for chlorine at the front end of the plant. Dosage of permanganate is calculated like any other dry chemical: pounds fed divided by million gallons, in turn divided by 8.34 will provide a dosage as milligrams per liter. Potassium permanganate has the unique quality of a color change mechanism, which signals to the operator its effectiveness and end point. A pink color is associated with water freshly dosed with permanganate. This color changes to straw or yellow at the end point. Operators can use this knowledge as an aid for dosage and demand predictions. Permanganate reactions are dependent largely on pH.

Like chlorine, there is a demand for permanganate. Therefore, operators can use jar testing techniques to calculate this demand.

Make a 1-mg/L stock solution each week by weighing 1 g potassium permanganate into a 1-L volumetric flask and diluting it to 1 L. This is a 1-mg/mL working stock. Obtain a raw water sample taken before the chemical addition point for potassium and pour it up to the 2-L mark in a jar test container at the jar test station. Add 1 mL of stock permanganate and agitate for 5 minutes at slow speed on the stirrer. This is a 0.5-mg/L dose. Add the appropriate amount (current operational amount) of stock alum solution and continue to agitate for another 20 minutes. Observe color change during that time.

Sample the settled water from the beaker and run a free chlorine residual test on it. Multiply the result by 0.89, which is the milligrams per liter of permanganate left in the sample. Subtract this result from 0.5 mg/L (dosage). The result is the amount that should be fed. For water of very little demand, adjustments may be necessary.

Coagulant Jar Testing

Preparation for Jar Testing Note: Following are generic procedures for jar testing. Because each water treatment plant is different, the procedures will have to be customized (calibrated) by the operators to fit the particular system.

The purpose of jar testing is to simulate full-scale plant operations. Successful jar testing depends on the ability of the operator to simulate the conditions of the water plant as closely as possible. Important considerations are those of proper dosage, proper chemical addition sequence, proper mixing times, and maintenance of operating temperatures. Also important is the need for good sampling and analysis techniques and application of the data produced to the full-scale operation.

Operators should determine the times and sequences of the jar testing procedure using knowledge of their particular plant. These sequences will provide for test results of higher quality if they are accurate. AWWA Manual M37, *Operational Control of Coagulation and Filtration Processes*, is an excellent resource for helping the operator determine times and sequences that should be used for testing. The main reason that jar testing fails to produce suitable results is that operators make the mistake of assuming that detention times for settling that are found at plant scale must be duplicated at the bench scale. In reality, bench-scale settling times should be adjusted, or scaled down, and M37 shows how to do that. Most importantly, the sedimentation basin overflow rate must be known because it determines the length of time that floc should be allowed to settle in the jars before sampling. Because overflow rates are a function of plant flow, jar test results can take on new meaning each time an operator makes a change in pump rate.

Starting with fresh stocks, fill weigh boats or pipettes with the proper amount of chemicals and place them next to each jar before proceeding. If using a programmable jar test apparatus, it should be preset for the times that are being used. Manual machines will have to be controlled by an operator at each step change.

Fill each jar test container with 2 L of fresh raw water just prior to testing. The raw water should be obtained from a point in the process before any chemical addition. Alternatively, if permanganate is being used and the testing being performed is for turbidity removal results only, raw water with the permanganate already in it can be used.

Jar Test Procedure A generic procedure for jar testing for coagulant-dosage control is provided in this section. Plant operators should alter these procedures and times to suit their particular needs. Consult AWWA Manual M37 for a more detailed jar testing procedure.

1. Begin the testing by setting the mixers to moderate speed. Operators should have calculated detention times of each sequence of their plant.

2. Add the permanganate (if desired) and allow mixing for a time equal to the travel time (baffling factor adjusted) to the rapid mix. If applicable, add chlorine stock.

3. Add coagulant and set mixers to full speed for the duration of the detention time that has been chosen. Be sure to add the coagulant to all jars simultaneously.

4. Turn down the speed of the mixers to simulate flocculation energy for the duration chosen.

5. After flocculation, turn the mixers to the lowest speed possible to simulate flow through the settling basin.

6. After 5 minutes, samples can be taken for analysis. The most common analysis is the test for turbidity (Figure 10-5).

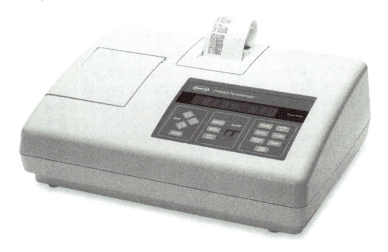

Figure 10-5 Bench-top turbidimeter. Unit is capable of turbidity measurements in a very low range.

Source: Hach Company.

Tips: For colder water temperatures, samples of raw water just prior to jar testing should be taken so that the operator will work with the coolest water possible. If the jar testing is being performed for TOC removals, be sure to wipe the mixer shafts with a clean cloth to remove any oils that may have seeped from the bearings of the mixer. If more than two jars are being tested, it is advisable that two operators perform the testing, which allows for the simultaneous application of chemicals to the jars. Use a clean syringe to take samples from the tops of the settled jars rather than using the attached hose. An excellent reference for jar test procedures is the *Partnership for Safe Water* manual, available from AWWA.

 WATCH THE VIDEO
Jar Testing (www.awwa.org/wsovideoclips)

Study Questions

1. Which type of sample should always be collected for determining the presence of coliform bacteria?
   a. Time composite
   b. Grab sample
   c. Proportional
   d. Composite

2. Samples to be tested for coliforms can be refrigerated for up to _____ hours before analysis, but should be done as soon as possible.
   a. 4
   b. 6
   c. 8
   d. 12

3. When a water sample is acidified, the final pH of the water must be
   a. <2.0.
   b. <2.5.
   c. <3.0.
   d. <3.5.

4. Which chemical is used to remove residual chlorine from water?
   a. $Na_2S_2O_3$
   b. Na_2SiO_3
   c. Na_2SiF_6
   d. NaOCl

5. Particle counters use the principle of light
   a. scattering.
   b. reflection.
   c. refraction.
   d. blockage.

6. Distribution system samples are intended to measure the quality of the water at
   a. a selected location in the system.
   b. various points in the system.
   c. points in the system designated by USEPA.
   d. the system's primary reservoir.

7. Water that contains high concentrations of _____ can cause the consumption of lime when using this method of softening.
   a. calcium bicarbonate
   b. fluoride
   c. free carbon dioxide
   d. iron and manganese

8. _____ is used as an oxidant for iron and manganese control, some organic precursor control, and taste-and-odor control.
   a. Potassium permanganate
   b. Chlorine
   c. Ferric chloride
   d. Alum

9. Which type of sample consists of multiple grab samples taken from the same location at different times and mixed together to make one sample?

10. Which type of coagulation testing is perhaps the least understood test but the most useful process tool available to operators?

Chapter 11
Corrosion Control

Many water systems must apply special chemical treatment because their source water either causes damaging **corrosion** or deposits scale (a process known as **scaling**) on pipelines and plumbing fixtures. The treatment process for controlling these problems is known as **stabilization**. Many more systems must provide corrosion control treatment under new federal and state regulations enacted to protect the public from the health dangers of lead and copper, and other trace metals in drinking water.

Water system operators are cautioned that complicated interactions often occur in the control of water corrosion and scaling. A seemingly simple change to improve one characteristic may have an adverse effect on some other water characteristic or treatment process. This chapter offers an overview of the need for corrosion and scaling control and their related processes. However, because of all the complex variables involved, it is not intended to be a guide for the best treatment method for any particular system. It is best to get professional guidance and state approval before beginning any new stabilization treatment.

Purposes of Corrosion and Scaling Control

Corrosion and scaling are controlled for the following reasons:

- Protect public health
- Improve water quality
- Extend the life of plumbing equipment
- Meet federal and state regulations

Protecting Public Health

Corrosive water can leach toxic metals from distribution and household plumbing systems. Lead and copper are the metals most likely to be a problem because they are commonly used in plumbing systems.

In addition, corrosion of cast-iron mains can cause the formation of iron deposits, called **tubercles**, in the mains. These deposits can protect bacteria and other microorganisms from chlorine, allowing them to grow and thrive. Changes in water velocity or pressure can then cause the microorganisms to be released, creating a potential for disease outbreaks. Some bacteria shielded by the tubercles can also accelerate the corrosion process.

corrosion

The gradual deterioration or destruction of a substance or material by chemical action. The action proceeds inward from the surface.

scaling

Metal deposits left in pipelines and plumbing fixtures.

stabilization

The water treatment process intended to reduce the corrosive or scale-forming tendencies of water.

tubercle

A knob of rust formed on the interior of cast-iron pipes as a result of corrosion.

221

Table 11-1 Estimated effect of scale on boiler fuel consumption

| Scale Thickness, | | Fuel Consumption, |
in.	mm	% increase
1/50	0.5	7
1/16	1.6	18
1/18	3.2	39

Improving Water Quality

Corrosive water attacking metal pipes can cause taste, odor, and color problems in a water system. Red-water problems occur when iron is dissolved from cast-iron mains by corrosive water. The iron will stain a customer's plumbing fixtures and laundry and make the water's appearance unappealing for drinking and bathing. The dissolved iron also acts as a food source for a group of microorganisms called *iron bacteria*, which can cause serious taste-and-odor problems. Corrosion of copper pipes can cause a metallic taste, as well as blue-green stains on plumbing fixtures and laundry.

Extending the Life of Plumbing Equipment

Unstable water can also result in significant costs to water systems and customers. Aggressive water can significantly reduce the life of valves, unprotected metal, and asbestos–cement (A–C) pipe. It can also shorten the service life and performance of plumbing fixtures and hot water heaters.

Buildup of corrosion products (a process known as *tuberculation*) or uncontrolled scale deposits can seriously reduce pipeline capacity and increase resistance to flow. This impaired flow in turn reduces distribution system efficiency and increases pumping costs. If scale deposits or tuberculation go unchecked, pipes can become completely plugged, requiring expensive repair or replacement. Scaling can also increase the cost of operation of hot water heaters by increasing their fuel consumption, as shown in Table 11-1.

Meeting Federal and State Regulations

As detailed later in Chapter 3, the Lead and Copper Rule enacted by the US Environmental Protection Agency (USEPA) in 1991 requires water systems to check if their water is corrosive enough to cause lead and copper corrosion products to appear in customers' water at levels exceeding the new action level. If the level is exceeded, the system is required to take action to reduce the corrosivity of the water.

Water System Corrosion

Corrosion can be broadly defined as the wearing away or deterioration of a material because of chemical reactions with its environment. The most familiar example is the formation of **rust** (oxidized iron) when an iron or steel surface is exposed to moisture. Corrosion is usually distinguished from erosion, which is the wearing away of material caused by physical causes, such as abrasion. Water that promotes corrosion is known as *corrosive* or *aggressive* water.

In water treatment operations, corrosion can occur to some extent with almost any metal that is exposed to water. Whether corrosion of a material will be extensive enough to cause problems depends on several related factors, such as

rust
Oxidized iron.

the type of material involved, the chemical and biological characteristics of the water, and the electrical characteristics of the material and its environment.

The relationships among these factors, as well as the process of corrosion itself, are quite complex. As a result, it is difficult to make general statements about what combinations of water and equipment will or will not have corrosion problems. The discussions in this chapter cover only basic principles; the operator faced with persistent corrosion in a given installation may require the assistance of corrosion-control specialists.

Chemistry of Corrosion

The chemical reactions that occur in the corrosion of metals are similar to those that occur in an automobile battery. In fact, corrosion generates an electrical current that flows through the metal being corroded. The chemical and electrical reactions that occur during *concentration cell corrosion* of iron pipes are illustrated in Figure 11-1.

As shown in Figure 11-1A, minor impurities and variations (present in all metal pipes) have caused one spot on the pipe to act as an electrical anode in relation to another spot that is acting as an electrical cathode. At the anode, atoms of iron (Fe^{+2}) are breaking away from the pipe and going into solution in the water. As each atom breaks away, it ionizes by losing two electrons, which travel through the pipe to the cathode.

In Figure 11-1B, it is shown that chemical reactions within the water balance the electrical and chemical reactions at the anode and cathode. Many of the water molecules (H_2O) have dissociated into H^+ ions and OH^- radicals. This is a normal condition, even with totally pure water. The Fe^{+2} released at the anode combines with two OH^- radicals from dissociated water molecules to form $Fe(OH)_2$, ferrous

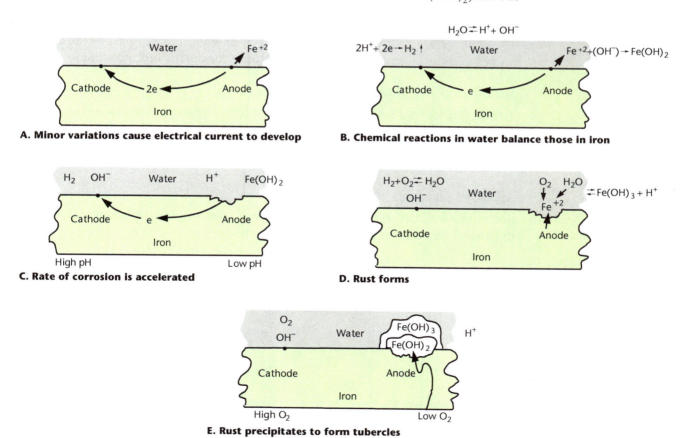

A. Minor variations cause electrical current to develop

B. Chemical reactions in water balance those in iron

C. Rate of corrosion is accelerated

D. Rust forms

E. Rust precipitates to form tubercles

Figure 11-1 Chemical and electrical reactions that occur during corrosion of iron pipe

hydroxide. Similarly, two H^+ ions from the dissociation of the water molecules near the cathode pick up the two electrons originally lost by the iron atom then bond together as H_2, hydrogen gas.

The formation of $Fe(OH)_2$ leaves an excess of H^+ near the anode, and the formation of the H_2 leaves an excess of OH^- near the cathode. This change in the normal distribution of H^+ and OH^- accelerates the rate of corrosion and causes increased pitting in the anode area (the concentration cell), as shown in Figure 11-1C.

If the water contains dissolved oxygen (O_2)—most surface water does—then $Fe(OH)_3$, ferric hydroxide, will form (Figure 11-1D). Ferric hydroxide is common iron rust. The rust precipitates, forming tubercles (Figure 11-1E). The existence of tubercles further concentrates the corrosion, increasing both pitting at the anode and growth of the tubercle. Tubercles can grow into large nodules (Figure 11-2), significantly reducing the carrying capacity of a pipe. During rapid pressure or velocity changes, some of the $Fe(OH)_3$ can be carried away, causing "red water."

 WATCH THE VIDEO
External Corrosion (www.awwa.org/wsovideoclips)

Factors Affecting Corrosion

The rate of corrosion depends on many site-specific conditions, such as the characteristics of the water and pipe material. Therefore, there are no established guidelines that determine the rate at which a pipe will be corroded.

Chemical reactions play a critical role in determining the rate of corrosion at both the cathode and the anode. Any factor that influences these reactions will also influence the corrosion rate.

Dissolved Oxygen

The concentration of dissolved oxygen (DO) in water is a key part of the corrosion process. As the concentration of DO increases, the corrosion rate will also increase.

Total Dissolved Solids

The total dissolved solids (TDS) concentration is important because electrical flow is necessary for the corrosion of metal to occur. Pure water is a poor conductor of electricity because it contains very few ions. But as the TDS is increased, water becomes a better conductor, which in turn increases the corrosion rate.

Alkalinity and pH

Both the alkalinity and the pH of the water affect the rate of chemical reactions. In general, as pH and alkalinity increase, the corrosion rate decreases.

Figure 11-2
Tuberculated pipe
Courtesy of Girard Industries.

Temperature

Because chemical reactions occur more quickly at higher temperatures, an increase in water temperature usually increases the corrosion rate.

Flow Velocity

The velocity of water flowing past a piece of metal can also affect the corrosion rate, depending on the nature of the water. If the water is corrosive, higher flow velocities cause turbulent conditions that bring DO to the corroding surface more rapidly, which increases the corrosion rate. However, if chemicals are being added to stabilize the water, the higher velocities will decrease the corrosion rate, allowing the chemicals (such as calcium carbonate) to deposit on the pipe walls more quickly.

Type of Metal

Metals that easily give up electrons will corrode easily. Table 11-2 lists metals commonly used in water systems, with those at the top being most likely to corrode. This listing is called the *galvanic series of metals*. Where dissimilar metals are electrically connected and immersed in a common flow of water, the metal highest in the galvanic series will immediately become the anode, the other metal will become the cathode, and corrosion will occur. This process is termed *galvanic corrosion* (as opposed to concentration cell corrosion).

The rate of galvanic corrosion will depend largely on how widely separated the metals are in the galvanic series. Widely separated metals will exhibit extremely rapid corrosion of the anode metal (the highest in the series), and the cathode metal will be protected from corrosion. A common example of galvanic corrosion occurs when a brass corporation cock is tapped into a cast-iron main and attached to a copper service line. The copper will be protected at the expense of the brass and cast iron.

Electrical Current

If electrical current is passed through any corrodible metal, corrosion will be accelerated. The two causes of electrical current in water mains are improperly grounded household electrical systems and electric railway systems.

Table 11-2 Galvanic series for metals used in water systems

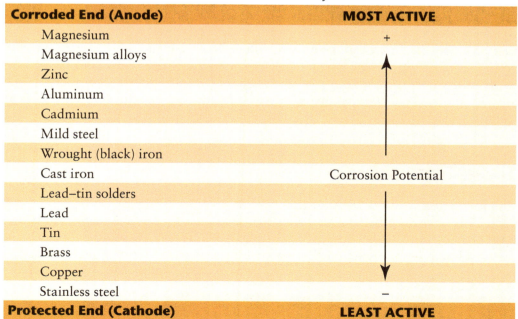

Corroded End (Anode)	MOST ACTIVE
Magnesium	+
Magnesium alloys	
Zinc	
Aluminum	
Cadmium	
Mild steel	
Wrought (black) iron	
Cast iron	Corrosion Potential
Lead–tin solders	
Lead	
Tin	
Brass	
Copper	
Stainless steel	–
Protected End (Cathode)	LEAST ACTIVE

Bacteria

Certain types of bacteria can accelerate the corrosion process because they produce carbon dioxide (CO_2) and hydrogen sulfide (H_2S) during their life cycles, which can increase the corrosion rate. These bacteria can also produce slime, which will entrap precipitating iron compounds and increase red-water problems and the amount of tuberculation.

Two groups of bacteria cause the most problems. Iron bacteria, such as *Gallionella* and *Crenothrix*, can form considerable amounts of slime on pipe walls, particularly if the water contains enough dissolved iron to allow them to survive. The iron present in the water can be naturally occurring in the source water or can be caused by corrosion of the pipe material. Beneath this slime layer, CO_2 production by the bacteria can significantly lower the pH, which will speed up the corrosion rate.

The periodic sloughing of these slime accumulations can cause other major problems, such as tastes and odors. The slimes can also prevent the effective deposition of a protective calcium carbonate ($CaCO_3$) layer by enmeshing it within the slime layer. As this layer sloughs away, it carries away the $CaCO_3$ and leaves the pipe surface bare.

Sulfate-reducing bacteria, such as *Desulfovibrio desulfuricans*, can accelerate corrosion when sulfate (SO_4^{-2}) is present in the water. They reduce the SO_4 under anaerobic conditions, which occur under the slime layer where oxygen is depleted by other bacteria. The products formed are iron sulfide (Fe_2S_3) and hydrogen sulfide (H_2S), which causes obnoxious odors and black-colored water. The CO_2 formed can also lower the pH of the water.

All of these factors interact with each other and with metal pipes, tanks, and various equipment in water plants, the distribution system, and the customer's plumbing. As the factors change, the corrosion rate will also change. The only factors over which the operator has significant control are pH, alkalinity, and the bacteriological content of the water. The techniques used to control these parameters are discussed later in this chapter.

Types of Corrosion

Corrosion in water systems can be divided into two broad classes: localized and uniform. Localized corrosion, the most common type in water systems, attacks metal surfaces unevenly. It is usually a more serious problem than uniform corrosion because it leads to a more rapid failure of the metal. Two types of corrosion that produce pitting are galvanic corrosion and concentration cell corrosion (discussed previously).

Uniform corrosion takes place at an equal rate over the entire surface. It usually occurs where waters having very low pH and low alkalinity act on unprotected surfaces.

Scale Formation

The formation of mild scale on the interior of pipes can protect the pipe from corrosion by separating the corrodible pipe material from the water. However, uncontrolled scale deposits can significantly reduce the carrying capacity of a distribution system. Figure 11-3 shows a pipe that is almost completely blocked by scale.

Figure 11-3 Scaling of pipe

Courtesy of Johnson Controls.

Chemistry of Scale Formation

Scale is formed when the divalent metallic cations associated with hardness, primarily magnesium and calcium, combine with other minerals dissolved in the water and precipitate to coat pipe walls. The most common form of scale is calcium carbonate ($CaCO_3$). Other scale-forming compounds include magnesium carbonate ($MgCO_3$), calcium sulfate ($CaSO_4$), and magnesium chloride ($MgCl_2$).

Factors Affecting Scale Formation

Water can hold only so much of any given chemical in solution. If more is added, it will precipitate instead of dissolve. The point at which no more of the chemical can be dissolved is called the saturation point. The saturation point varies with other characteristics of the water, including pH, temperature, and TDS.

The saturation point of calcium carbonate ($CaCO_3$) depends primarily on the pH of the water. For example, if water with a certain temperature and TDS concentration can maintain 500 mg/L of $CaCO_3$ in solution at pH 7, then the same water will hold only 14 mg/L of $CaCO_3$ in solution if the pH is raised to 9.4.

Temperature also affects the saturation point, although not as dramatically as pH. The solubility of $CaCO_3$ in water decreases as temperature increases. The most common example is when the higher temperature in hot water heaters and boilers causes scale to precipitate out of the water and build up on pipe and tank walls. Because the presence of other minerals in the water affects the solubility of $CaCO_3$, the TDS concentration must be known in order to determine the $CaCO_3$ saturation point. As the TDS concentration increases, the solubility of $CaCO_3$ increases.

Corrosion and Scaling Control Methods

The following are the basic methods used for stabilizing water to protect against the problems of corrosion or scaling:

- Adjustment of pH and alkalinity
- Formation of a calcium carbonate coating
- Use of corrosion inhibitors and sequestering agents

The selection of which method or methods are finally used on any water system depends on both the chemical characteristics of the raw water and the effects of other treatment processes being used.

The type of source water, the number of sources, and the hydraulics and flow patterns of the system can also have a bearing on the choice of corrosion-control measures that should be taken. Systems that have multiple sources with different chemistry may have particularly complex problems. The pH of the water is sometimes critical in the use of chemical corrosion-control measures; some measures work properly only within a narrow pH range. The principal treatment techniques for corrosion and scaling control are summarized in Table 11-3.

Table 11-3 Summary of treatment techniques for controlling corrosion and scaling

Treatment	Application	Effectiveness	Comments or Problems
To Prevent Corrosion			
Lime alone or lime with sodium carbonate or sodium bicarbonate	Increase pH Increase hardness Increase alkalinity	Most effective in water with low pH and hardness Excellent protection for copper, lead, and asbestos–cement pipe in stabilized waters Good protection for galvanized and steel pipe	May be best overall treatment approach Oversaturation may cause calcium deposits
Sodium hydroxide	Increase pH	Most effective in waters with sufficient hardness and alkalinity to stabilize water May provide adequate protection against lead corrosion in low-alkalinity, soft waters	Should not be used to stabilize waters without the presence of adequate alkalinity and hardness May cause tuberculation in iron pipes at pH 7.5–9.0
Sodium hydroxide and sodium carbonate or sodium bicarbonate	Increase pH Adjust alkalinity	Most effective in water with low pH and sufficient hardness Excellent protection for lead corrosion in soft waters at pH 8.3	Combination of high alkalinity and hardness with low pH is more effective than combination of high pH with low hardness and alkalinity
Inhibition with phosphates (primarily sodium zinc phosphate and zinc orthophosphate)	Form a protective film on pipe surfaces	Effective at pH levels above 7.0 Good protection for asbestos–cement pipe Addition of lime may increase effectiveness of treatment for copper, steel, lead, and asbestos	May cause leaching of lead in stagnant waters May encourage the growth of algae and microorganisms May cause red water if extensive tuberculation is present May not be effective at low pH levels
Inhibition with silicates	Form a protective film on pipe surfaces	Most effective in waters having low hardness and pH below 8.4 Good protection for copper, galvanized, and steel pipe	May increase the potential of pitting in copper and steel pipes May not be compatible with some industrial processes
To Prevent Scale Formation			
Carbon dioxide or sulfuric acid	Decrease pH Decrease alkalinity	Effective with high-pH, high-alkalinity water such as lime-softened water	Overfeeding can cause low pH and corrosion
Sequestering with phosphates (primarily sodium hexametaphosphate and tetrasodium polyphosphate)	Sequester scale-forming ions	Effective in controlling scale formation from lime-softened waters and iron in the source water	Can loosen existing deposits and cause red-water complaints Compounds lose sequestering ability in hot water heaters, causing precipitation of calcium carbonate or iron

Adjustment of pH and Alkalinity

In general, soft waters that have a pH of less than 7 and are slightly buffered will be corrosive to lead and copper. Water that has too much alkalinity can also be quite corrosive. Water that is nominally corrosive naturally can also have the corrosivity increased by the addition of other water treatment chemicals. For instance, gaseous chlorine will reduce pH levels.

Usually, a moderate increase in pH and alkalinity levels can reduce corrosion, and a decrease can prevent scale formation. The formation of a protective film on the interior of lead and copper pipes is also usually aided by increasing pH.

Lime is normally used to increase both pH and alkalinity because it is less expensive than other chemicals having the same effect. Soda ash (sodium carbonate) can be added along with the lime to further increase the alkalinity. Sodium bicarbonate is sometimes used instead of sodium carbonate because it will also increase alkalinity without as much of an increase in pH. The increased alkalinity buffers the water against pH changes in the distribution system. This has proven particularly effective in controlling corrosion of lead and copper service pipes. Instead of lime, caustic soda with soda ash or sodium bicarbonate can be used to increase pH and alkalinity.

Lime-softened water can cause severe scale problems if it is not stabilized. Stabilization after softening is accomplished by the addition of carbon dioxide (a process called *recarbonation*) or sulfuric acid. Both chemicals lower the pH so that calcium carbonate will not precipitate in the distribution system.

Formation of a Calcium Carbonate Coating

Because corrosion attacks the surface of a pipe, a protective coating on the pipe surface can inhibit corrosion. A coating of cement, plastic, or asphaltic material is commonly applied to the interior of pipes and tanks and on metal equipment to protect them from corrosion. Although these coatings form an effective barrier against corrosion, very aggressive water can sometimes attack the coatings. In addition, if there are any breaks in the coating, corrosion will be particularly severe at these points.

As a result, many systems apply an additional protective coating by controlling the chemistry of the water. A common protective-coating technique is to adjust the pH of the water to a level just above the saturation point of calcium carbonate. When this level is maintained, calcium carbonate will precipitate and form a protective layer on the pipe walls. This process must be closely controlled. A pH that is too low may result in corrosion, and a pH that is too high may result in excessive precipitation, which will cause a clogging of service pipes and a restriction of flow in the distribution system. Lime, soda ash, sodium bicarbonate, or sodium hydroxide can be used to raise the pH level. Lime is often used because it also adds needed calcium (hardness) and alkalinity.

Use of Corrosion Inhibitors and Sequestering Agents

Some waters do not contain enough calcium or alkalinity to make the formation of calcium carbonate coatings economical. Water obtained primarily from snowmelt is an example, having alkalinity and calcium concentrations as low as 2 mg/L as calcium carbonate. In this event, other chemical compounds can be used to form protective coatings.

The most common compounds are polyphosphates and silicates. The chemical reactions by which these compounds combine with corrosion products to form a protective layer are not completely understood; however, the chemicals have proven successful in many water systems.

Some polyphosphates can also be used as sequestering agents to prevent scale formation. These compounds sequester, or chemically tie up, the scale-forming ions of calcium and magnesium so that they cannot react to form scale. Because these compounds remain in solution, they are eventually ingested by consumers. Therefore, any sequestering agent selected must be suitable for use in drinking water.

Polyphosphates also sequester iron, whether it is dissolved in water from the source or from corrosion of the system. This prevents the precipitation of the iron compounds, so red water will not result. However, this effect does not prevent corrosion; it merely prevents the corrosion by-products from being noticed.

Corrosion and Scaling Control Facilities

Whatever method or methods are used to control corrosion and scaling, the facilities required for the process consist primarily of selected chemicals and the equipment required to store, handle, and feed these chemicals. A wide range of chemicals is available for stabilization. Some of these are primarily intended for industrial or other nonpotable applications. Therefore, all chemicals used to stabilize drinking water must be approved for such use. Chemicals must also comply with NSF International and American Water Works Association standards to ensure quality. Table 11-4 lists the chemicals commonly used for stabilizing potable water.

Chemical Storage and Handling

Chemicals should be purchased in quantities that will maintain a 30-day minimum supply at all times. This practice guards against interruptions of service because of temporary shortages and other unforeseen events. In determining the quantity of chemical to be stored at any one time, the operator should consider

Table 11-4 Chemicals commonly used for stabilization of potable water

Treatment Method	Chemical Name	Chemical Formula
Increase pH and alkalinity	Unslaked lime (quicklime)	CaO
	Slaked lime (hydrated lime)	$Ca(OH)_2$
	Sodium bicarbonate	$NaHCO_3$
	Sodium carbonate (soda ash)	Na_2CO_3
	Sodium hydroxide (caustic soda)	$NaOH$
Decrease pH and alkalinity	Carbon dioxide	CO_2
	Sulfuric acid	H_2SO_4
Form protective coatings	Sodium silicate (water glass)	$Na_2O(SiO_2)n$*
	Sodium hexametaphosphate (sodium polyphosphate, glassy)	$(NaPO_3)n \bullet Na_2O$†
	Sodium zinc phosphate	$(MPO3)n \bullet M_2O$‡
	Zinc orthophosphate	$Zn_3(PO_4)_2$
Act as sequestering agents	Sodium hexametaphosphate (sodium polyphosphate, glassy)	$(NaPO_3)n \bullet Na_2O$†
	Tetrasodium pyrophosphate	$Na_4P_2O_7 \bullet 10H_2O$

*Typically $n = 3$.
†Typically $n = 14$.
‡M = Na and/or ½ Zn, typically $n = 5$.

storage space available, discounts for purchasing large quantities, and length of time the chemical can be held without losing potency or caking so that it becomes unusable.

Lime

Lime is available in either unslaked or slaked form. Unslaked lime, also called *quicklime* or *calcium oxide* (CaO), is available in a variety of sizes from a powder to pebble form. Although available in bags, it is considerably more economical and easier to handle in bulk. As a result, quicklime is normally used only by treatment plants that use large quantities. It should not be stored for more than three months because it can deteriorate. Quicklime is noncorrosive in the dry form and can be stored in steel or concrete bins. A minimum of two bins should be available to provide for maintenance and ease in unloading. The bulk deliveries are transferred from the railcars or trucks to the bins by mechanical or pneumatic conveyors.

For treatment plants that do not use large quantities of lime, hydrated (or slaked) lime—calcium hydroxide, $Ca(OH)_2$—is more cost-effective. Hydrated lime is a finely divided powder available in 50- or 100-lb (23- or 45-kg) bags and in bulk shipments. The bulk form is unloaded and stored like quicklime and should be used within three months. The bagged hydrated lime can be stored for up to one year without serious deterioration. Both types should be stored in a dry, well-ventilated area. This is particularly important for quicklime because moisture will start the slaking process, generating a tremendous amount of heat, which could start a fire if combustible materials are nearby.

Soda Ash

Soda ash (sodium carbonate, Na_2CO_3) is a white, alkaline chemical. It is available as granules and powder in bulk or in 100-lb (45-kg) bags. Because soda ash absorbs moisture and will readily cake, it should be stored in a dry, well-ventilated area. Dry soda ash is not corrosive, so it can be stored in steel or concrete bins. Sodium carbonate should not be stored near products containing acid. If acid contacts the chemical, large quantities of carbon dioxide can be released, creating a safety problem.

Sodium Bicarbonate

Sodium bicarbonate ($NaHCO_3$) is also commonly known as *baking powder*. It is a white, alkaline chemical available in either powdered or granular form. It can be purchased in 100-lb (45-kg) bags or in barrels up to 400 lb (180 kg). It must be stored in a cool, dry, well-ventilated area because it decomposes rapidly as the temperature nears 100°F (38°C).

As with sodium carbonate, sodium bicarbonate should not be stored near products containing acid. Corrosion-resistant materials must be used for storing and transporting sodium bicarbonate solution.

Sodium Hydroxide

Sodium hydroxide (NaOH) is commonly called *caustic soda*. It is available as a liquid or in dry form in flakes or lumps. Regardless of the form, it must be handled carefully because of the many hazards involved. As a liquid, caustic soda is available in solution concentrations of 50 percent or 73 percent NaOH. The 50 percent solution will begin to crystallize if its temperature drops below 54°F (12°C). The 73 percent solution will crystallize as its temperature drops below 145°F (63°C). Special storage facilities are therefore required to maintain the temperature of the solution high enough to prevent crystallization. This can be accomplished by further dilution or by installing heaters in the storage tanks.

The manufacturer's recommendations should be followed closely to prevent problems. The storage tanks should be constructed of nickel–cadmium or nickel–alloy steel, lined with caustic-resistant material, such as rubber or polyvinyl chloride (PVC). Tanks made of PVC or fiberglass can also be used.

If the dry form of caustic soda is used, it is normally dissolved immediately on delivery in a storage tank. The dilution concentration is at a point where crystallization will not be a problem at local temperatures.

With either the dry or the liquid form, the addition of water for dilution generates a considerable amount of heat. Therefore, the rate of dilution must be carefully controlled so that boiling or splattering of the solution does not occur. A source of flushing water should be readily available in case of spills.

Sulfuric Acid

Sulfuric acid (H_2SO_4) is a corrosive, dense, oily liquid available in strengths containing 62 percent, 78 percent, or 93 percent H_2SO_4. It can be purchased in 55- or 110-gal (210- or 420-L) barrels or in bulk in tank cars or trucks. The acid must be stored in corrosion-resistant tanks.

If diluted, the acid must always be added to the water (not the water to the acid). If the water is added to the acid, there will be splattering and rapid generation of heat.

Sodium Silicate

Sodium silicate is also frequently called *water glass*. It is an opaque, syrupy, alkaline liquid, available in barrels and in bulk. The barrels can be stored, or the liquid can be added immediately to storage tanks. Bulk deliveries are unloaded directly into storage tanks. The tanks can be constructed of steel or plastic because sodium silicate is not corrosive.

Carbon Dioxide

Carbon dioxide (CO_2) is a colorless, odorless gas that can be generated on-site or purchased in bulk in a liquified form. On-site generation involves producing the CO_2 in a furnace or in an underwater burner.

Liquified CO_2 is stored in insulated, refrigerated pressure tanks ranging in capacity from 6,000 to 100,000 lb (2,700 to 45,000 kg). The liquified CO_2 must be kept at about 0°F (–18°C) and 300 psig (2,070 kPa [gauge]) so that it will remain in liquid form. Carbon dioxide vapor forms above the liquid surface and is withdrawn and piped to the application point. Each tank has a built-in vaporizer, which helps maintain a constant vapor supply. The storage tank should be located as close as possible to the application point.

Phosphate Compounds

A number of phosphate compounds can be used for stabilization. Three of the most common are sodium hexametaphosphate (sodium polyphosphate, glassy), sodium zinc phosphate, and zinc orthophosphate. Sodium hexametaphosphate is available in bags or drums in a form that looks like broken glass. The other two compounds are generally available in dry form in 50-lb (23-kg) bags and in liquid form in drums or tank trucks. The concentrated liquid solutions are slightly acidic, so they should be stored in corrosion-resistant tanks constructed of fiberglass, PVC, or stainless steel.

Chemical Feed Equipment

Feeding Lime

The type of equipment needed for lime feeding depends on the type of lime being used. If slaked lime is used, a gravimetric or volumetric dry feeder adds the lime to a solution chamber, where water is added to form a slurry (Figure 11-4). When unslaked lime is used, the dry feeder adds lime to a slaker, where the lime and water are mixed together to slake the lime and form a slurry ($CaO + H_2O \rightarrow Ca(OH)_2$). This slurry is often called *milk of lime*. Slakers are described in more detail in Chapter 12. The hoppers or bins supplying the dry feeders must be equipped with agitators or vibrators, because lime does not flow freely and will pack and "bridge" during storage.

Regardless of the type of lime used, the slurry must be added to the water. It is best to minimize the distance from the feeders to the application point because lime slurry will cake on any surface. Open troughs or flexible feed lines should be used to carry the slurry to the application point because they must frequently be cleaned and flushed. Solution feeders used with the slurry should be designed for easy cleaning.

Feeding Soda Ash and Sodium Bicarbonate

Soda ash and sodium bicarbonate can be fed with gravimetric or volumetric dry feeders. Agitators should be provided on the hoppers to prevent the chemical from bridging, which will block the flow.

Soda ash solutions can be fed using conventional solution feeders and lines. Because soda ash goes into solution slowly, it is necessary to provide larger than usual dissolving chambers to provide proper detention time, as well as good mixers to achieve thorough mixing. Sodium bicarbonate solutions are caustic. Therefore, metering pumps, dissolving chambers, and pipes should be constructed or lined with caustic-resistant materials, such as stainless steel or PVC.

Feeding Sodium Hydroxide

Sodium hydroxide solutions can be fed by metering pumps designed for handling corrosive and caustic solutions. All piping should be of caustic-resistant material, such as PVC. Valves and fittings should not contain any copper, brass, bronze, or aluminum because these materials will deteriorate rapidly.

Figure 11-4 Lime-feeding equipment

Feeding Sulfuric Acid

Sulfuric acid can be fed directly from the shipping container by a corrosion-resistant metering pump and piping. The diluted acid solution is even more corrosive than the pure acid and must be handled carefully.

Feeding Sodium Silicate

Sodium silicate can be fed directly from the shipping container or from a day tank, by using a conventional metering pump and piping.

Feeding Carbon Dioxide

If carbon dioxide is generated on-site, the feeding system consists of the generation equipment and related piping. A typical generation system consists of a combustion unit located beneath the surface of the water in a reaction basin. The carbon dioxide is produced and mixed by the same unit. The submerged combustion unit is also discussed in Chapter 12.

Another common system uses gas drawn from a carbon dioxide storage tank. The equipment necessary is shown in Figure 11-5. The gas can be fed to a recarbonation basin or directly to a pipeline for pH control.

Feeding Phosphate Compounds

Phosphate compounds purchased in liquid form are often fed directly from the shipping container; they can also be diluted in a day tank. If dry compounds are used, they are normally dissolved in the storage tanks by placing them in stainless-steel baskets suspended in the tanks. Metering pumps must be designed to handle corrosive materials, and piping, valves, and fittings should be made of PVC or stainless steel.

Limestone Contactors

A limestone bed contactor is a treatment device in which water flows through a container packed with crushed limestone and thereby dissolves calcium carbonate. The dissolving of the calcium carbonate increases the pH and alkalinity of the water. This treatment has been shown to reduce the corrosion of lead and copper in water systems having nominally corrosive water.

The contactors are simple, are low cost, and require minimal maintenance, making them especially suitable for corrosion control in small water systems.

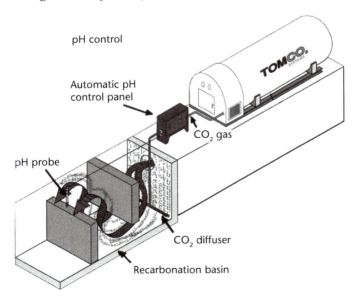

Figure 11-5 Liquid CO_2 recarbonation system

Courtesy of TOMCO Equipment Co.

Study Questions

1. Which index determines the calcium carbonate deposition property of water by calculating the saturation pH, where a negative value indicates corrosive water and a positive value indicates depositing water?
   a. Baylis curve
   b. Langelier saturation index
   c. Marble test
   d. Ryzner index

2. Which of the following composes the majority of scale in pipe?
   a. $Ca(HCO_3)_2$
   b. $CaCO_3$
   c. $MgCO_3$
   d. $CaSO_4$

3. The corrosion process can be accelerated by certain bacterial organisms because they produce which chemical?
   a. N_2
   b. CO_2
   c. $MgCO_3$
   d. $CaCO_3$

4. Which chemical would be best to use for corrosion inhibition if the water is very low in alkalinity and calcium concentration?
   a. Sodium carbonate
   b. Sodium bicarbonate
   c. Sodium hydroxide
   d. Polyphosphates

5. Stabilization refers to the treatment process used to control
   a. turbidity.
   b. temperature fluctuations.
   c. corrosion.
   d. pH imbalances.

6. In water treatment operations, corrosion can occur to some extent with
   a. both plastics and metals alike.
   b. almost any metal that is exposed to water.
   c. all metals except cast iron.
   d. all metals except copper and stainless steel.

7. Which metal is the most likely to corrode?

8. What are the two broad classes into which corrosion in water systems can be divided?

9. List two chemicals commonly used to increase both pH and alkalinity.

Chapter 12
Lime Softening

Lime Softening Chemical Reactions

The following equations represent the basic reactions involved in the lime–soda ash process. The down arrow (↓) indicates precipitates formed by the reactions.

First, although carbon dioxide does not cause hardness, it reacts with and consumes the lime added to remove hardness. Therefore, it must be considered when the lime dosage is being determined. Sufficient lime must be added to convert carbon dioxide (CO_2) to calcium carbonate ($CaCO_3$), as shown in Equation 12-1. This reaction is complete when a pH of 8.3 is reached. Some systems use aeration to remove some of the carbon dioxide, thereby reducing the lime requirement.

$$CO_2 \quad + \quad Ca(OH)_2 \quad \longrightarrow \quad CaCO_3\downarrow \quad + \quad H_2O \tag{12-1}$$

carbon   lime   calcium   water
dioxide   carbonate

The minerals that cause carbonate hardness are then precipitated out of the water either as calcium carbonate (Equation 12-2) or as magnesium hydroxide (Equations 12-3 and 12-4). In Equation 12-2, enough lime is added to raise the pH to 9.4, at which point the calcium precipitate is formed. A similar process is shown in Equations 12-3 and 12-4, where excess lime is added to elevate the pH above 10.6 to form the magnesium hydroxide precipitate, $Mg(OH)_2$.

$$Ca(HCO_3)_2 \quad + \quad Ca(OH)_2 \quad \longrightarrow \quad 2CaCO_3\downarrow \quad + \quad 2H_2O \tag{12-2}$$

carbon   lime   calcium   water
bicarbonate   carbonate

$$Mg(HCO_3)_2 + Ca(OH)_2 \longrightarrow CaCO_3\downarrow + MgCO_3 + 2H_2O \tag{12-3}$$

magnesium   lime   calcium   magnesium   water
bicarbonate   carbonate   carbonate

$$MgCO_3 \quad + \quad Ca(OH)_2 \quad \longrightarrow \quad CaCO_3\downarrow \quad + \quad Mg(OH)_2 \tag{12-4}$$

magnesium   lime   calcium   magnesium
carbonate   carbonate   hydroxide

The minerals that cause noncarbonate calcium hardness are precipitated out of the water by the addition of soda ash. The chemical reactions differ slightly, based on the types of noncarbonate calcium compounds causing the hardness. Two examples of these chemical reactions are as follows:

$$CaSO_4 \quad + \quad Na_2(CO)_3 \quad \longrightarrow \quad CaCO_3\downarrow \quad + \quad Na_2SO_4 \tag{12-5}$$

calcium sulfate | soda ash | calcium carbonate | sodium sulfate

$$CaCl_2 \quad + \quad Na_2(CO)_3 \quad \longrightarrow \quad CaCO_3\downarrow \quad + \quad 2NaCl \tag{12-6}$$

calcium chloride | soda ash | calcium carbonate | sodium chloride

The minerals that cause noncarbonate magnesium hardness are precipitated out of the water by the addition of lime (Equations 12-7 and 12-9). However, this process also results in the formation of noncarbonate salts (such as $CaSO_4$ and $CaCl_2$) that cause noncarbonate hardness. Therefore, soda ash must be added to the water to react with these salts to form calcium carbonate (Equations 12-8 and 12-10).

$$MgCl_2 \quad + \quad Ca(OH)_2 \quad \longrightarrow \quad Mg(OH)_2\downarrow \quad + \quad CaCl_2 \tag{12-7}$$

magnesium chloride | lime | magnesium hydroxide | calcium chloride

$$CaCl_2 \quad + \quad Na_2CO_3 \quad \longrightarrow \quad CaCO_3\downarrow \quad + \quad 2NaCl \tag{12-8}$$

calcium chloride | soda ash | calcium carbonate | sodium chloride

$$MgSO_4 \quad + \quad Ca(OH)_2 \quad \longrightarrow \quad Mg(OH)_2\downarrow \quad + \quad CaSO_4 \tag{12-9}$$

magnesium sulfate | lime | magnesium hydroxide | calcium sulfate

$$CaSO_4 \quad + \quad Na_2CO_3 \quad \longrightarrow \quad CaCO_3\downarrow \quad + \quad Na_2SO_4 \tag{12-10}$$

calcium sulfate | soda ash | calcium carbonate | sodium sulfate

Because the softened water has a pH close to 11 and a high concentration of calcium carbonate, it must be stabilized so that the calcium carbonate will not precipitate on the filter media, on filter underdrains, or in the distribution system. The water is stabilized through a process called recarbonation, which involves adding carbon dioxide to the softened water (Equation 12-11).

When carbon dioxide is added, soluble calcium bicarbonate is formed (resulting in a small amount of hardness), and the pH is reduced to about 8.6, or to a level at which the water is stabilized to prevent scale formation or corrosion.

Equation 12-12 shows that any residual $Mg(OH)_2$ will also be converted to a soluble compound, $MgCO_3$.

$$CaCO_3 \quad + \quad CO_2 \quad \longrightarrow \quad H_2O \quad + \quad Ca(HCO_3)_2 \tag{12-11}$$

calcium carbonate | carbon dioxide | water | calcium bicarbonate

$$Mg(OH)_2 \quad + \quad CO_2 \quad \longrightarrow \quad MgCO_3\downarrow \quad + \quad H_2O \tag{12-12}$$

magnesium hydroxide | carbon dioxide | magnesium carbonate | water

Sulfuric acid is sometimes used to stabilize water. However, this process adds noncarbonate hardness ($CaSO_4$), and the acid is more difficult to handle. Therefore, carbon dioxide is more frequently used for stabilizing softened water.

Lime Softening Facilities

The lime–soda ash softening process involves the following components:

- Chemical storage facilities
- Chemical feed facilities
- Rapid-mix basins
- Flocculation basins
- Sedimentation basins and related equipment
- Solids-contact basins (often used in place of the rapid-mix, flocculation, and sedimentation basins)
- Pellet reactors (may be used in place of rapid-mix, flocculation, and sedimentation basins)
- Sludge recirculation, dewatering, and disposal equipment
- Recarbonation facilities
- Filtration facilities

A discussion of these components follows.

Chemical Storage Facilities

The size and number of facilities used for storing chemicals depend on the type and quantity of chemicals used. Lime can be purchased in two forms:

- Calcium hydroxide, $Ca(OH)_2$, which is called hydrated lime or *slaked lime*
- Calcium oxide, CaO, which is called quicklime or *unslaked lime*

Both calcium hydroxide and calcium oxide are dry chemicals, varying in texture from a light powder to pebbles (Figure 12-1). Although calcium oxide requires special feeding equipment, large treatment plants typically use it because it is less expensive. Smaller plants, however, usually find calcium hydroxide less expensive. Soda ash (sodium carbonate, Na_2CO_3) is available in either powdered or granular form. All of these chemicals are available in bags, drums, or bulk truck or train carloads.

Large quantities of chemicals are used in the water softening process. For example, to reduce the hardness of 1 mgd (4 ML/d) of water from 200 mg/L to 50–85 mg/L hardness, more than 1,200 lb (540 kg) of soda ash is required. Chemical storage bins (Figure 12-2) are used to store the large quantities of chemicals that are required.

It is normally good practice to store a minimum of one month's supply of each chemical, so chemical storage space requirements for a plant must be carefully calculated. Calcium oxide and soda ash are most often delivered by hopper-bottom trucks and are transferred by mechanical or pneumatic conveyors to storage bins (Figure 12-3). Calcium hydroxide is usually delivered and stored in bags. When liquid caustic soda (50 percent NaOH) is used, it is stored and fed from lined-steel or plastic tanks.

hydrated lime
Another name for calcium hydroxide, $Ca(OH)_2$, which is used in water softening and stabilization. Also called slaked lime.

quicklime
Another name for calcium oxide, CaO, which is used in water softening and stabilization. Also called unslaked lime.

Figure 12-1 Pebble quicklime

Courtesy of the National Lime Association, Arlington, Virginia.

Figure 12-2 Chemical storage bins

Courtesy of the National Lime Association, Arlington, Virginia.

Figure 12-3 Conveyor used to transport calcium oxide and soda ash

Courtesy of the National Lime Association, Arlington, Virginia.

Chemical Feed Facilities

Slaked lime and soda ash can conveniently be fed by conventional dry feeders, which are usually located directly beneath the storage bins. Dispensing calcium oxide is a more complex process. In addition to a dry feeder, a lime slaker (Figure 12-4) and a solution feeder are required. The dry feeder dispenses the chemical into the slaker, where the calcium oxide and water are mixed together to form a slurry known as **milk of lime** [$CaO + H_2O \rightarrow Ca(OH)_2$]. The milk-of-lime slurry is then fed into the treatment system by a solution feeder.

The slurry will cake on any surface, so all feed equipment must be cleaned regularly. Open troughs or troughs that have removable covers are normally used to transport the slurry to allow for easy cleaning.

Tremendous heat is generated when lime is slaked, so adequate ventilation must be provided to the area where slakers are located. Provisions must also be made to remove the grit produced during slaking. The grit consists of undissolved lime and impurities. The quantity of grit produced depends on the amount of lime slaked. Small plants can often collect the grit in a wheelbarrow, but larger plants are usually designed to load the grit directly to dump trucks, which haul it to a disposal site daily.

milk of lime
The lime slurry formed when water is mixed with calcium hydroxide.

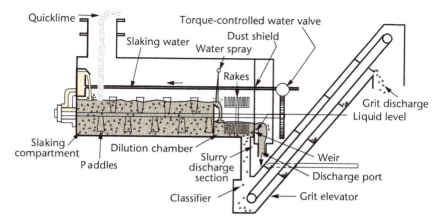

Figure 12-4 Lime slaker
Courtesy of US Filter/Wallace
& Tiernan.

Conventional dry feeders and solution feeders are also used to dispense coagulants (such as alum) or coagulant aids (such as organic polymers) if necessary. Whenever lime, soda ash, or other dry powdered chemicals are used, chemical dust can pose a health hazard. Therefore, dust control equipment should always be used while working with the chemicals.

Rapid-Mix Basins

Rapid-mix basins used in lime–soda ash softening are similar to those described in Chapter 4 for coagulation and flocculation. They serve the following two purposes:

1. They facilitate thorough mixing of lime and soda ash with water.
2. They allow mixing to continue long enough for the chemicals to dissolve. This extended mixing time is especially critical for lime because it dissolves rather slowly.

Rapid-mix basins with high-speed, mechanical mixers are preferred over baffled basins because the mixing rate can be controlled independently of the flow rate. This makes it possible to mix the chemicals more vigorously. Some rapid-mix basins are also equipped to receive sludge that is recirculated from the settling basin in order to speed up the chemical reactions.

Although coagulants can be added to rapid-mix basins, better results are usually obtained if they are added upstream in a separate rapid-mix basin.

Flocculation Basins

Flocculation is the process by which impurities are gathered together into solid particles, known as *floc*, that form and grow after coagulants have been added to the water. Flocculation basins used in the lime–soda ash softening process are similar to those used in conventional water treatment. They are equipped with gentle mixing devices, such as rotating paddle wheels, that control the mixing of chemicals. Variable-speed drives on the mixers provide a means of varying the mixing speed required for the most effective flocculation under the varying conditions of raw-water quality and flow rate.

Sedimentation Basins

Sedimentation is the process by which suspended floc particles are removed by gravity (i.e., settle out of water). Sedimentation basins used in the lime–soda ash softening process can be identical to those described in Chapter 5. However, because of the larger amount of sludge formed, mechanical equipment to collect and remove the sludge in the basins is essential.

Solids-Contact Clarifiers

The solids-contact basin (or clarifier), described in Chapter 5, is sometimes used in the conventional treatment process. The basin is a single unit in which the processes of coagulation–flocculation, sedimentation, sludge collection, and sludge recirculation are performed (Figure 12-5). Several designs of solids-contact basins are available. All have two major zones:

1. A *mixing zone*, which includes the reaction flocculation area in the basin. In this zone, lime (or lime with soda ash) is mixed with water and also with some of the previously formed slurry, which is recirculated to enhance the chemical reactions and formation of precipitates.
2. A *settling or clarification zone* where the precipitates are allowed to settle as the clarified water passes upward through the sludge blanket to the effluent troughs.

A portion of the settling solids is returned to the mixing zone. The remainder is allowed to settle completely and is periodically withdrawn to maintain a relatively fixed slurry concentration in the basin.

The major advantage of the solids-contact basin is that it saves on construction costs by handling many softening processes in one component.

Pellet Reactors

A pellet reactor is a conical tank in which the softening reactions take place quite rapidly as the water passes upward through the unit (Figure 12-6). In operation, the tank is half-filled with fine granules of calcium carbonate. Raw water enters the bottom through a nozzle placed at an angle so that an upward swirling action is produced. At the same time, a dose of lime or lime and soda ash is introduced through another opening.

The chemicals mix with the water as it moves upward, and the softening action takes place. The calcium carbonate granules gradually grow in size, with the heaviest sinking to the bottom, where they are periodically drawn off. At the same time, additional finer particles are added to make up for the loss. One advantage of the process is that the equipment does not require very much space, and the initial cost is relatively low. The detention time is only about 5–10 minutes. In addition, the "waste" from the unit is a granular material (Figure 12-7), which is much easier to handle than conventional lime-softening sludge.

pellet reactor
A conical tank, filled about halfway with calcium carbonate granules, in which softening takes place quite rapidly as water passes up through the unit.

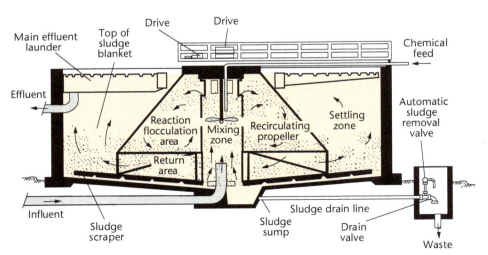

Figure 12-5 Solids-contact basin
Courtesy of General Filter Company, Ames, Iowa

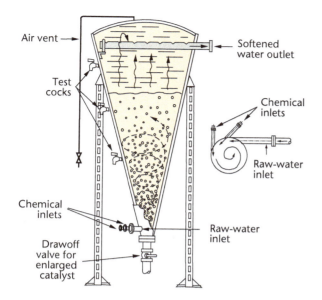

Figure 12-6 Design of a pellet reactor

Courtesy of the National Lime Association, Arlington, Virginia.

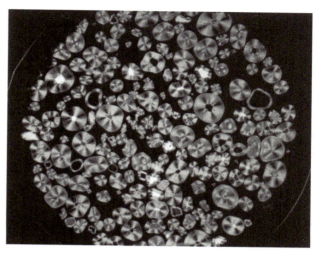

Figure 12-7 Thin section of water softening pellets from a pellet reactor. Concentric buildup of calcium carbonate on the seed particles is shown.

Photograph by D. J. Wiersma, The Netherlands.

A principal disadvantage is that magnesium removal is difficult in the units, so pellet reactors should not be considered if the raw-water magnesium content is high.

Sludge Pumps

The flocculation and settling of calcium carbonate and magnesium hydroxide precipitates occur more quickly in the presence of previously formed floc particles. For this reason, some settled sludge is commonly recirculated to a rapid-mix basin. Whether or not sludge recirculation is practiced, sludge must be removed from the settling basin. Although sludge removal can be accomplished by gravity drains, it is rare that the plant can be set up for this to be possible. As a rule, special sludge-handling pumps must be provided for both sludge recirculation and removal. The handling and disposal of sludge are discussed further in Chapter 5.

Pumps for handling sludge are usually located in rooms adjacent to the settling basins. Centrifugal pumps with open impellers designed to prevent plugging are typically used for sludge disposal. Use of slow-speed centrifugal pumps or positive-displacement pumps minimizes the breakup of the flocculated solids, which helps the sludge retain good settling characteristics.

Sludge recirculation equipment also includes flow-measuring devices, such as a magnetic or ultrasonic meter that will not become clogged. A progressing-cavity, variable-speed metering pump is also sometimes used to both pump and measure the returned sludge.

Recarbonation Facilities

Recarbonation is usually accomplished by adding carbon dioxide gas to softened water to stabilize the water before it enters the distribution system. Recarbonation is performed in a basin that normally provides 15–30 minutes of detention time.

There are several ways to obtain carbon dioxide gas. In large, older plants, it is collected from furnace exhaust gases. It is then scrubbed by spraying water through it to remove impurities, moved by a compressor to the recarbonation basin, and diffused into the water. This technique is no longer recommended because

recarbonation
The reintroduction of carbon dioxide into the water, either during or after lime–soda ash softening, to lower the pH of the water.

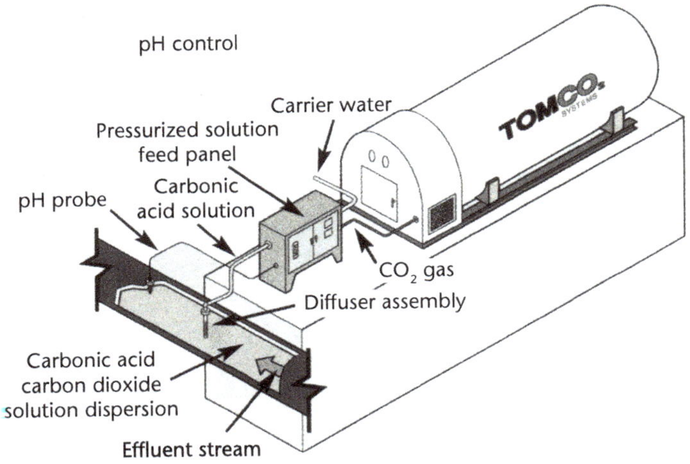

Figure 12-8
Submerged-combustion
recarbonation system

Courtesy of TOMCO
Equipment Co.

sulfur and phenolic materials in the gas may cause taste-and-odor problems in the finished water. In addition, the equipment required for this system requires considerable maintenance because moist carbon dioxide is very corrosive.

In large, newer plants, carbon dioxide is produced by (1) burning a mixture of natural gas and air in a submerged combustion chamber (Figure 12-8), or (2) burning a mixture of natural gas and air in a forced-draft generator on the surface. Natural gas is used in these processes because it has relatively few impurities. The carbon dioxide produced is released near the bottom of the recarbonation basin, and the amount of gas generated is controlled by regulating the amount of gas burned. This equipment also produces carbon monoxide, so good ventilation is essential to prevent the level from reaching dangerous concentrations in low areas around the basin.

Small treatment plants and plants requiring only small amounts of carbon dioxide can purchase carbon dioxide in the form of either dry ice or liquid. Some principal advantages to purchasing the gas are the elimination of the costs of purchasing and maintaining combustion equipment and compressors and the danger of carbon monoxide being present in the plant.

If carbon dioxide is purchased in the form of dry ice or liquid, an evaporator is used to convert it to a gas. The gas then flows through a pressure-regulating valve, and the flow is controlled by a rotameter, similar to the type used on gas chlorinators. The gas is then fed into the recarbonation basin through diffusers. Because the gas generated from dry ice or liquid carbon dioxide is pure carbon dioxide, pipes for the diffusers can be much smaller than those needed for carbon dioxide generated on-site by combustion.

Carbon dioxide gas is heavier than air. A leak can result in a hazardous buildup of the gas in low areas of the treatment plant. Therefore, a good ventilation system is needed wherever carbon dioxide is used.

Filtration Facilities

The filters used in a lime–soda ash softening plant are identical to those used in a conventional rapid sand or multimedia filter plant.

Regulations

There are no federal regulations that require the softening of water. It should be noted that addition of lime–soda ash softening could affect a system's compliance with some regulations.

Trihalomethane Control

The formation of trihalomethanes (THMs) or other disinfection by-products can be increased by softening because of the high pH required to operate the process. If this is found to be a problem, other methods can be considered for reducing the THMs sufficiently to meet the requirements. One method is to simply change disinfectants. If this effort is not sufficient, a preliminary treatment step may have to be added to remove precursors before the softening process. Although THMs can also be removed after other treatment is completed by use of granular activated carbon, this is a rather expensive alternative.

TOC Removal Requirements

As stated in Chapter 3, the Disinfection By-products Rule (DBP Rule) requires the removal of total organic carbon (TOC) levels based on source water alkalinity (see Table 3-3 in Chapter 3). Softening plants that cannot meet these TOC removal percentages can comply by meeting two other alternative compliance criteria:

1. Softening that results in lowering the treated-water alkalinity to less than 60 mg/L (as $CaCO_3$), measured monthly and calculated quarterly as an annual running average
2. Softening that results in removing at least 10 mg/L of magnesium hardness (as $CaCO_3$), measured monthly and calculated quarterly as an annual running average

Meeting Lead and Copper Rule Requirements

The federal Lead and Copper Rule requires that samples from customers' taps be analyzed for the presence of lead and copper. Excessive lead and copper are normally the result of corrosion in lead and copper piping. Soft water is generally more aggressive than hard water, so systems that have softening treatment may find it necessary to take additional steps to stabilize their water to meet the Lead and Copper Rule requirements (as discussed further in Chapter 11).

Meeting Surface Water Treatment Rule Requirements

The Surface Water Treatment Rule (SWTR) requires that all water systems using a surface source, as well as certain groundwater systems designated by the state, must maintain specified removal and disinfection levels to ensure microbiological safety of the finished water. Although the lime–soda ash process is known to be effective in removing viruses, the amount of "credit" that should be given to a softening plant for overall microbiological removal or control has not yet been determined. Systems required to comply with SWTR requirements should consult with state authorities on how the requirements must be met by their system.

WATCH THE VIDEO
Lime Softening (www.awwa.org/wsovideoclips)

Study Questions

1. The minerals that cause noncarbonate magnesium hardness are precipitated out of water by the addition of
   a. carbon dioxide.
   b. chlorine.
   c. lime.
   d. calcium bicarbonate.

2. To reduce noncarbonate hardness, soda ash must be added to water to form
   a. calcium bicarbonate.
   b. carbon dioxide.
   c. magnesium carbonate.
   d. calcium carbonate.

3. In lime softening facilities, it is normally good practice to store a minimum of _____ supply of each chemical.
   a. 1 year's
   b. 1 week's
   c. 1 month's
   d. 1 day's

4. What is the term for a conical tank in which the softening reactions take place quite rapidly as the water swirls upward through the unit?
   a. Rapid-mix basin
   b. Pellet reactor
   c. Solids-contact clarifier
   d. Flocculation basin

5. When magnesium hardness is treated with lime what is precipitated?

6. Why is soda ash needed to remove noncarbonate hardness?

7. What happens when quicklime is added to water?

8. What is carbon dioxide used for in lime softening?

9. What federal regulation requires that samples from customers' taps be analyzed for the presence of lead and copper?

Chapter 13

Ion Exchange

Ion Exchange Softening

When ion exchange units are used to soften water, they use a resin that releases the nonhardness-causing ion attached to it in favor of the hardness-causing ions present in the raw water. The resin holds the unwanted ion temporarily and releases it when a regenerant solution is used to restore the resin to its original form. The process of regeneration allows the resin to be reused.

The regeneration process using sodium chloride, or salt, as the regenerant involves the following chemical reactions:

$$CaX + 2NaCl \longrightarrow CaCl_2 + Na_2X$$

$$MgX + 2NaCl \longrightarrow MgCl_2 + Na_2X$$

The ion exchange softening process does not alter the pH or alkalinity of the water. However, the stability of the water is altered by the removal of calcium and by an increase in total dissolved solids (TDS). For each 1 mg/L of calcium removed and replaced with sodium, the TDS increases 0.15 mg/L. For each 1 mg/L of magnesium removed and replaced with sodium, the TDS increases by 0.88 mg/L.

The measurements commonly used to express water hardness in the ion exchange softening process are different from those in the lime–soda ash process. Hardness in the lime–soda ash process is commonly expressed in terms of milligrams per liter as calcium carbonate ($CaCO_3$). In the ion exchange process, hardness is expressed in grains per gallon, which is simply referred to as *grains*. Currently, both grains/gallon and mg/L as $CaCO_3$ are used. The following conversion factors show the relationship between mg/L and grains:

$$1 \text{ grain} = 17.12 \text{ mg/L}$$

$$1 \text{ grain} = 0.142 \text{ lb per } 1,000 \text{ gal}$$

$$7,000 \text{ grains} = 1 \text{ lb per gal}$$

Early ion exchangers were natural silica compounds called *zeolites*. Today, synthetic organic exchange materials are available with improved properties and expanded uses.

Facilities

The ion exchange process requires the following basic components:

- Ion exchange materials (resins)
- Ion exchange units
- Salt storage tanks

> **ion exchange**
> A process used to remove hardness from water that depends on special materials known as resins. The resins trade nonhardness-causing ions (usually sodium) for the hardness-causing ions, calcium and magnesium. The process removes practically all the hardness from water.
>
> **regeneration**
> The process of reversing the ion exchange softening reaction of ion exchange materials. Hardness ions are removed from the used materials and replaced with nontroublesome ions, thus rendering the materials fit for reuse in the softening process.

Ion Exchange Resins

The majority of commercially available ion exchange **resins** are made by the copolymerization of the organic polymers, styrene and divinylbenzene (DVB). Styrene provides the basic matrix of the resin. DVB is used to cross-link the resin. Cross-linking provides insolubility and toughness to the resin. The degree of cross-linking determines the internal pore structure, which in turn affects the movement of ions into and out of the resin.

Figure 13-1 shows that a synthetic resin has a "whiffle ball," skeleton-like structure. It is insoluble in water, with electrically charged exchange sites holding ions of opposite charge at the exchange sites.

Synthetic resins are available in bead form and range in size from 20 mesh (0.84 mm in diameter) to 325 mesh (0.044 mm in diameter). Most ion exchange applications in water use resins in the 20- to 50-mesh size range. Figure 13-2 shows typical resin beads.

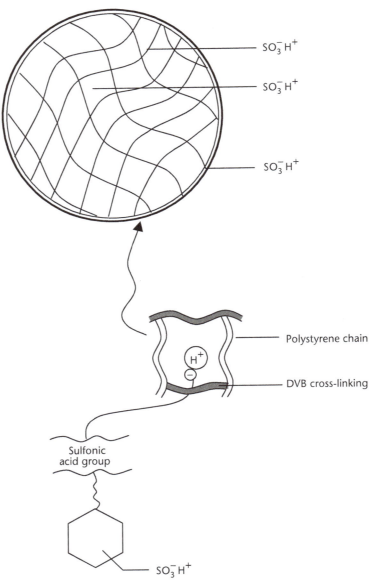

Figure 13-1 Cation exchange resin structure

Courtesy of Pall Corp.

resin

In water treatment, the synthetic, bead-like material used in the ion exchange process.

Figure 13-2 Beads of polystyrene resin
Courtesy of US Filter/Permutit.

Each ion exchange resin has its own order of exchange preference. A bumping order based on valence exists. In general, trivalent is preferred over divalent, which is preferred over monovalent. Aluminum ions are preferred over calcium, and calcium is preferred over sodium. A bumping order, based on atomic number, exists for ions with the same valence (i.e., sodium is preferred over hydrogen). These properties means that an ion exchanger can be used in either the sodium or calcium cycle to remove aluminum ions.

Cation Exchangers

Cations are positively charged ions, such as calcium and magnesium, that migrate toward the cathode. Cation exchange resins are used to exchange unwanted positively charged cations with cation species, such as sodium or hydrogen.

Cation exchangers are available with different properties. The most common cation exchangers are called *strong-acid resins*.

Strong-acid cation exchangers exchange one cation for another and operate over the entire pH range. Their operating capacity (i.e., the practical usable portion or ion exchange capacity of an ion exchange resin bed) is less than stoichiometric. Strong-acid cation resins must be regenerated more frequently than weak-acid resins. Most strong-acid resins used in water treatment applications contain sulfonic acid functional groups (see Figure 13-1). These resins differ mainly in their DVB content or structure. Water softening applications (i.e., removal of calcium and magnesium ions) typically use strong-acid resins.

Weak-acid cation exchangers are not highly dissociated. They operate only above pH 4. The functional group is the organic compound carboxylic acid. Weak-acid cation exchangers remove more cations per unit volume than strong-acid resins and generate almost stoichiometrically. They are sometimes used in conjunction with a strong-acid polishing resin. This combination allows for economic operation in terms of regenerant requirements but also produces a treated water of quality comparable to the use of just a strong-acid resin.

Anion Exchangers

Anions are negatively charged ions, such as nitrate and sulfate, that migrate toward the anode. Anion exchange resins are used to exchange unwanted anions with anion species, such as chloride or hydroxide. If cations are exchanged for

cation
A positive ion.

cation exchange
Ion exchange involving ions that have positive charges, such as calcium and sodium.

anion
A negative ion.

anion exchange
Ion exchange involving ions that have negative charges, such as chloride.

hydrogen and the anions are exchanged for hydroxide, the result is HOH, or demineralized water.

Anion exchangers were developed almost exclusively as synthetic resins. Organic exchangers were among the earliest ion exchange resins produced. The first patents issued for anion exchangers were for resins having weak-base amino groups. Later, resins with strong-base quaternary ammonium groups were prepared.

Strong-base anion exchangers operate over the entire pH range, but their capacity is less than stoichiometric. Like strong-acid resin cation exchangers, they must be regenerated more frequently than weak-base resins, which exhibit much higher capacities and regenerate almost stoichiometrically. A problem with strong-base resins is that they tend to irreversibly sorb humic acid substances, losing capacity. Activated carbon or a weak-base organic trap is typically used to prevent resin fouling.

Weak-base anion exchange resins behave much like their weak-acid counterparts. They do not remove anions above a pH of 6. They regenerate with a nearly stoichiometric amount of base (with the regeneration efficiency possibly exceeding 90 percent) and are resistant to organic fouling. They do not remove carbon dioxide or silica, and they have capacities about twice as great as that of strong-base exchangers. They are useful following strong-acid exchangers to save the cost of regenerant chemicals. They act as organic "traps" to protect strong-base exchangers and to remove color. Weak-base resins have a higher capacity for the removal of chlorides, sulfates, and nitrates. For waters containing organic contaminants (humic and fulvic acids), macroreticular weakly basic anion resins are preferred.

Ion Exchange Units

The vessels used in ion exchange systems (Figure 13-3) resemble those used in pressure filters. A major difference is that the interior of the tanks are coated with a special lining to protect them from the corrosion caused by the brines (high salt concentration) used in regeneration.

Figure 13-4 is a cutaway view of a vertical-downflow ion exchange unit designed with the following components:

- Hard-water inlet
- Soft-water outlet
- Wash-water inlet and collector
- Brine inlet and distribution system
- Brine and rinse-water outlet
- Rate-of-flow controllers
- Sampling taps
- Underdrain system, which also serves to distribute backwash water
- Graded gravel to support the ion exchange resins

Both upflow and downflow ion exchange units are available. Units designed for the water to flow downward are more commonly used because they can be operated at higher flow rates during the softening cycle. All units are equipped with automatic controls.

The size of the ion exchange unit and the volume of resin required are determined by the concentration of ions to be removed (e.g., the hardness of the raw water and the desired length of time between regenerations). The minimum recommended depth is 24 in. (0.6 m). The resin is supported either by an underdrain system or by 15–18 in. (0.38–0.46 m) of graded gravel.

Figure 13-3 Ion exchange pressure tanks

Courtesy of Ionics, Inc.

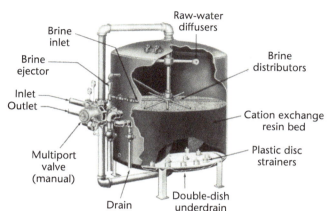

Figure 13-4 Vertical-downflow ion exchange unit

Courtesy of US Filter/Permutit.

In downflow units, a regenerant distribution system is used to direct the flow of regenerant downward and evenly through the unit during regeneration so that all resin comes into contact with it. The rinse water can also be distributed through the same system.

Regenerant Storage Tanks (Softening)

Sodium chloride and potassium chloride are the most common salts used in the water softening process to form brine, which is used to regenerate the resin. The amount of salt used in creating the brine ranges from 0.25 to 0.45 lb (0.11 to 0.20 kg) for every 1,000 grains of hardness removed (7,000 grains = 1 lb, 1 grain/gal = 17.1 mg/L).

The salt is normally stored in tanks large enough to hold brine for a 24-hour period of operation or for three regenerations, whichever is greater. Salt brines are corrosive to salt attacks and wear away concrete and steel and stainless steel. Tanks are coated with a salt-resistant material.

The salt used for resin regeneration must meet standards for purity. Rock salt or pellet-type salt is the best for preparing the brine. Road salt is not suitable because it often contains impurities. Block salt is sometimes used for home softeners, but it is not suitable for larger installations because its small surface area does not allow it to dissolve fast enough. Fine-grained salt, such as table salt, is not suitable because it packs tightly and does not dissolve easily.

Salt storage tanks are covered to prevent dirt and foreign material from entering. The access holes in the cover should have raised lips so that any dirty water on the cover does not flow into the hole and contaminate the salt.

To fill a salt storage tank and prepare the brine, water is added to the tank and the tank is filled with rock salt. More water is then added to submerge the salt so that it will dissolve. The water fill line must allow for an air gap above the top of the tank to prevent the possibility of brine siphoning back into the water supply. An excess of undissolved salt should be kept in the tank to ensure that a concentrated solution is achieved.

Because brine is heavier than water, the highest concentration of brine will be in the lower part of the storage tank. The brine used for regeneration is therefore usually pumped from the bottom of the tank.

Operation of Ion Exchange Processes

As shown in Figure 13-5, there are four basic cycles in the ion exchange water softening process:

1. Backwash
2. Regeneration
3. Rinse—both a slow and a fast rinse
4. Service (i.e., softening)

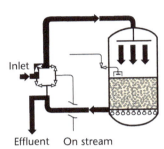

Softening. Influent water passes through the bed of ion exchange material to the effluent.

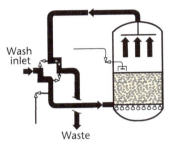

Backwash. Influent water is passed upward through bed of ion exchange material to loosen the bed and remove suspended solids that may have been deposited in the bed during operation.

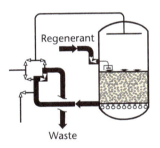

Regeneration. Regenerant solution is passed through to waste at a controlled concentration and flow.

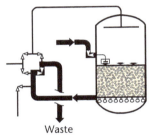

Slow rinse. Water is passed through the bed to displace the regenerant solution to waste.

Figure 13-5 Four cycles in the ion exchange process

Courtesy of Infilco Degremont Inc., Richmond, Virginia.

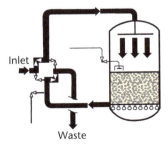

Fast rinse. Influent water is passed through the bed to waste to remove the last traces of regenerant chemicals.

Operating procedures will vary somewhat depending on the type of equipment used, so it is important to follow the manufacturer's specific instructions. General guidelines for the process operation follow.

Backwash

The function of the backwash cycle is to loosen the resin that has compacted during the service cycle, randomly mix the resin, and remove any silt, dirt, precipitated iron, or other accumulated insoluble matter.

Water is introduced through the bottom of the ion exchange column at a rate sufficient to expand the bed by 50–75 percent.

Typical flow rates are 5–8 gpm/ft² (0.20–0.33 kLm/m²). Hydraulic expansion data are provided by the resin manufacturer and provide bed expansion information as a function of temperature and flow rate.

Regeneration

Backwash and rinsing are actually steps in the regeneration process. These steps require 5–15 percent of the product water. The regeneration step uses only 5 percent of the water treated. It concentrates the ions removed during the service cycle by as much as 20 times.

Regeneration follows backwash and is the displacement of the ions held on the resin sites. These ions were removed from the process feedwater during the service cycle. If the column contains a strong-acid resin regenerated with an acid, for example, hydrogen ions (H^+) are exchanged onto the resin in place of those released. This resin is now in the hydrogen form or hydrogen cycle. If the column is a strong-base resin and is regenerated with sodium hydroxide, hydroxyl ions (OH^-) are exchanged on the resin. It is then in the hydroxyl form or cycle. The regeneration step can be accomplished in either a downflow or upflow mode. When softening water or using the column in the sodium cycle, the regeneration step is often referred to as *brining*. In most cases, concentrated sodium chloride is passed through the bed to drive the calcium and magnesium (softening) or other metals (industrial uses) off the resin.

Rinse

The purpose of the rinse cycle is to remove excess regenerant from the resin bed. Raw water is typically used for all rinsing operations. Softening operations usually involve a slow rinse and a fast rinse. Raw water is introduced through the brine distribution line. Product water is then sometimes used for the slow and fast rinse until the effluent regenerant concentration equals the influent.

Service

The service cycle follows rinsing. It is the operational cycle—that is, the process involving softening, demineralization, and removal of the unwanted contaminant from the water.

The cycle is complete when a predetermined concentration of contaminant is measured in the effluent (breakthrough) ion. A flowmeter can be adjusted to sound an alarm when the volume of water corresponding to the unit's rated capacity has passed through the unit. The loading rate for units that use polystyrene resin is in the range of 10–15 gpm/ft² (0.41–0.61 kLm/m²).

Regenerant Disposal

An important technical consideration of the ion exchange process is the disposal of the wastewaters generated during the backwash, regeneration, and rinse cycles. Proper disposal options must be evaluated before the ion exchange process is selected. The costs involved in disposing of the brine usually influence the process selected for softening and can make other softening processes more cost-effective. The method allowed will depend primarily on the requirements of the state pollution control agency.

The total amount of spent brine varies from 1.5 to 7 percent of the volume of water softened. The wastewater contains calcium chloride, magnesium chloride, sodium chloride, and other cations and anions present in the raw water and in the water used to prepare the regenerant solution. If there is a high concentration of radium in the source water, there will also be a high concentration of radium in the wastewater. Even with the dilution provided by the backwash and rinse water, the total concentration of dissolved solids in the wastewater is usually between 35,000 and 45,000 mg/L.

Wastewater with this concentration of dissolved solids can, with time, cause pipe corrosion and upset the biological processes in a wastewater treatment plant.

Brines should not be discharged to any waterway where there is not adequate dilution. If brine is not properly applied to a land surface, it can make the soil unusable for agricultural purposes. If discharged to a seepage bed, brine could eventually contaminate an underground aquifer.

Operating Problems

Several conditions cause operating problems in the ion exchange softening process, including the following:

- Resin deterioration
- Iron fouling
- Turbidity, organic color, and bacterial-slime fouling
- Removal of arsenic, barium, radium, nitrate, and uranium
- Unstable water

Resin Deterioration

Ion exchange systems are designed and casted under the assumption that polystyrene resins have an operational life of 15–20 years. Certain operating conditions cause the resin to degrade. Many resins are susceptible to attack to some degree by oxidants.

In most cases, ion exchange resins are considered as capital equipment. Expected life of cation exchange resins used in softening operations is between 2,000,000 and 10,000,000 gallons of water treated per cubic foot of resin.

Cation exchange resins deteriorate for many reasons. Each situation is unique and either a single effect or combination of effects must be considered: (1) physical degradation to smaller particles, (2) decross-linking and subsequent swelling, (3) loss of breakthrough capacity, (4) loss of acidity of functional groups, (5) poisoning of functional groups, and (6) fouling due to precipitation in or on the exchanger surface.

The particle size distribution becomes too small and causes an unacceptable increase in head loss when the resin assumes a jelly-like consistency from oxidative attack.

Iron Fouling

Iron is present in all water supplies in concentrations ranging from nondetectable to as high as 40 mg/L. Iron is present in waters in many forms, including ferrous bicarbonate, colloidal iron, and soluble iron complexed with organic acids. Uncomplexed soluble iron is cationic in nature and is removed by cation exchangers in the hydrogen or sodium cycle. Iron complexed with organic matter is anionic in chemistry and is not removed by cation exchange.

The property of iron to form insoluble hydroxides when exposed to air and other oxidants can result in the formation of precipitates in and on cation exchange resins, which causes decreased exchange capacity and channeling and leakage.

The best solution to the iron problem is to remove the iron prior to the iron exchange operation (see Chapter 8 on iron and manganese removal). Whenever pretreatment is not possible or practical, a number of precautions are effective, including (1) the addition of a strong reducing agent, such as sodium hydrosulfite, during regeneration to prevent precipitation, and (2) the use of a strong reducing agent in combination with polyphosphates for fouled resins.

Turbidity, Organic Color, and Bacterial-Slime Fouling

Turbidity, color caused by humic substances, and bacterial slimes as measured by heterotrophic plate count (HPC) in raw water will coat an ion exchange resin and cause loss of ion exchange capacity and excessive head loss. Although backwashing helps to remove some of these materials, it will not remove large volumes because the materials become tightly held on the resins.

Highly colored water, highly turbid water, and water with a high HPC require pretreatment.

Unstable Water

Water softened by the ion exchange process is corrosive, as tested by the Langelier saturation index. Blending raw and softened water so that the finished water is only nominally soft usually helps. If the water is still corrosive, chemicals may have to be added to provide corrosion control. Additional information on corrosion control is included in Chapter 11.

Ion Exchange for Removal of Arsenic, Barium, Radium, Nitrate, TOC, and Uranium

Arsenic

Arsenic exists as arsenite (AsO_3^{-3}) and arsenate (AsO_4^{-3}). Soluble forms of arsenic can be removed using strong-base anion exchange in the chloride form when waters are low in divalent anions, such as sulfate. The arsenic is first oxidized to the arsenate species.

Barium and Radium

Barium and radium are found frequently in groundwater in several locations in the United States, and they exceed the maximum contaminant level (MCL) in many wells. Barium and radium are effectively removed by strong-acid cation exchangers in the sodium cycle or hydrogen cycle.

Nitrate

Nitrate is most often found in shallow wells in rural areas, primarily as a result of agricultural contamination. It is also present in excess of the MCL in some surface sources.

Drinking water regulations in the United States limit the amount of nitrate in public water supplies to an MCL below 10 mg NO_3^-–N/L (or 45 mg NO_3^-/L). (NO_3^-–N/L stands for nitrate nitrogen per liter; NO_3^-/L stands for nitrate per liter.) The European Community limits nitrate to 11.3 mg NO_3^-–N/L (or 50 mg NO_3^-/L). Some local agencies, such as the Municipal Water District of Southern California, limit nitrate to 80 percent of the national MCL.

The strong-base anion exchange process, which uses a single bed in the chloride cycle, is the most widely used ion exchange process for nitrate removal in the United States. The effluent is sometimes blended with untreated surface water to meet regulations without treating the entire stream. Weak-base resins require a more complicated process scheme and are currently not preferred for the municipal market.

Total Organic Carbon

Total organic carbon (TOC) in source water acts as the precursor to disinfection by-product production. Removal of TOC prior to chlorine addition is an excellent strategy when ion exchange is used because it can reduce the expense of downstream coagulant use and the resultant production of residuals from coagulation. Naturally occurring organic matter such as TOC typically has a negative charge and therefore anionic exchange resins will work to remove the contaminant. Because resins are subject to deterioration (fouling) from particulate material in the source water, a standard fixed ion exchange bed employed to remove TOC would have to be situated downstream of conventional filtration and might be exposed to chlorine. The media might still be clogged eventually in spite of accepting filtered water if the water weren't of consistently low turbidity. Using the MIEX system (which derives its name from the term magnetic ion exchange), the magnetically enhanced media are added to the source water prior to coagulation and filtration, thus avoiding those problems.

The MIEX process uses a "dual stage" strategy that follows these steps:

1. Resin is fed into the first stage from a slurry.
2. The incoming water and added resin are mixed in the first stage for a prescribed period of time to keep the resin suspended and in intimate contact with the water to be treated (greater contact time allows more time for ion exchange to occur).
3. Resin particles agglomerate and settle quickly in the clarifier (second stage) due to the magnetic nature of the resin.
4. Resin is recycled to the first stage, repeating steps 2 through 4.
5. Five to 10 percent of the resin slurry stream is diverted to the regeneration step.

This process has been shown to lower the amount of coagulant needed. One study showed that it can be added in a separate stage prior to conventional coagulation with alum, or downstream of alum coagulation at the conventional settling stage, and still be expected to reduce alum use by significant amounts. In this same study, the use of alum alone did not remove as much TOC as did alum plus MIEX.

Uranium

Both cation and anion exchange resins have been used successfully to remove uranium from contaminated water. Strong-base anion exchange in the chloride cycle is reported to exhibit better performance than cation exchange and offers cost-effective treatment to small communities.

Activated Alumina Fluoride Removal Process

The fluoride levels in some source waters exceed the regulatory limits and, thus, must be reduced. Both surface and groundwaters may be subject to fluoride removal. Fluoride can be removed by several processes including: reverse osmosis, electrodialysis, adsorptive media, and anion exchange. Activated alumina is a selective ion exchange material that is often the method of choice for fluoride removal.

Activated alumina (AA) is a granular filter media that is primarily aluminum oxide (Al_2O_3). It has been activated by heat and washing with sodium hydroxide (caustic soda). Fluoride ions in water are attracted to the surfaces of the AA. This is a pH-sensitive process where the optimum is a pH of 5.5. Fluoride can be removed to below 0.5 mg/L at this pH.

The capacity of the AA filters varies depending on the concentration of fluoride in the feed water. As the capacity of the filter is reached, the media can be regenerated by washing with a caustic solution. The wash waste can be recycled and the AA media can be used again to remove fluoride from the water.

Arsenic (III and V) is also removed from water by AA. If present in the water, arsenic will compete with fluoride for adsorption on AA. This is not often a problem because arsenic is usually present in very low concentrations when compared to fluoride. Therefore, it does not materially affect the capacity of the AA for fluoride removal. A modification of the AA treatment process can be employed in the rare case where arsenic is present in high concentration. Arsenic held on the AA requires higher concentrations of caustic for regeneration than does fluoride.

A typical treatment system consists of two downflow pressure vessels in series. Fluoride removal is accomplished in the first vessel initially. As more water is treated, the fluoride-containing area of the AA in the vessels migrates down to the second vessel. The first vessel is then removed from service and regenerated. The second vessel continues to process water containing fluoride. When the first vessel regeneration is completed, it is returned to service in the lag position. In this way, continuous treatment is accomplished.

Adsorptive Media

Fluoride, arsenic, uranium, and other anions are often removed using adsorptive media. Specific materials used can be AA, modified AA, iron sorbents, and other proprietary adsorptive materials. The process operates similarly to the AA process

activated alumina (AA)

The chemical compound aluminum oxide, which is used to remove fluoride and arsenic from water by adsorption.

described above. The pH or other operating details may be different from the optimized AA process for fluoride removal. The system manufacturer should be contacted for operating specifics. Extensive pilot testing should be conducted to verify the optimum operating conditions.

Study Questions

1. Which of the following is the best type of salt to use in the regeneration of ion exchange softener resin?
   a. Fine-grained salt
   b. Block salt
   c. Block or road salt
   d. Rock salt or pellet-type salt

2. Once the ion exchange resin can no longer remove hardness, it is said to be
   a. wasted.
   b. consumed.
   c. expended.
   d. exhausted.

3. When backwashing an ion exchange unit, how much should the bed expand from water introduced from the bottom of the ion exchange column?
   a. 25–35%
   b. 35–50%
   c. 50–75%
   d. 75–85%

4. Ion exchange processes can typically be used for direct groundwater treatment as long as turbidity and _____ levels are not excessive.
   a. calcium carbonate
   b. iron
   c. carbon dioxide
   d. sodium sulfate

5. Weak-base anion exchange resins behave much like their weak-acid counterparts; they do not remove anions above a pH of
   a. 6.
   b. 7.
   c. 8.
   d. 9.

6. Block salt is sometimes used for home softeners, but it is not suitable for larger installations because
    a. it packs tightly and does not dissolve easily.
    b. the blocks required would be too heavy for operators to lift.
    c. it often contains impurities.
    d. its small surface area does not allow it to dissolve fast enough.

7. What are cations?

8. What is another term for the service cycle?

9. What element in source water acts as the precursor to disinfection by-product production?

Chapter 14
Activated Carbon Adsorption

The Principle of Adsorption

Adsorption works on the principle of adhesion. In the case of water treatment, organic contaminants are attracted to the adsorbing material. They adhere to its surface by a combination of complex physical forces and chemical action.

For adsorption to be effective, the adsorbent must provide an extremely large surface area on which the contaminant chemicals can adhere. If the process is to be economical to build and operate, the total surface area of adsorbent required must be contained in a tank of reasonable size.

Porous adsorbing materials help achieve these objectives. Activated carbon is an excellent adsorbent because it has a vast network of pores of varying size to accept both large and small contaminant molecules. These pores give activated carbon a very large surface area. Just 1 lb (0.45 kg) of activated carbon has a total surface area of about 150 acres (60 ha). This amount of activated carbon can, for example, trap and hold over 0.55 lb (0.25 kg) of carbon tetrachloride.

Figure 14-1 is a photograph taken through an electron microscope showing the large number of pores in a grain of activated carbon. These pores are created during the manufacturing process by exposing the carbon to very high heat in the presence of steam. This process is known as *activating* the carbon. Activation oxidizes all particles on the surface of the carbon, leaving the surface free to attract and hold organic substances. Figure 14-2 shows the structure of the carbon after

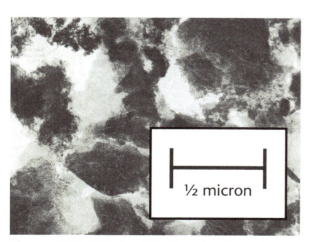

Figure 14-1 Details of the fine structure of activated carbon

½ micron

Courtesy of Calgon Carbon Corp.

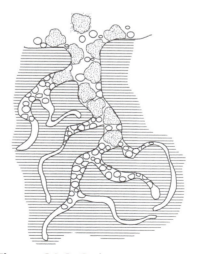

Figure 14-2 Carbon structure after activation showing small and large adsorbed chemical molecules

adsorption
The water treatment process used primarily to remove organic contaminants from water. Adsorption involves the adhesion of the contaminants to an adsorbent, such as activated carbon.

adsorbent
Any material, such as activated carbon, used to adsorb substances from water.

261

activation and indicates that different sizes of chemical molecules can be adsorbed within the pores.

Once the surface of the pores is covered by adsorbed material, the carbon loses its ability to adsorb. The spent carbon can then be reactivated by essentially the same process as the original activation, or it can be discarded and replaced with fresh carbon.

Adsorption Facilities

Activated carbon can be made from a variety of materials, such as wood, nutshells, coal, peat, and petroleum residues. One of the principal sources of activated carbon used in water treatment is bituminous or lignite coal. The coal is slowly heated in a furnace without oxygen so that it will not burn, thus converting the coal to carbon. The carbon is then activated through exposure to a steam–air mixture. The activated carbon is then crushed and screened to obtain the desired particle size.

The two common forms of activated carbon used in water treatment are powdered activated carbon (PAC) and granular activated carbon (GAC). Table 14-1 compares the general properties of types of PAC and GAC. Figure 14-3 shows the difference in particle size.

Table 14-1 Comparison of GAC and PAC properties

| Property | Types of Carbon | |
	GAC	PAC
Density, lb/ft³	26–30	20–45
(g/cm³)	(0.42–0.48)	(0.32–0.72)
Surface area, m²/g	650–1,150	500–600
Mean particle diameter, mm	1.2–1.6	Less than 0.1

Figure 14-3 Difference in particle size between GAC and PAC

Powdered Activated Carbon

Powdered activated carbon (PAC) is typically available in 50-lb (23-kg) bags or in bulk truck or railroad car shipments. It can be fed in dry form or as a slurry with water.

Dry Feed Systems

Treatment plants that only periodically need to use PAC normally use the dry feed method. The PAC can be stored in bags or in bulk form in steel tanks. The tanks should be located so that they can feed directly into the hoppers of dry feeders by using gravity only.

When PAC is used in dry form, chemical feeders specifically designed to handle carbon should be used. The feeders must be capable of operating over a wide range of feed rates. In some instances, only a very light application may be desired to eliminate slight tastes or odors in the water. In other cases, very heavy dosages may be required if there are serious tastes or odors or chemical contamination of the raw water.

The feeder must feed to a tank or ejector where the carbon will quickly be mixed with water to form a thin slurry and then discharged to the application point (Figure 14-4). The hopper above the feeder must have walls slanting at a 60-degree angle, as well as a vibrator mechanism to keep the carbon flowing and to prevent it from arching.

Systems that feed PAC only occasionally find that a relatively low-cost method is to use a helix-type feeder positioned over an eductor, as illustrated in Figure 14-5. Although the feed rate may not be very accurate, the dosage of carbon required during a temporary taste or odor episode is generally an estimate based on past experience, so precise feed is only a secondary consideration.

Carbon dust is extremely fine and light. If not carefully confined and handled, it will easily float throughout a plant and cover floors and surfaces with a fine,

> **powdered activated carbon (PAC)**
> Activated carbon in a fine powder form. It is added to water in a slurry form primarily for removing those organic compounds causing tastes and odors.

Figure 14-4 Dry carbon feeder

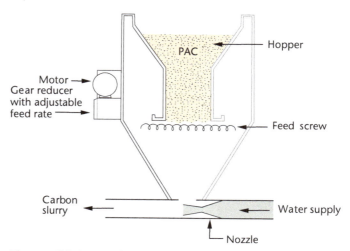

Figure 14-5 PAC feed using a helix feeder

black coating. To keep dust to a minimum, dry carbon feeders should be located in a confined room of the treatment plant.

Slurry Feed Systems

Because of the handling problems encountered with dry PAC, many plants that consistently use PAC have a slurry feed system of the type shown in Figure 14-6. If delivered in bulk form, the PAC is removed from the truck or railcar via an eductor or pneumatic system directly to a storage tank. Plants usually have two storage tanks so that a shipment can be placed in one before the other is empty. The tanks are normally made of steel or concrete and have a capacity about 20 percent greater than the maximum load that will be delivered. The tanks should have an epoxy or other type of lining to protect them from corrosion. If PAC is delivered in bags, it can be added directly to the storage tank.

The tanks are usually provided with two-speed mixers. If dry carbon is added directly to the water in the tank, the mixer must initially be operated at high speed to create a slurry; otherwise, the carbon just tends to lie on top of the water. After all of the carbon has been thoroughly wetted to form a slurry, the mixer can be set to a slow speed to maintain the proper slurry condition. If power to the mixer is ever turned off, perhaps because of an electrical outage, it may be necessary to operate the mixer at the high speed again for a duration to reestablish a uniform slurry.

The slurry is usually pumped to a day tank that holds the quantity to be used for a day or shift. The tank should be plastic or lined steel and equipped with a mixer to keep the slurry in suspension. The slurry is then fed from the day tank by a volumetric feeder.

Unless the feeder is located at exactly the feed point, there is a danger that the carbon will settle out in the feed line because of the low flow velocity. To prevent this, the feeder can discharge onto an eductor that maintains a high flow velocity in the feed line. The feed line should slope downward all the way to the application point, and provisions should be made for flushing any carbon that may settle out and clog the pipe. It sometimes happens that the PAC will settle and form a difficult-to-remove mass when insufficient slope is available or if the slurry is too concentrated. Most PAC systems are designed to create and feed a slurry with a concentration of 1 pound of PAC per gallon. Many operators have

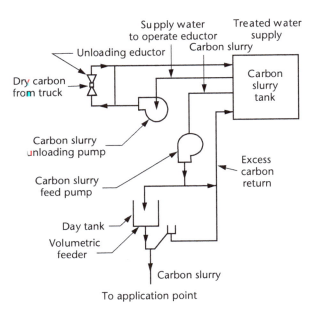

Figure 14-6 Slurry feed system for handling PAC

learned that this is too concentrated to be practical, and they prefer to operate with a slurry concentration of 0.25 lb/gal to avoid difficult maintenance. As a result, they need to make day tank slurries four times as often (or they need day tanks four times as big) as they might at design conditions, but they believe this is a worthwhile effort.

A PAC feed system should be kept in service at all times to prevent clogging. Often, PAC feed systems that are shut down when not dosing are not ready to go when needed. Many operators feel that it is better to operate the machinery at a very low dose so that when the PAC is needed, the operators only need to increase the speed of the feeder.

The type of pipe used for the feed line must be both corrosion- and erosion-resistant material, such as rubber, plastic, or stainless steel. Day tanks are typically made of fiberglass.

PAC is usually added to the water before the normal coagulation–flocculation step, but multiple application points are desirable. PAC must not be fed too near the chlorine application points. The carbon must be removed from the water to prevent it from moving through the plant into the distribution system. Most of the PAC becomes part of the sedimentation basin sludge, so it is not practical to recover and reuse it.

Granular Activated Carbon

Granular activated carbon (GAC) is used principally where continuous removal of organics is required. Typical uses include the following:

- Removal of organics following surface water treatment
- Removal of organics from groundwater
- Special-purpose removal of certain inorganics or radionuclides

GAC Used as a Filter Medium

GAC has been installed in place of sand or anthracite in open filters (Figures 14-7 and 14-8). The GAC can be either a capping layer over other media or a complete replacement. The material then acts as both an adsorbent and a filtering medium.

The decision to use this approach must be based on a study of how long the adsorption qualities of the GAC will last, how much it will cost to remove

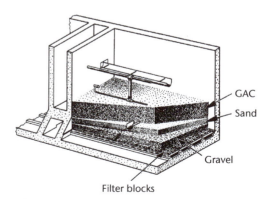

Figure 14-7 Partial replacement with GAC

Courtesy of Calgon Carbon Corp.

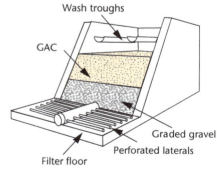

Figure 14-8 Complete replacement with GAC

Courtesy of Calgon Carbon Corp.

granular activated carbon (GAC)
Activated carbon in a granular form, which is used in a bed, much like a conventional filter, to adsorb organic substances from water.

exhausted material, and how much it will cost to have the old material either reactivated or replaced with new material. The effective life of the GAC in typical installations has been anywhere from a few months to 3 or 4 years, depending not only on the organics concentration, but also on the types of substances being removed. For instance, GAC will generally last much longer for removal of tastes and odors than it will for removal of synthetic organic chemicals (SOCs).

There are design and operational concerns associated with using GAC as a filtering medium, and for these reasons operators may prefer GAC post filtration contactors. Some of the concerns of using GAC as a filtering medium are as follows:

- It is lighter than other filter materials such as anthracite, and so is more easily washed away in an aggressive backwash sequence. Its specific gravity is about 1.4 compared to anthracite at about 1.6. Therefore the designer needs to specify a larger effective size for the anthracite, and that may alter the filtering characteristics of the bed.

- It is softer than anthracite and so is more easily broken up into smaller pieces by repeated washing, resulting in an abundance of fines in the filter. They in turn will get washed away prematurely.

- Removal of organics is dependent on the empty bed contact time (discussed later in the chapter) needed for each organic species that is to be removed. A high empty bed contact time might be impossible to achieve in the filter.

- If chlorine is used in the pretreatment scheme, the GAC will remove the chlorine residual in the incoming settled water and therefore create a higher demand for that disinfectant, and exhaust the GAC prematurely.

- If the GAC cap becomes fouled by adsorbing an unexpected pollutant, it may render the filter unusable until it can be replaced or regenerated.

Many or all of these problems can be avoided if space is available to install post-filtration GAC contactors.

GAC Contactors

GAC contained in closed pressure tanks (GAC contactors) has been used to treat the effluent of surface water systems after full conventional filtration treatment (Figure 14-9). This treatment may be provided for taste-and-odor control or to reduce the level of SOCs not removed by other treatment processes. Where the total organics load of the raw-water source is relatively high, using GAC after other treatment is typically found to be more economical than using it as a filter medium.

GAC contactors can also be used when a concentration of trihalomethanes (THMs) or other disinfection by-products (DBPs) cannot be prevented from forming by an adjustment of the disinfectants or other treatment processes. In these cases, GAC can be used as a final treatment to remove the DBPs to an acceptable level before the finished water enters the distribution system.

The principal use of GAC contactors on groundwater systems has been for removal of SOCs from wells that have been contaminated. In most cases where contamination of groundwater has been identified, it is most cost-effective to abandon the well if another good-quality source is available. The second choice is typically to use aeration if the problem contaminants are volatile, which means they diffuse freely from the water. Where there are no other choices, GAC will

Figure 14-9 GAC contactors installed as the final treatment at a large surface water treatment plant

Courtesy of the Greater Cincinnati Water Works.

Figure 14-10 Skid-mounted GAC contactors

Courtesy of Calgon Carbon Corp.

normally work for chemical removal, but the operating cost will probably be relatively high. Groundwater has also been treated by passing the water through open beds, rather than tanks, of GAC. This has a few advantages, such as ease in monitoring and replacement of the GAC, but enclosed tank contactors are usually preferred.

GAC contactors have also been used extensively as an emergency method for treating a wide variety of water contamination problems. Truck or skid-mounted GAC contactors can be set up quickly at a site to treat a contaminated water supply on an interim basis (Figure 14-10). In these cases, the capacity of the GAC unit is usually sufficient to give the water system enough time to examine alternatives and to design either a change in water source or a permanent treatment system before the GAC is exhausted.

New GAC is available in 60-lb (27-kg) bags or in bulk form delivered by trucks or railcars. GAC is usually placed in filters or contactors in a slurry form by an eductor, both to facilitate handling and to reduce dust.

GAC Regeneration

Depending on the type and concentration of organic compounds in the water being treated, GAC gradually loses its adsorption ability over a period of time, ranging from a few months to several years. The old GAC must then be removed and replaced with fresh carbon. The used GAC can be reactivated and reused, as illustrated in Figure 14-11. Reactivation consists of passing the spent carbon through a regeneration furnace, where it is heated to 1,500–1,700°F (820–930°C) in a controlled atmosphere that oxidizes the adsorbed impurities. About 5 percent of the carbon is lost during the process, so more new carbon must be added when the GAC is replaced in the filter or contactor.

Reactivation can be done on-site, but it is generally practical only for a relatively large installation. It is usually not cost-effective to ship exhausted GAC to a distant location to be reactivated and then return it to the treatment plant.

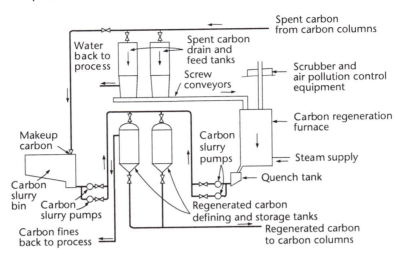

Figure 14-11 Schematic of GAC reactivation

Regulations

The use of activated carbon in water treatment has increased in recent years as a result of new regulations and health advisories limiting public exposure to various SOCs. Most of the uses have involved groundwater systems where well contamination exceeding a maximum contaminant level has been identified. These systems are usually directed by state authorities to stop using the well or to add treatment for contaminant removal.

In many cases, GAC is used as an emergency treatment measure because it will effectively remove almost all SOCs of concern; treatment units can be delivered to the site and made operational on very short notice.

Operating Procedures for Adsorption

The operating procedures for adsorption differ depending on whether PAC or GAC is used.

Application of PAC

PAC is used primarily to help control those organic compounds responsible for tastes and odors. It has also been used to remove compounds that will form THMs or other DBPs. In some instances, massive doses of PAC have been used for temporary treatment of a water source that has been contaminated by a chemical spill.

In addition to adsorption of organic compounds, the carbon particles will often aid coagulation by providing a center, or nuclei, on which floc will form. This is a benefit where a water source has naturally low turbidity.

PAC can be applied in the treatment plant at almost any point before filtration, but the following considerations should be made in the selection of an application point:

- The contact time between the PAC and the organics is important and depends on the ability of the carbon to remain in suspension. At least 15 minutes of contact time is typically considered advisable.
- The surfaces of the PAC particles lose their capacity to adsorb if coated with coagulants or other water treatment chemicals.

■ PAC will adsorb chlorine. If PAC and chlorine are fed at the same time, the chlorine will reduce the effectiveness of the PAC. At the same time, more chlorine than normal must be fed in order to provide the required disinfection.

These factors indicate that the raw-water intake is the most advantageous point for PAC application if adequate mixing and retention facilities are available. This allows organic compounds to be removed before the application of chlorine. When chlorine reacts with organic compounds, new compounds are often produced that may not be as easily removed by the carbon.

If the raw-water intake is not a possible application point, PAC can be used at other points, but the dosages normally must be higher to account for shorter contact times and the interference by other chemicals.

If PAC can be fed only to the effluent from the sedimentation basins or as the water enters the filters, particular care must be taken in filter operation because finely divided carbon particles can pass through filters. If it does pass through the filter, some of the PAC will settle to the bottom of clearwells, where it can be stirred up and pumped to the distribution system at a time when the pumping rate is unusually high. The system operator will then most certainly receive "black water" complaints from customers. If PAC does pass into the clearwells, the clearwells should be cleaned as soon as possible.

A particularly effective way of applying PAC is to use two or more application points. Some carbon is added to the raw water, and smaller doses are added before filtration to remove any remaining taste- and odor-causing compounds.

The PAC dosage depends on the types and concentrations of organic compounds present. Common dosages range from 2 to 20 mg/L, but they can also be as high as 100 mg/L to handle severe problems.

Application of GAC in Conventional Filters

When GAC is used as a medium in a conventional filter, an eductor like the one shown in Figure 14-12 is typically used to remove the carbon when it has become exhausted. If the GAC has been placed over a layer of sand, it is difficult to remove the GAC without also removing some of the sand.

Except for very large installations, it is generally not cost-effective to regenerate the GAC when it has become exhausted. The old GAC is therefore discarded and replaced with new material. It is usually easiest and least troublesome in terms of dust to replace the new material in the filter as a slurry, as shown in Figure 14-13.

Figure 14-12 Eductor used for filter media removal

Figure 14-13 Placement of GAC slurry

Courtesy of Calgon Carbon Corp.

Once the carbon has been placed, the filters should be backwashed to remove trapped air and small particles of carbon, or carbon fines. It will probably be noticeable for several weeks that fines are still being washed out during backwashing.

Contact Time

The time that water is in contact with GAC as it passes through the bed is usually termed the **empty bed contact time (EBCT)**. The EBCT is equal to the volume of the bed, which is based on the medium's depth and the dimensions of the filter compartment, divided by the flow rate through the bed. Filters with GAC as a medium are normally operated in the same manner as regular rapid sand filters, with a filtration rate of about 2 gpm/ft² (1.4 mm/sec). This provides an EBCT of about 7.5 to 9 minutes, which is sufficient to remove many organic compounds.

Backwashing

A filter bed containing GAC is backwashed using the same general procedures as for conventional filters. The backwash rate depends to some extent on the GAC particle diameter and density, the water temperature, and the amount of bed expansion desired. A 50 percent bed expansion is recommended for GAC. Operational curves are available from GAC manufacturers to help select the proper backwash rate. It is also recommended that surface washers be used to ensure adequate cleaning of the carbon layer and to prevent the formation of mudballs within the bed.

It is important to have precise control of the backwash rate in order to achieve good cleaning of the bed without losing too much of the filter medium over the wash-water troughs. The particle density of GAC is about 1.4 g/cm³, which is much less than that of sand, which is about 2.65 g/cm³. An appropriate adjustment must therefore be made in the backwash rate to prevent loss of the medium if GAC is used to replace sand.

Carbon Loss

The major causes of carbon loss are entrapment within the bed and excessive backwash rates. Before backwashing, the filter should be drained to a minimal distance beneath the backwash troughs. Surface washers should be turned on for about 2 minutes to break up any mat that may have formed on the surface and to clean floc particles from the carbon. If polymers are used as a coagulant or filter aid, it is particularly important to provide thorough surface washing because the floc will be very sticky and hard to remove. The agitation created by the surface washers also helps to remove any air entrapped in the bed.

Backwashing should begin at 2–3 gpm/ft² (1.4–2.0 mm/sec) to allow any remaining air to be released slowly, so that it does not cause carbon loss into the wash-water troughs. The rate should then be increased to 5–6 gpm/ft² (3.4–4 mm/sec) for about 1 minute, and then gradually increased to the rate necessary to achieve the desired bed expansion.

Backwashing should continue until the flow into the wash-water troughs is clear. The time required for backwashing and the lengths of filter runs depend primarily on the amount of floc and other suspended matter in the water passing through the filters.

Some loss of carbon will occur during backwashing. It is important to keep track of the bed depth; this way, more GAC can be added when the depth becomes

empty bed contact time (EBCT)

The volume of the tank holding an activated carbon bed, divided by the flow rate of water. The EBCT is expressed in minutes and corresponds to the detention time in a GAC filter bed.

insufficient. One method of monitoring bed depth is to take routine measurements of the distance from the top of the wash-water troughs to the top of the carbon bed. Another method is to place a marker, such as a stainless-steel plate, on the filter walls to indicate the proper carbon level. It is not unusual for about 1 in. (25 mm) of carbon to be lost from filters over a period of a year.

Carbon Life

The life of a GAC bed depends primarily on the concentrations and types of organic compounds being removed. For typical taste- and odor-causing compounds, bed life may be as long as 3 years. However, the bed life for removing organic compounds, such as chloroforms, can be as short as 1 month. In general, as the influent concentration increases, the bed life decreases.

One major advantage of GAC beds is that they are not all used up at once. Breakthrough of the contaminants takes place over a long period of time, rather than suddenly. Figure 14-14 indicates a typical breakthrough pattern for a 30-in. (760-mm) GAC bed. This pattern is important because it allows the replacement of the GAC to be phased, so that only one filter needs to be out of service at a given time.

Application of GAC in Contactors

When GAC is used in closed tanks, the containers are known as contactors or *adsorbers*. An advantage of contactors is that they can be manufactured to the size necessary to provide the desired EBCT. For example, if the contact time of GAC in a filter bed is not sufficient, there is little that can be done about it. However, if a contactor is used, the size and flow rate can be designed to provide any contact time desired. Table 14-2 shows the effect of these factors in a hypothetical water treatment situation. The design of contactors for a water system is normally based on pilot tests of the water to be treated.

In most installations, two or more contactors are operated in parallel. The units are started up at different times so that the beds will not all be exhausted at the same time. The beds are usually backwashed periodically to remove suspended matter and carbon fines. Contactors are backwashed when there is a noticeable pressure drop between the inlet and outlet sides of the unit.

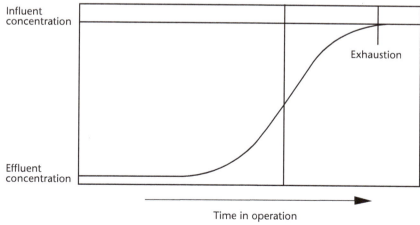

NOTE: Vertical line denotes point at which the effluent concentration is approximately one half of the influent concentration and GAC is replaced.

Figure 14-14 Typical breakthrough pattern for a GAC bed

contactor
A vertical, steel cylindrical pressure vessel used to hold the activated carbon bed.

Table 14-2 Example of GAC adsorption characteristics for various bed depths

Bed Depth, ft	(m)	EBCT, min	Average Influent Chloroform Concentration mg/L	Time to Bed Exhaustion,* weeks
2.5	(0.8)	6.2	67	3.4
5.0	(1.5)	12.0	67	7.0
7.5	(2.3)	19.0	67	10.9
10.0	(3.0)	25.0	67	14.0

*Bed is exhausted when effluent concentration is equal to influent concentration.

Samples of the effluent from each contactor must be collected and analyzed periodically for the presence of the organic chemicals that are to be removed. When monitoring indicates that the chemicals are passing through the contactor, the GAC is exhausted and must be replaced.

Operating Problems

Problems Associated With Powdered Activated Carbon

The most common operating problem with PAC is handling it. It is a fine powder, so dust can be a major problem. If it is used continuously or if large quantities are used periodically, a slurry system should be considered.

PAC passing through the filters and entering the distribution system can result in complaints of "black water." Carbon passing through the filters is usually caused by inadequate coagulation–sedimentation, resulting in carbon carryover to the filters. It can also be caused by adding heavy doses of PAC just before the filters.

Occasionally, taste-and-odor problems persist, in some cases regardless of the PAC dosage. This problem is usually caused by adding the PAC and chlorine in stages too close to one another. The chlorine reacts with the organic compounds to produce additional compounds that are more difficult to adsorb. One solution to this problem is to add the PAC so that it has at least 15 minutes of contact time before any chlorine is added.

Problems Associated With Granular Activated Carbon

When GAC is used as a filter medium, it is important that the coagulation, flocculation, and sedimentation processes be operated continuously to maximize the removal of suspended matter going to the filters. The adsorption capacity of the carbon can quickly be lost if the carbon becomes coated with floc, resulting in shorter bed life and higher operating costs.

In addition, filter rates should be carefully controlled so that they do not fluctuate rapidly. Fluctuations can drive previously deposited material through the filter and can form channels through the bed. The channels will reduce the filtering capacity of the bed and greatly reduce the contact time of water passing through the filter medium.

Proper backwashing is essential for effective filtration and adsorption. The rate must be sufficient to achieve a bed expansion of about 50 percent for proper cleaning. If the bed is not adequately cleaned, both filtration and adsorption capacity will be lost, and mudballs will begin to form. This is particularly a problem where polymers are used as coagulants and filter aids.

GAC is lighter than other filter media, so it can easily be washed away during backwashing. If an excessive amount of GAC is being lost, backwash rates may have to be reduced. The water pressure to surface washers may also have to be decreased if it is found that they are contributing to excessive GAC loss. If carbon loss is over 2 in. (50 mm) per year, the backwashing procedures should be reviewed to determine how the loss can be reduced. If the height of the wash-water troughs above the filter beds is not adequate, the troughs may have to be raised to prevent further carbon loss.

A special problem with GAC filters is the rapid growth of bacteria within the bed. The carbon removes and holds organic compounds, and bacteria can feed and thrive on this material. The carbon also adsorbs chlorine, so any chlorine added ahead of the filters to control bacterial growth is ineffective. As a result, the bacterial concentration in the filter effluent may be thousands of times higher than in the filter influent.

Research indicates that the high bacterial populations in the bed actually enhance organics removal because the bacteria break down complex compounds to simpler products that are more easily adsorbed by the GAC. Some treatment plants backwash with heavily chlorinated water to help control the bacterial levels. In general, it is not good practice to allow high concentrations of bacteria to pass through the filters. The final chlorination must be closely controlled to ensure that the bacteria are destroyed before they enter the distribution system.

If GAC contactors are used after conventional filters, it is essential that only a minimum of suspended material come through the filters. If the length of time between backwashing a contactor begins to decrease, or if the bed life is much shorter than it should be, floc carryover from the filters may be a problem. The turbidity of filter effluent should be monitored continuously to prevent reducing the life of GAC in the contactors. Bacterial growth in contactors can also be a problem, but it can be handled in the same manner as that for GAC in conventional filters.

Control Tests

Testing of the effluent from carbon adsorption systems to determine if there is a breakthrough of organic chemicals requires sophisticated water quality analyses using a gas chromatograph. In general, only very large water systems have the equipment and expertise to run these analyses themselves. Other systems should contract with a convenient private laboratory that will perform the analyses.

Determining PAC Dosage Rates

The required dosage of PAC can be approximated by modifying the jar test. The stirring apparatus, as well as all glassware, should be cleaned with an unscented detergent and rinsed thoroughly with odor-free water. (The method for producing odor-free water is described in *Standard Methods for the Examination of Water and Wastewater*. The standard jar test and taste-and-odor testing are described in Chapter 10.)

One-liter samples of the raw water are then dosed with varying amounts of a well-shaken stock PAC solution—for example, 5, 10, 20, and 40 mL. The stock solution is prepared by adding 1 g of PAC to 1 L of odor-free water. Each milliliter of this solution, when added to a 1-L sample of raw water, is equivalent to a dosage of 1 mg/L.

The four dosed samples, and a fifth sample to which no PAC is added, are stirred for a period that approximates the contact time the PAC will have with the water as it passes through the plant. At the end of that time, each sample is filtered through glass wool or filter paper to remove the PAC. The first 200 mL of each sample through the filter is discarded, and the remainder is subjected to the threshold odor test to arrive at a threshold odor number (TON) for each sample.

The results can then be plotted, as illustrated in Figure 14-15, to obtain an optimal PAC dosage. In the example, the dosage needed to reduce the TON to 3 (which is recommended by USEPA's secondary drinking water regulations) would be 29 mg/L. Experience has shown that plant-scale PAC application is more efficient than indicated by jar tests. However, a plant should begin with the dosage indicated by the jar test result and then gradually reduce it while sampling the plant effluent. To maintain the optimum PAC dosage, the threshold odor test should be conducted by each shift, or at least daily, on the raw and treated water while PAC is being fed.

If PAC is used to remove organic compounds other than those causing tastes and odors, different procedures are needed to determine the proper dosage.

Tests of GAC Used in Filters

GAC is used in a conventional gravity filter for both filtration and adsorption, so most of the control tests are the same as those used for filtration, as described in Chapter 6. The condition of the filter bed should be checked frequently, particularly during and after backwashing, to determine if there are problems with cracks, shrinkage, channeling, or mudballs.

The distance between the top of the carbon and the top of the wash-water troughs (or other reference mark) should be measured at least every 3 months to determine the rate of carbon loss. Samples of backwash water can also be collected and tested to see if they contain excessive amounts of carbon. Excessive loss may indicate a need for changes in the backwashing procedures.

Core samples of the carbon bed should be taken at the time of installation and at least every 6 months to determine both the amount of bed life remaining and the condition of the medium. The sample should represent the carbon from top to bottom of the bed. GAC manufacturers can provide the necessary information on sampling, testing, and interpreting results.

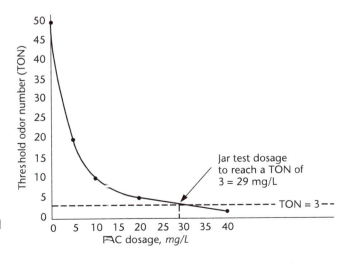

Figure 14-15 Example determination of optimal PAC dosage

If the filter is designed to remove taste- and odor-causing compounds, the threshold odor test should be conducted routinely on the raw and finished water as a check on the effectiveness of the filter. If the filter is intended to remove other organic compounds, special analyses of samples will periodically have to be performed for these compounds. Taste-and-odor testing cannot be used as an indicator of the removal of other organic compounds. In general, the holding capacity of GAC for other organic compounds is much less than it is for taste- and odor-causing compounds.

Because bacteria can thrive in GAC filters, standard plate count analyses should be performed each day on the filtered water and the water after final chlorination. This will help determine the required final chlorination dosage and will indicate whether chlorination of the backwash is needed. The plate count should be kept well below 500 organisms/mL.

Tests of GAC Used in Contactors

Because GAC contactors are typically used for removing organic compounds other than those causing tastes and odors, more sophisticated control tests are usually needed. There is normally a specific compound or group of compounds that are to be removed, so periodic analyses of the contactor effluent must be made to check on the effectiveness of the treatment and to determine when the carbon is exhausted.

Head loss through the contactors should also be monitored and recorded continuously so that backwashing can be performed at the proper time. Core samples of the carbon bed should be taken at least every 3 months so that the approximate remaining bed life can be determined. Turbidity of the effluent should also be monitored and recorded continuously to determine if carbon fines or other suspended matter are passing through the GAC bed. Standard plate count analyses of the contactor effluent before and after chlorination will also help determine what chlorine dosages are needed to control bacterial growth.

Safety Precautions

Bags of PAC and GAC should be stored on pallets in a clean, dry place so that air can circulate underneath. The bags should be stacked in single or double rows with access aisles around every stack to allow easy handling and fire inspection. They should never be stored in stacks over 6 ft (2 m) high.

Although activated carbon is not considered explosive, it will burn like charcoal without producing smoke or flame, and it glows with the release of intense heat. A carbon fire is difficult to detect and to extinguish. This is another advantage of using slurry storage rather than dry feed.

The storage area for dry carbon should be fireproof and have self-closing fire doors separating it from other storage areas. Storage bins for dry bulk carbon storage should be fireproof and equipped for fire control. Smoking must be prohibited during the handling and unloading of carbon and in the storage and feeding areas.

Burning carbon should not be doused with a large stream of water because doing so will only cause carbon particles to fly in all directions and spread the fire. A fine mist of spray from a hose or a chemical foam extinguisher is most effective in controlling a fire of this type.

Activated carbon must never be stored near gasoline, mineral oils, or vegetable oils. When mixed with carbon, these substances will slowly oxidize until the ignition temperature of 600–800°F (316–427°C) is reached.

Figure 14-16 Safety clothing to be worn in the handling of carbon

Stored carbon should also be kept well away from chlorine compounds and potassium permanganate because spontaneous combustion can occur when they are mixed. Carbon is an electrical conductor, so explosion-proof light fixtures and electrical wiring should be used in all carbon storage and feeding areas. As an added precaution, the electrical equipment should be cleaned frequently.

All tanks receiving dry carbon should be vented and provided with dust-control equipment, such as bag-type dust collectors. Even slurry storage tanks should be so equipped. If a PAC dry feed system is used, the hoppers and feeders should be enclosed in a separate room so that dust will be confined.

Oxygen is removed from air in the presence of wet activated carbon. As a result, slurry tanks or other enclosed spaces containing carbon may have seriously reduced oxygen levels. Personnel must be careful when entering these spaces to ensure that adequate oxygen is available. Devices to indicate the amount of oxygen present should be used before anyone enters this type of closed space. Personnel entering a tank or other enclosure should also have attached safety belts and another worker standing by to pull them from danger if necessary.

Personnel who unload or handle carbon should be provided with dust masks, face shields, gauntlets, and aprons (Figure 14-16). In addition, shower facilities should be provided for workers to use if they get unduly dirty from handling carbon.

Record Keeping

Good record keeping can help prevent taste- and odor-causing compounds from reaching consumers, while preventing waste of PAC. Suggested records include the following:

- The type of odor in the water (such as musty, septic, or rotten egg) and the taste (such as sweet or bitter)
- Dates on which PAC was added
- The calculated dosage based on jar tests
- The actual plant dosage that was effective
- The TON values for the raw and treated water

This information may give some clue to the source of the problems, how long they may last, how they may be prevented, and when they may occur again.

Records of GAC in filters should include the following:

- Hours of filter operation for computing the quantity of water processed
- Monthly measurements of the carbon level to monitor GAC loss
- Dates when GAC is replaced so that bed life can be monitored
- Results of periodic threshold odor tests

Records of GAC contactor operations could include the following:

- Quantity of water treated
- Times when the contactor is backwashed and the amount of backwash water used
- Dates of GAC replacement for computing the bed life
- Results of analyses of raw-water and finished-water organics concentrations

Study Questions

1. Powdered activated carbon is primarily used to control
   a. disinfectant by-products.
   b. organic compounds responsible for tastes and odors.
   c. synthetic organic chemicals.
   d. humic and fulvic acids.

2. GAC contactors can be used when formation of trihalomethanes or other disinfection by-products _____ an adjustment of the disinfectants or other treatment processes.
   a. can easily be controlled by
   b. can be quickly reduced by
   c. cannot be prevented by
   d. will likely be produced by

3. Which of the following happens at the surfaces of the PAC particles if they are coated with coagulants or other water treatment chemicals?
   a. They emit toxic by-products.
   b. They lose their capacity to adsorb.
   c. They dissolve into the surrounding water.
   d. They harden and crack.

4. How should burning carbon be handled?
   a. It should be allowed to cool without any interference.
   b. It should be covered in aqueous film-forming foam.
   c. It should be covered with a fine mist of spray from a hose or a chemical foam extinguisher.
   d. It should be doused in buckets of water.

5. In a slurry feed system, where should the feeder be located to ensure there is no danger that the carbon will settle out in the feed line because of the low flow velocity?

6. In a slurry feed system, day tanks are typically made of what material?

Chapter 15
Aeration

Aeration is a gas transfer process used in water treatment. As a contaminant removal process, aeration is used to transfer volatile substances from water into the air. Examples of *air stripping* or *desorption* include removal of organic compounds, odorous substances, carbon dioxide, and radon.

Aeration is also used to add oxygen from the air to water. This process increases the dissolved oxygen (DO) level that can be used to oxidize certain pollutants and enhance removal. Examples of oxidation include removal of sulfide and iron. Higher levels of DO are sometimes needed to increase biological activity desired for improved operation of subsequent treatment processes and to avoid undesirable tastes and odors.

The general categories of aerators are based on the two main aeration methods. Water-into-air aerators produce small drops of water that fall through air. Air-into-water aerators create small bubbles of air that rise through the water being aerated. Both categories of aerators are designed to create extensive contact area between the air and the water. Within the two main categories, there is a wide variety of equipment. The more common types of aerators are discussed in this chapter.

Water-Into-Air Aerators

Cascade Aerators

A cascade aerator is a series of steps that may be designed like a stairway or stacked metal rings, as shown in Figures 15-1 and 15-2. In all cascade aerators, aeration occurs in the splash areas. Splash areas on the inclined cascade are created by placing riffle plates (blocks) across the incline. Cascade aerators can be used to oxidize iron and to partially reduce dissolved gases.

Cone Aerators

Cone aerators are similar to ring-type cascade aerators. Air portals draw air into stacked pans, mixing that air with the falling water. Water enters the top pan through a vertical center feed pipe. The water fills the top pan and begins cascading downward to the lower pans through specially designed, cone-shaped nozzles in the bottom of each pan, as shown in Figures 15-3 and 15-4. Cone aerators are used primarily to oxidize iron, although they are also used to partially reduce dissolved gases.

> **aeration**
> The process of bringing water and air into close contact to remove or modify constituents in the water.

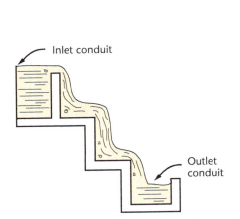

Figure 15-1 Stairway-type cascade aerator

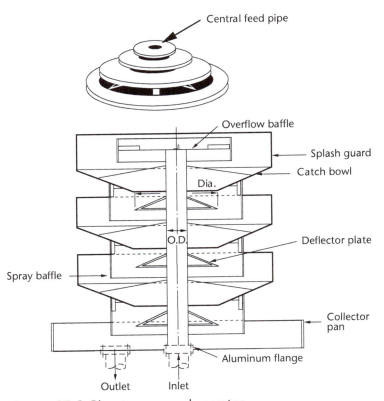

Figure 15-2 Ring-type cascade aerator
Courtesy of US Filter-General Filter Products.

Figure 15-3 Cone aerator
Courtesy of ONDEO Degremont.

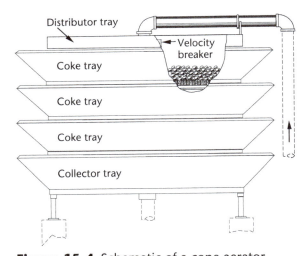

Figure 15-4 Schematic of a cone aerator
Courtesy of US Filter-General Filter Products, Ames, Iowa.

Slat-and-Coke-Tray Aerators

As shown in Figure 15-5, slat-and-coke-tray aerators usually consist of three to five stacked trays that have spaced wooden slats, usually made of redwood or cypress. Early models of this type of aerator were usually filled with about 6 in. (150 mm) of fist-sized pieces of coke. The media used in more recent years may also be rock, ceramic balls, limestone, or other materials.

The aerator is usually constructed to have sloping sides, called *splash aprons*, which are used to protect the splash from wind loss and freezing. Water is

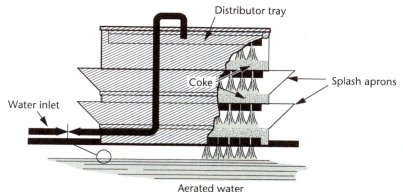

Figure 15-5 Slat-and-coke-tray aerator

Courtesy of *Power Magazine*, McGraw-Hill Inc.

introduced into the top tray (the distributing tray) and moves down through the successive trays, splashing and taking in oxygen each time it hits a layer of media. This type of aerator is commonly used to oxidize iron and, to a limited extent, to lower the concentration of dissolved gases.

Draft Aerators

A draft aerator is similar to those already discussed except that it also has an air-flow created by a blower. There are two types of draft aerators: the positive draft type and the induced draft type.

As shown in Figure 15-6, a positive draft aerator is composed of a tower of tiered slats (wooden in the illustration) and an external blower that provides a continuous flow of air. The water is introduced at the top of the tower, and as it flows and splashes down through the slats, it is subjected to the high-velocity airstream from the blower. The impact caused by the free-falling droplets of water hitting the slats aids in releasing dissolved gases. The high-velocity airflow rushes past the water droplets, carrying away carbon dioxide, methane, or hydrogen sulfide, and constantly renews the supply of DO needed to oxidize iron and manganese.

The induced draft aerator, shown in Figure 15-7, differs in that it has a top-mounted blower, which pulls an upward flow of air from vents located near

Figure 15-6 Positive draft aerator

Courtesy of US Filter-General Filter Products.

Figure 15-7 Induced draft aerator

Courtesy of US Filter-General Filter Products.

the bottom of the tower. Draft aerators are much more efficient than the previously discussed types for dissolved gas removal and for oxidizing iron and manganese.

Spray Aerators

A spray aerator consists of one or more spray nozzles connected to a pipe manifold. Moving through the manifold under high pressure, the water leaves each nozzle in a fine spray and falls through the surrounding air, creating a fountain effect. When relatively large high-pressure nozzles are used, the resulting fountain effect can be quite attractive. This type of aeration device is sometimes located in a decorative setting at the entrance to the water treatment plant.

Spray towers (Figure 15-8) are also used for aeration because the structure protects the spray from windblown losses and reduces freezing problems. Spray aerators are also sometimes combined with cascade and draft aerators, as shown in Figure 15-9, to capture the best features of each, depending on the application.

In general, spray aeration is successful in oxidizing iron or manganese and is very successful in increasing the DO level of water.

Packed Towers

The use of **packed towers** (or *air strippers*, as they are commonly called) for aeration of drinking water is a relatively new development. They have been designed principally for removal of less-volatile compounds, such as volatile organic chemicals, from contaminated water.

A typical packed tower consists of a cylindrical tank containing a packing material. Water is usually distributed over the packing at the top, and air is forced in at the bottom. The very large surface area of the packing material provides for considerably more liquid–gas transfer compared with other aeration methods.

The packing material is of two general types. The type commonly referred to as *dumped packing* consists of shaped pieces of ceramic, stainless steel, or plastic that are randomly dumped into the tower. Figure 15-10 illustrates examples of commercially available packing pieces. Plastic packing is normally used in water treatment air strippers. Fixed packing is also available as prefabricated sheets that are placed in the tower. The manufacturer can provide the operational transfer efficiency rates for these sheets.

packed tower

A cylindrical tank containing packing material, with water distributed at the top and airflow introduced from the bottom by a blower. Commonly referred to as an air stripper.

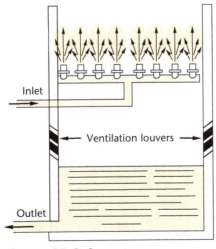

Figure 15-8 Spray tower

Courtesy of ONDEO Degremont Inc., Richmond, Virginia.

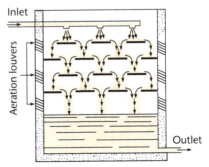

Figure 15-9 Spray aerator combined with cascade and draft aerators

Courtesy of ONDEO Degremont Inc., Richmond, Virginia.

Figure 15-10 Typical plastic packing pieces
Source: US Environmental Protection Agency.

The principal parts of a packed tower with *loose packing* are shown in Figures 15-11 and 15-12. The redistributors placed at intervals in the column direct water that is flowing along the column wall back toward the center. The demister at the top of the column removes moisture from the air as it leaves to prevent objectionable clouds of moisture from coming off the column.

Airflow is provided by a centrifugal blower driven by an electric motor. Small towers designed to remove relatively volatile compounds may require only a 5-hp (3,700-W) or smaller motor. Other columns require much larger and more powerful blowers. Formulas are available for computing the air-to-water ratio required for removing various volatile chemicals. Care must be taken in protecting the blower intake to prevent contaminants from being blown into the tower.

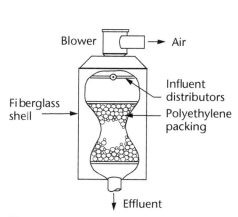

Figure 15-11 Cutaway view of interior of a packed tower

Figure 15-12 View of an installation of a packed tower
Source: US Environmental Protection Agency.

The relative volatility of a substance is expressed in terms of a factor known as *Henry's constant*. In general, compounds with a factor over 100 can feasibly be removed by a packed tower. For values from 10 to 100, removal may or may not be feasible, depending on what other contaminants are in the water and other factors. In most cases, the state will require that pilot-plant tests be run on the water if there is any question of how well a contaminant would be removed.

Factors that must be considered in the design of a packed tower unit include the following:

- The height and diameter of the unit
- The air-to-water ratio required
- The packing depth
- The surface loading rate

The minimum air-to-water ratio at peak flow in a packed tower should normally be at least 25:1, and the maximum should be no more than 80:1. The design of a tower should also make allowance for potential fouling of the packing by calcium carbonate, precipitated iron, and bacterial growth. If any of these are expected to be a problem, pretreatment may have to be provided. If the problem is not expected to be too serious, provisions can be made for periodic chemical backwashes to clean the packing.

Air-Into-Water Aerators

Diffuser Aerators

A typical diffuser aeration system (Figure 15-13) consists of an aeration basin or tank constructed of steel or concrete. The basin is equipped with compressed air piping, manifolds, and diffusers. The piping is usually steel or plastic, and the individual diffusers may be plastic or metallic devices, porous ceramic plates, or simply holes drilled into the manifold pipe.

The air diffuser releases tiny bubbles of compressed air into the water, usually near the bottom of the aeration basin. The bubbles rise slowly but turbulently through the water, setting up a rolling-type mixing pattern. At the same time, each air bubble gives up some oxygen to the surrounding water. This form of aeration is used primarily to increase the DO content in order to prevent tastes and odors.

The essential piece of equipment in every diffused air system is the air compressor or blower. There are two types of blowers: centrifugal and positive-displacement. A typical unit of the more common type, the centrifugal blower, is shown in Figure 15-14; the cutaway view in Figure 15-15 identifies the major component parts.

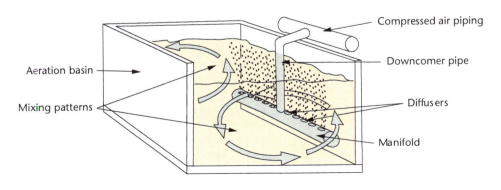

Figure 15-13 Diffuser aeration system

Figure 15-14 Centrifugal blower

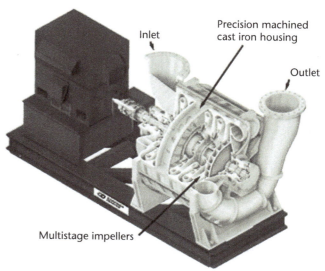

Figure 15-15 Cutaway view of a centrifugal blower
Courtesy of Gardner, Denver Water.

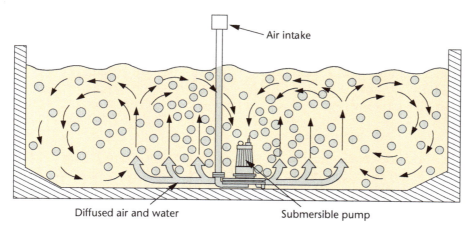

Air intake

Diffused air and water

Submersible pump

Figure 15-16 Draft-tube aerator

Draft-Tube Aerators

The draft-tube aerator is a submersible pump equipped with a draft tube (air intake pipe), as illustrated in Figure 15-16. A partial vacuum is created at the eye of the spinning turbine impeller, causing air to enter through the draft tube and water to enter through the water intake. The air and water are mixed by the turbine impeller and then discharged into the aeration basin. The draft-tube aerator is a convenient, low-cost method for adding aeration to an existing basin or tank.

Combination Aerators

Mechanical Aerators

A mechanical aerator consists of a propeller-like mixing blade mounted on the end of a vertical shaft driven by a motor. By a rapid rotation of the mixing blade in the water, air and water are violently mixed.

Figure 15-17 is a side-by-side comparison of the following four types of mechanical aerators:

1. Surface aerators (water-into-air type).
2. Submerged aerators (air-into-water type).
3. Combination mechanical aerators (combination type).
4. Draft-tube surface aerators (water-into-air type).

Surface Aerators

Surface aerators draw water into the blade of the aerator and throw the water into the air in tiny droplets, so that the water can pick up oxygen. The mixing pattern of a typical surface aerator is shown in Figure 15-18. Violent mixing is necessary for efficient oxygen transfer and release of unwanted gases, tastes, and odors.

Submerged Aerators

Submerged aerators usually consist of two components: a submerged air diffuser (called a *sparger*) and a submerged blade that mixes the air into the water. Figure 15-19 shows the mixing pattern of a submerged turbine aerator with air entering the water just below the propeller blade. Note that the mixing pattern is opposite to the surface aerator pattern. Figure 15-20 shows more clearly how the air is introduced by the sparger and how the submerged turbine distributes the air.

A submerged turbine aerator produces relatively calm water at the surface compared with surface aerators. Because of this relative calm, submerged aerators are best used to increase DO levels, rather than surface aerators, whose greater turbulence removes unwanted gases and incidentally acts as a very effective mixer.

Combination Mechanical Aerators

Combination mechanical aerators, like the one shown in Figure 15-17C, offer the features of both the surface and submerged types. They can be used to oxidize iron and manganese and to remove unwanted gases, tastes, and odors.

Draft-Tube Aerators

Draft-tube aerators, like the one shown in Figure 15-17D, are used to ensure better mixing when surface aerators are installed in deep aeration basins. The draft

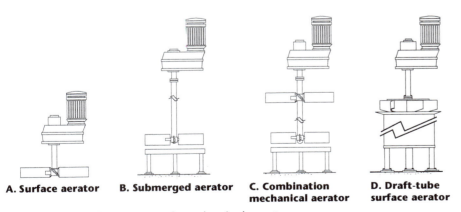

A. Surface aerator B. Submerged aerator C. Combination mechanical aerator D. Draft-tube surface aerator

Figure 15-17 Four types of mechanical aerators
Courtesy of Philadelphia Mixers Corporation.

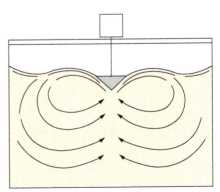

Figure 15-18 Mixing pattern of surface aerator
Courtesy of Dorr-Oliver EIMCO.

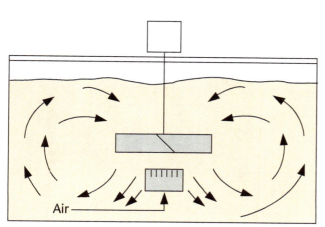

Figure 15-19 Mixing pattern of submerged turbine aerator

Courtesy of Dorr-Oliver EIMCO.

Figure 15-20 Functioning submerged turbine aerator

Courtesy of Philadelphia Mixers Corporation.

tube is open at both ends. When the surface aerator is spinning, water is drawn from the very bottom of the basin into the bottom of the draft tube. The water rises up the tube into the impeller and is thrown out onto the water surface. In this way, the draft tube improves the bottom-to-top mixing and turnover.

Pressure Aerators

There are two basic types of pressure aerators. The type diagrammed in Figure 15-21 consists of a closed tank continuously supplied with air under pressure. The water to be treated is sprayed into the high-pressure air, allowing the water to pick up DO quickly. Aerated water leaves through the bottom of the tank and moves on to further treatment. Aerators of this type are used primarily to oxidize iron and manganese for later removal by filtration.

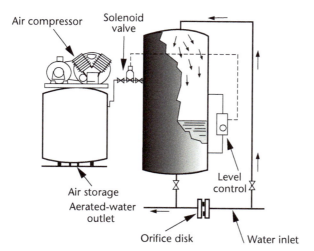

Figure 15-21 Pressure aerator with pressure vessel

Courtesy of *Power Magazine*, McGraw-Hill Inc.

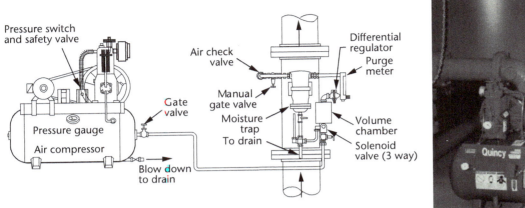

Figure 15-22 Pressure aerator with air diffused directly into pressure pipeline

Courtesy of US Filter-General Filter Products

The second type of pressure aerator is diagrammed in Figure 15-22. This type has no pressure vessel; instead, air is diffused directly into a pressurized pipeline. The diffuser inside the special aeration pipe section distributes fine air bubbles into the flowing water. As in any pressure aerator, the higher the pressure, the more oxygen that will dissolve in water. The more oxygen there is in solution, the quicker and more complete the oxidation of the iron and manganese will be.

Study Questions

1. Which type of aerator causes aeration to occur in splash areas and can be used to oxidize iron and to partially reduce dissolved gases?
   a. Slat-and-coke-tray aerator
   b. Packed tower
   c. Cascade aerator
   d. Cone aerator

2. Which type of aerator is much more efficient in removing dissolved gases and for oxidizing iron and manganese?
   a. Packed tower
   b. Slat-and-coke-tray aerator
   c. Cascade aerator
   d. Draft aerator

3. Which of the following is the normal air-to-water ratio at peak flow in a packed tower aerator?
   a. 10:1
   b. 15:1
   c. 25:1
   d. 35:1

4. _____ aerators draw water into the blade of the aerator and throw the water into the air in tiny droplets, so that the water can pick up oxygen.
   a. Surface
   b. Submerged
   c. Draft-tube
   d. Mechanical

5. Which type of aerator consists of a propeller-like mixing blade mounted on the end of a vertical shaft driven by a motor?
   a. Surface aerator
   b. Submerged aerator
   c. Draft-tube aerator
   d. Mechanical aerator

6. In a slate-and-coke-tray aerator, the aerator is usually constructed to have sloping sides, called _____, which are used to protect the splash from wind loss and freezing.
   a. splash slats
   b. coke trays
   c. splash aprons
   d. distribution trays

7. List five components of a ring-type cascade aerator.

8. In general, spray aeration is successful in oxidizing iron or manganese and is very successful in increasing what characteristic of water?

Chapter 16
Membrane Treatment

Microfiltration membranes can enhance and, in some cases, substitute for conventional municipal water treatment processes. They enable water treatment plants using this technology to meet more stringent regulations. Microfiltration membranes, for example, can produce treated water that meets the log removal requirements of the Surface Water Treatment Rule for *Cryptosporidium* and *Giardia* and the ultra-low turbidity goals of some systems. Microfiltration is also often used as effective pretreatment for spiral-wound nanofiltration and reverse osmosis membrane systems. Reverse osmosis and nanofiltration membranes are capable of removing minerals and a broad spectrum of possible contaminants. However, the operating costs of these energy-driven processes is higher than conventional water treatment systems.

Microfiltration Facilities

This section describes the equipment and chemicals associated with hollow-fiber microfiltration. Figure 16-1 shows a microfiltration installation.

Characteristics of Microfiltration Membranes

Microfiltration membranes are available with pore sizes ranging from 0.03 to 1.2 microns. The most robust are those ranging from 0.03 to 0.2 microns and made from polyvinylidene fluoride (PVDF), which has excellent chemical resistance to oxidants, such as chlorine, ozone, and permanganate; acids; and bases. These membranes have a long service life ranging from 5 to 10 years. Polypropylene and polysulfone membranes are also available.

Figure 16-1 7.8-mgd hollow-fiber system at San Patricio, Texas

Courtesy of Pall Corporation.

Configuration of Microfiltration Membranes

Hollow-fiber microfiltration membranes are organic polymeric tubes (fibers) usually less than a millimeter in diameter and are enclosed in a module. Figure 16-2 shows a magnified cross section of a hollow fiber. Figure 16-3 is a cutaway of a typical hollow-fiber module. The fibers are sealed at the bottom end of the filter module in such a manner as to direct flow streams to the outside or shell side of the fiber. This is shown in Figure 16-4. The fibers are sealed at the top end of the filter module to allow filtered water to exit from the inside of the fiber (lumen side). The water to be treated is pumped into the module and exits from the open ends of the lumens. A vertical configuration allows the use of gravity to separate air and water during flux recovery processes. The number of hollow fibers housed in a module can range from several hundred to several thousand. Module length usually ranges from 1 to 2 meters (3–6 ft). A pump upstream of the module pressurizes the shell side of the membrane. Feedwater contacts the shell (outside) of the hollow fiber, and product water collects on the fiber lumen (inside).

Operation of Microfiltration Membranes

Hollow-fiber-configured microfiltration membranes are operated in either a direct flow or cross-flow mode. Flux ranges from 35 to 50 gal/ft^2 (1,426–2,037 L/m^2) of membrane area (specified as outside area or inside area) per day (gfd [m^3/m^2/d]) on surface water.

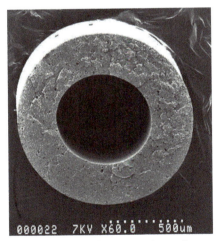

Figure 16-2 Cross section of hollow-fiber microfilter

Courtesy of Pall Corporation.

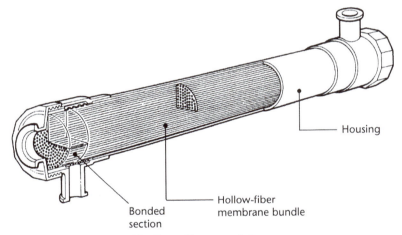

Figure 16-3 Typical hollow-fiber module

Courtesy of Pall Corporation.

Figure 16-4 Hollow-fiber module inlet

Fluxes in excess of 75 gfd (3 m³/m²/d) computed for the outside-in configuration have been documented using a 0.1-micron, PVDF microfiltration system. At these fluxes, operating pressures range from 5 psi (35 kPa) for clean membranes to 30 psi (207 kPa) and higher, at which point membranes require chemical cleaning. Process recovery is typically 95 percent for surface water and can be as high as 97 percent. When used to treat surface water, microfiltration membranes are typically cleaned every one to three months. Chemical cleaning of fouled membranes depends on the membrane's resistance to oxidants. PVDF membranes can be cleaned with strong acids; strong bases; chelating agents, such as citric acid; and oxidants, such as chlorine or peracetic acid.

Figure 16-5 is a schematic that shows the operation of a hollow-fiber microfiltration system in the filtration, reverse filtration, air-scrubbing, and clean-in-place modes. These operations are described in the following sections.

Reverse Flow (Backwash)

Fouling species are removed from the PVDF membrane surface by reversing the direction of flow across the membrane. The efficiency of the process is affected by the velocity, volume, and duration of the reversed fluid flow at the membrane surface.

Reverse flow (RF) is generated using water or water with oxidants, acids, or compressed air, or both. In both the air and liquid backwash, water molecules trapped in the pores of the membrane will dislodge the retained material. The fluid velocity at the membrane surface is greater during air scrub. The reverse flow duration is usually short when combined with air scrub. Material dislodged from the fiber bundle can be flushed from the module with feedwater. Oxidants

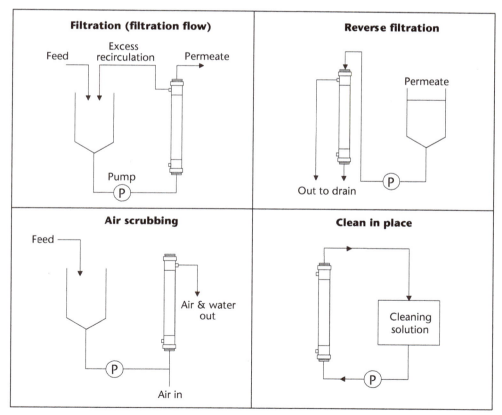

Figure 16-5 Membrane system operations
Courtesy of Pall Corporation.

are sometimes added to the reverse flow water to enhance the efficiency of liquid backwash. Strong oxidants, such as free chlorine and chlorine dioxide, can disrupt the structure of the retained material, which facilitates its removal from the membrane surface. Alternatively, the recirculation of small air bubbles on the membrane surface can scour and disrupt the retained material.

Chemical Cleaning

Reverse flow and air scrub provide a short-term strategy for removing fouling species. Although an effective reverse flow–air scrub strategy will retard the rate of membrane fouling, eventually (4–6 weeks) the membrane must be chemically cleaned to restore its initial transmembrane pressure.

Cleaning protocols for microfiltration membranes vary according to the membrane tolerance to changes in pH and resistance to oxidation. In general, a low pH solution (pH 2–3) removes cationic species, and a high pH solution (pH 11–12) removes organic material. The action of the caustic cleaning solutions can be enhanced by the addition of nonionic surfactants, provided the surfactants can be rinsed to meet regulatory requirements. The surfactants help to disperse the organic particles without binding to the membranes and chelating agents, which are particularly useful for the disruption of bridging in biologically fouled membranes. Chlorine dioxide, free chlorine, or peracetic acid can be used with synthetic polymer membranes, which are resistant to strong oxidants.

Recovery

Unlike reverse osmosis and nanofiltration, the recovery of the microfiltration process is not limited by the formation of scale or the precipitation of salts. **Recovery** is defined as the ratio of the volume of product water, or permeate produced, to raw water treated. The microfiltration process recovery is a function of the filtrate flow rate, the length of the filtration cycle, and the volume of water used in a backwash. For surface waters, the filtration cycle is usually 20 to 30 minutes with a process recovery of 90–97 percent.

Pleated Membrane Facilities

This section describes the equipment and chemicals associated with pleated membrane microfiltration.

Characteristics of Pleated Membranes

The characteristics of a pleated membrane technology designed to serve as a *Cryptosporidium* barrier for very high-quality waters with low fouling potential, such as the effluent from conventional water treatment plants and groundwater under the influence, are summarized in Table 16-1. A flat sheet of polyethersulfone membrane with a 1-micron absolute pore size is folded into a pleated arrangement and backed by a mesh support material to provide maximum surface area for filtration and to provide structural support for the membrane (Figure 16-6).

The pleated membrane is integrated with a polypropylene support and housing to form a module, 9 in. (229 mm) in diameter and 40 in. (1,016 mm) in length. Each module provides 150 ft^2 (14 m^2) of membrane surface. The system is designed to operate at high hydraulic loading rates (referred to as *product water flux* in membrane applications). The nominal design flux of 430 gfd (17.5 m^3/m^2/d) results in a nominal treatment capacity of 45 gpm (170 L/min) per module.

recovery
The ratio of the volume of product water, or permeate produced, to raw water treated.

Table 16-1 Septra CB system characteristics

Characteristic	Description
Membrane configuration	Pleated flat sheet
Rated pore size (absolute)	1 μm
Driving force	Pressurized feedwater
Module dimensions	6-in. outside diameter, 40-in. length
Module housing dimensions	9-in. outside diameter, 55-in. length
Module surface area (external)	150 ft²
Membrane material	Polyethersulfone
Membrane support medium, core, end caps, and outer wrap material	Polypropylene
Module housing material	Polyvinyl chloride
Maximum inlet pressure	150 psig
Module clean water pressure drop	3.5 psid at 50 gpm
Terminal head loss	35 psid
Free chlorine compatibility	Up to 5 mg/L (continuous)
Chloramine compatibility	Up to 8 mg/L (continuous)
Allowable pH range	5–9
Allowable temperature range	1–32°C
Integrity test method	Manual air pressure hold (bubble point)

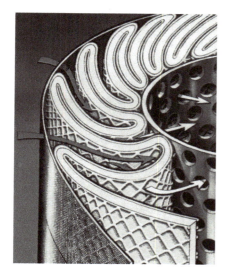

Figure 16-6 Pleated membrane, cartridge cutaway showing membrane, mesh support, and center product water line
Courtesy of Pall Corporation.

Figure 16-7 Disposable pleated membrane cartridge and module
Courtesy of Pall Corporation.

Modules

The pleated membrane assembly is housed in a polyvinyl chloride (PVC) housing with raw water and finished water piping connections (Figure 16-7). The basic system design uses a skid-mounted "rack" assembly containing a manifold with 10 modules. Each rack can process approximately 450 gpm (1,703 L/min) and is approximately 1 ft wide by 12 ft long (0.30 m by 3.7 m) (Figure 16-8). Multiple racks operate in parallel to meet system capacity requirements. Figures 16-9 and 16-10 illustrate how these membranes are configured to treat a surface water and groundwater source.

The membrane module in the *Cryptosporidium* barrier system is designed to be disposable. The system does not employ backwash and thus produces no wastewater. The system is applicable only to very high-quality, low-turbidity waters with no iron or manganese, and low total organic carbon. The modules are replaced when pressure drop becomes excessive. Membrane service life depends on the solids loading (cumulative pounds of solids per square foot [square meter] of membrane area) and the presence of any constituents in the water that may cause or facilitate fouling. Measurement of feedwater turbidity alone may not

Figure 16-8 Septra CB rack assembly, 10 modules, 450-gpm nominal capacity
Courtesy of Pall Corporation.

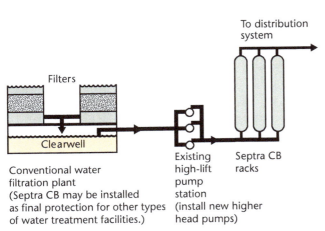

Figure 16-9 Pleated guard filter following conventional treatment
Courtesy of Pall Corporation.

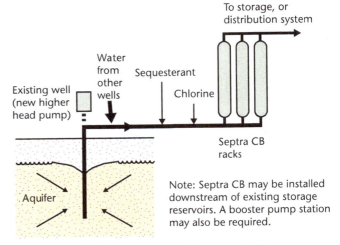

Figure 16-10 Pleated guard filter applied to groundwater under the direct influence of surface water
Courtesy of Pall Corporation.

be a good predictor of membrane life. The system is intended for use on waters where the total suspended solids is less than 0.3 mg/L and the silt density index is less than 5.0.

Backwashable Pleated Membranes

Patented pleated membrane filters have been designed specifically for removal of iron from groundwaters.

A large-diameter, coreless, single, open-ended, pleated cartridge with an outside-to-inside flow pattern has been developed with backwash capability. The filter provides a high surface area for filtration in a compact cartridge design. The cartridge traps all debris and particles 0.2 microns and larger within the filter matrix. The system, operating in the direct flow mode, can operate at high pressure. This allows a municipality to use wellhead pumps to drive the flow through the membrane and into the distribution system without the need to break head in a clearwell or other storage vessel. Flux maintenance of the system is designed to occur every 4 hours for 30 seconds when air and the water in the filter are used to remove any iron or particulate particles trapped in the filter matrix.

Recovery, defined as the ratio of the volumes of product water, or filtrate produced, to raw water treated, is greater than 99 percent. Process recovery is a function of the filtrate flow rate, the length of the filtration cycle, and the volume of water used in flux maintenance. Over 99 percent of the water treated is recovered, and less than 1.0 percent must be handled as waste. The high-flow capacity and high-pressure properties of the filter allow the unit to operate without excessive head loss.

Reverse Osmosis Facilities

This section describes the equipment and chemicals associated with reverse osmosis (RO). Note that nanofiltration is a pressure-driven process analogous to RO. It can remove multivalent cations and anions but at lower pressures than RO (50–150 psi vs. 300–600 psi). Nanofiltration uses many of the same system components and operates in the same way as RO.

Figure 16-11 shows a schematic for a simple RO installation. Figure 16-12 shows a typical package RO unit.

Characteristics of Reverse Osmosis Membranes

The two membranes most commonly used for water treatment are cellulose acetate and polyamide–composite. Each type has specific operating characteristics, such as efficiency of salt rejection, best pH operating range, resistance to degradation if exposed to chlorine or other oxidants, susceptibility to biological attack, and resistance to hydrolysis.

Over time, the performance of all membranes will change, mainly as a result of compaction and fouling. Compaction, which is similar to plastic or metal "creep" under compression, gradually closes the pores in the membrane. The compaction rate increases when membranes are operated at higher pressures and higher temperatures. Membrane fouling can be traced to substances in the feedwater, such as calcium carbonate or calcium sulfate scales, fine colloids, iron or other metal oxides, and silica. Pretreatment of water before RO is usually necessary to prevent premature fouling of the RO membrane.

In some cases, fouled membranes can be restored by periodic cleaning with acid. Otherwise, they must be replaced.

reverse osmosis (RO)
A pressure-driven process in which almost-pure water is passed through a semipermeable membrane. Water is forced through the membrane and most ions (salts) are left behind. The process is principally used for desalination of sea water.

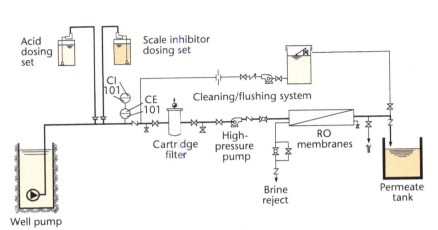

Figure 16-11 Typical reverse osmosis flow schematic
Courtesy of American Engineering Services, Inc., Tampa, Florida.

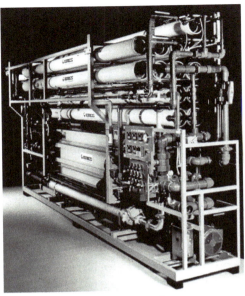

Figure 16-12 Typical package reverse osmosis treatment unit
Courtesy of Ionics, Inc.

Four membrane configurations are currently available: spiral wound, hollow fiber, tubular, and plate and frame. Only the first two configurations are used in commercial drinking water treatment plants.

Spiral-Wound Membranes

As illustrated in Figure 16-13, spiral-wound membranes consist of two flat sheets of membrane material separated by porous sheets. These layers are sealed on three sides to form an envelope. Feedwater enters at one end, and the open side of the envelope is attached to a plastic tube that collects the product water.

 WATCH THE VIDEO
Membrane Technology (www.awwa.org/wsovideoclips)

Hollow-Fiber Membranes

As illustrated in Figure 16-14, hollow-fiber membranes consist of a compact bundle of thousands of fibers that surround the feedwater distribution core. Each fiber is laid in a U-shape in the bundle, and both ends are encapsulated in an end sheet.

Feedwater Concerns

The term feedwater describes raw water that has undergone pretreatment (acidification or scale inhibitor addition) prior to entering the membrane arrays. Feedwater quality and pressure are important issues for the RO process. Feedwater treatment is almost always necessary.

Feedwater Quality

The quality of feedwater passing through RO units is critical for prolonging the life of membranes. Improperly treated feedwater can rapidly cause irreversible damage to the membrane. Surface water usually requires more pretreatment and closer monitoring because its quality is not as stable as groundwaters. To

feedwater
Raw water that has undergone pretreatment (acidification or scale inhibitor addition) prior to entering the membrane arrays.

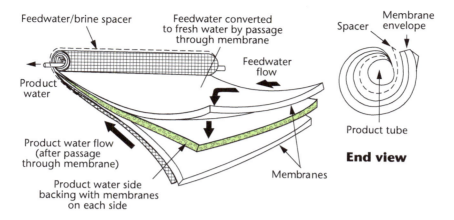

Sectional view

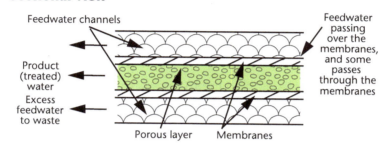

Figure 16-13 Details of a spiral-wound membrane
Courtesy of the Dow Chemical Company.

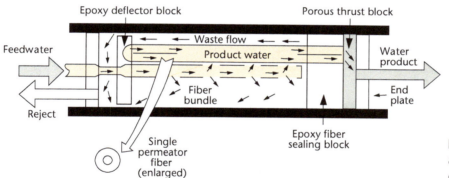

Figure 16-14 Details of a hollow-fiber module
Courtesy of US Filter/Permutit.

maximize membrane life, pretreatment of RO feedwater may be needed for any one or more of the following purposes:

- Turbidity reduction
- Iron or manganese removal
- Stabilization to prevent scale formation
- Microbial control
- Chlorine removal
- Hardness reduction

The types of materials used in feedwater piping and the pressure pump can adversely affect membrane life or performance by releasing trace amounts of metals; therefore only noncorrosive materials, such as stainless steel, PVC, and fiberglass, are preferred. To prevent air ingress, either vertical-turbine types that have mechanical seals or submersible pumps are used.

Feedwater Treatment

In addition to any special pretreatment for contaminant control, as listed above, sulfuric acid is usually added to the feedwater to prevent calcium and magnesium carbonate scaling of the membrane. Cartridge filters are also commonly installed immediately ahead of RO units for final removal of suspended matter down to the 5- or 10-mm level. This is necessary regardless of whether the supply is surface water or groundwater. Cartridge filter containers are furnished with inlet and outlet pressure gauges to monitor head loss. Cartridges should be replaced when the difference in pressure between the gauges reaches 15 psi (100 kPa).

Feedwater Pressure

The pressure differential applied to RO membranes depends on the desired recovery. Although the process can be designed to operate with a pressure as low as 50 psi (340 kPa), commercial installations are commonly operated at 300 psi (2,100 kPa) or higher, because the osmotic pressure increases as the recovery is pushed to higher levels.

Posttreatment

Additional treatment following RO treatment typically includes the following:

- Degasification for removal of carbon dioxide, hydrogen sulfide, or other undesirable gases if they are present
- pH adjustment to minimize corrosion in the distribution system
- Disinfection

Reverse Osmosis Membrane Cleaning

As membranes foul, water production rates and removal efficiencies decrease and higher pressure differentials are required. The following are some indications that a membrane needs cleaning:

- The passage of salt through the membrane increases by 15 percent.
- The pressure drop through the unit increases by 20 percent.
- Feed pressure requirements increase by 20 percent.
- Product water flow drops or increases by 5 percent.
- Fouling or scaling is evident.

Each membrane manufacturer has recommended procedures and chemicals that should be used to rejuvenate its membranes. These guidelines should be carefully followed to prevent damage to the membranes. Cleaning chemicals are usually corrosive, so all cleaning system components must be made of stainless steel or other noncorrosive materials. Obviously, the discharge of these cleaning chemicals is another disposal issue that requires careful planning.

Reject Water

The amount of reject water resulting from the operation of a commercial RO plant can range from 20 to 50 percent of the feedwater to an RO unit. The amount of reject water depends on the number of stages in which the membranes are configured and the feed pressure.

Reject water presents two potentially significant problems. The first is that the water source must be capable of supplying up to twice the amount of water needed by the system. In areas with limited groundwater availability, other treatment processes that do not waste as much water may need to be considered, even if those processes are more expensive.

The other problem is waste disposal. One acceptable method is discharge to the local waste treatment system. However, the relatively large quantity of wastewater produced and the high mineral content of the reject water may overtax the local treatment system. If the source water is subject to agricultural runoff high in pesticides and fertilizer concentrations, or it is well water with high arsenic levels or fluoride, or even if the source water is salty sea water, the concentrated reject water may limit the disposal options. If disposal to the local wastewater system is out of the question, other common disposal methods are deep-well injection (Class V injection wells) and evaporative ponds. With ponds, care must be taken to protect any aquifers, and the utility needs to know that this approach has only delayed the eventual disposal. Discharge to surface waters is possible if a National Pollutant Discharge Elimination System (NPDES) permit can be obtained. The amount of reject water from an RO installation can be reduced to a limited extent by an increase in the feedwater pressure. However, this increased pressure usually results in a shorter membrane life and increased operating costs.

A two-stage system is shown in Figure 16-15. The two first-stage modules purify 50 percent of the water fed to the system. The reject water from the first stage is then processed by the second-stage unit, which purifies it another 50 percent. The final flow ends up being 75 percent purified water and 25 percent reject water.

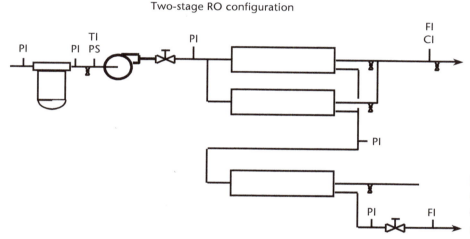

Two-stage RO configuration

Figure 16-15 Schematic of a two-stage reverse osmosis treatment system
Courtesy of US Filter.

Study Questions

1. Which of the following is *not* a component of a typical hollow-fiber module?
   a. Reverse osmosis membrane
   b. Hollow-fiber membrane bundle
   c. Bonded section
   d. Housing

2. Process recovery is typically _____ percent for surface water and can be as high as _____ percent.
   a. 50; 100
   b. 95; 97
   c. 10; 25
   d. 80; 90

3. In a typical reverse osmosis flow schematic, which of the following components most closely follows the cartridge filter?

   a. High-pressure pump
   b. Reverse osmosis membranes
   c. Water pump
   d. Permeate tank

4. What process is defined as the ratio of the volume of product water, or permeate produced, to raw water treated?

5. What term describes raw water that has undergone pretreatment (acidification or scale inhibitor addition) prior to entering the membrane arrays?

Chapter 17
Plant Waste Treatment and Disposal

Residuals are comprised of all of the unwanted material that has been removed from the water and all of the compounds that are created by the addition of treatment chemicals (floc, lime solids, etc.). Alum sludge is the most common form of residual material produced in water plant sedimentation basins, but ferric sludge also is common. Both of these wastes come from the filter backwash process or the sedimentation basins at low concentrations by weight. Disposal of this dilute sludge is inefficient, and so water plant operators must use techniques that dewater and concentrate it.

Removal of Sludge from Conventional Sedimentation Processes

Manual Removal

Older sedimentation basins, and newer installations where sludge accumulation is not expected to be excessive, are designed to have the sludge removed manually. Many plants remove the sludge twice a year—in spring, just before heavy summer water use is expected, and in the fall, after high use has ended. Slimes that accumulate on the walls of basins harbor pathogens, which are removed using high-pressure water hoses to blast the walls. In most cases, fire hoses are used to wash the sludge to the drain, but smaller hoses can be used for smaller basins. Basin floors are usually sloped toward the inlet end because most of the solids settle closer to the inlet. If sludge is not removed regularly from horizontal-flow tanks, the tank must be designed with enough depth to allow for sludge accumulation so that the sedimentation efficiency remains unaffected. Operating staff may need to follow safe entry and working conditions if the basins are considered to be confined spaces.

Mechanical Removal

Sludge can be removed frequently by mechanical sludge scrapers that sweep the sludge to a hopper at the end of the basin. The hopper is then periodically emptied hydraulically. For rectangular basins, one of the following types of collectors is typically used to remove sludge:

- A *chain-and-flight collector*, which consists of a steel or plastic chain and redwood- or fiberglass-reinforced plastic flights (scrapers). A motor drives the chain, which pulls the flights along the basin bottom.

- A *traveling-bridge collector*, which consists of a moving bridge that spans one or more basins. The mechanism has wheels that travel along rails mounted on the basin's edge. In one direction, the scraper blade moves the sludge to the hopper. In the other direction, the scraper retracts, and the mechanism skims any scum from the water's surface.

- A *floating-bridge siphon collector* (Figure 17-1), which uses suction pipes or a submersible pump to withdraw the sludge from the basin. The pipes are supported by foam plastic floats, and the entire unit is drawn along the length of the basin by a motor-driven cable system.

Circular and square basins usually are equipped with scrapers or plows that slant downward toward the center of the basin and sweep sludge to the sludge hopper or pipe (Figure 17-2).

If sludge scrapers are installed, it is not necessary to interrupt plant operations to take basins out of service for manual cleaning. Sludge scrapers also

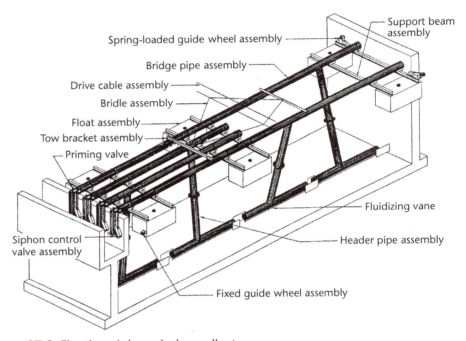

Figure 17-1 Floating siphon sludge collector

Courtesy Leopold, Inc. (from *Water Treatment Operator Handbook*).

Figure 17-2 Sludge scrapers

Courtesy Philadelphia Water Department (from *Water Treatment Operator Handbook*).

reduce personnel requirements and reduce the problems of decomposing sludge that occur under some conditions. The basins will still need to be dewatered and hose-blasted at least yearly to remove slimes that accumulate. If the sedimentation sludge is sent to the local wastewater treatment plant, it is best to deliver a small quantity of sludge every few days rather than large quantities less frequently. This part of the operation may be subject to a "permit" from the receiving entity.

Continuous mechanical sludge collection is also desirable when tube or plate settlers are used because more sludge is created in a smaller basin. If manual cleaning is used with tube or plate settlers, then the frequency of cleaning must be increased.

Softening Sludge Handling, Dewatering, and Disposal

Lime–soda softening plants generate large volumes of sludge. A good rule of thumb is that a typical plant will generate 2.5 lb (1.1 kg) of dry solids daily for each pound (0.5 kg) of commercial lime used in the softening process.

When land is available, lagooning has proved the most economical solution to disposal. The cheapest way to transport the sludge to the lagoons is via pipeline.

If land is not available or the contaminants in the sludge preclude lagooning, the sludge is dewatered. The process of sludge dewatering reduces the amount of water in the sludge, thereby reducing the bulk and making it easier to handle. Dewatering is the process of removing water from the thin sludge, allowing the solid material to remain (thickening) for ultimate disposal.

Most plants dispose of sludge by trucking it to a land disposal site. In general, the drier the sludge, the lower the disposal cost will be. Figure 17-3 shows the relationship between lime sludge volume and percent solids.

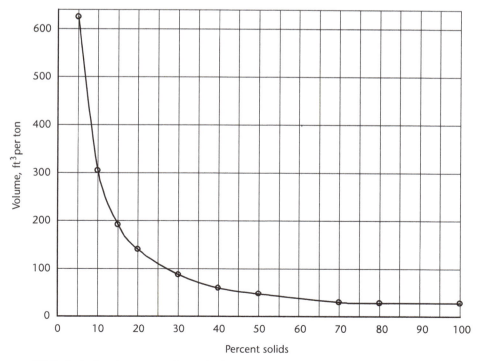

Figure 17-3 Relation between lime sludge volume and percent solids concentration

lagooning
The disposal of sludge by transporting it to a lagoon.

dewatering (of sludge)
A process to remove a portion of water from sludge.

At 10 percent solids, a ton of lime sludge occupies a volume of 300 ft³ (8.5 m³). At 40 percent solids, the sludge volume is reduced to 75 ft³ (2.1 m³). The more commonly used sludge-dewatering devices include the following:

- Drying beds
- Lagoons
- Sludge thickeners
- Vacuum filters
- Belt filters
- Filter presses
- Centrifuges

Amounts of sludge produced by the lime–soda softening process can be calculated using the formula

$$S = 8.34(Q)(2.0 \text{ Ca} + 2.6 \text{ Mg}),$$

where S is sludge produced (in lb/d), Q is plant flow (in mgd), and Ca and Mg are the calcium and magnesium hardness removed as mg/L of calcium carbonate.

Solids Separation Technologies

Sand Drying Beds

A drying bed is a layer of sand placed over graded gravel or stones and an underdrain system. The sand is usually 4–12 in. (100–230 mm) deep. It is placed on top of 8–18 in. (200–460 mm) of graded aggregate.

The aggregate rests on top of an underdrain system consisting of perforated pipe or pipe laid with open joints. The underdrain collects the water and either returns it to the water treatment plant or pipes it to a waste disposal site. Solids separation is accomplished through the mechanism of drainage and evaporation. Therefore, uncovered drying beds perform best in regions having clear sunny days, warm weather, and little rain and snow. It is useful, however, to try pilot testing of beds to determine if they will work in any given climate (Figure 17-4). Freeze-drying in cold climates may be useful. Some utilities have shown surprising results with pilot testing of sand beds in northern climates lacking ideal climate conditions.

Lagoons

Lagoons are large holding ponds that normally provide the most economical way to dewater sludge if sufficient land area is available within a reasonable distance of the treatment plant. They are usually about 10 ft (3 m) deep and can be 1 acre (0.4 ha) or more in size. Sludge is piped directly into the lagoon, where the solids settle out. The water on top (referred to as **decant**) is usually piped back to the treatment plant. More than one lagoon is always required, so that as the settled solids are being cleaned from one, the sludge from the plant can be diverted to the other.

Sludge Thickeners

Two types of sludge thickeners are currently used for softening sludges: conventional gravity thickeners and mechanical gravity belt thickeners.

decant
To draw off the liquid from a basin or tank without stirring up the sediment in the bottom.

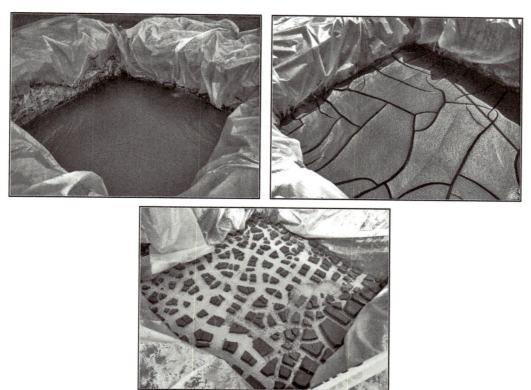

Figure 17-4 Pilot-scale sand drying beds showing three stages of drying

The purpose of a sludge thickener is to remove sufficient water such that the dry solids content of the thickened sludge that remains is constant in the range of 5–10 percent. A thickened sludge enhances the performance of traditional dewatering technologies.

Gravity thickening is accomplished in a tank equipped with a slowly rotating rake mechanism that breaks the bridge between sludge particles, thereby increasing settling and compaction.

Gravity sludge thickeners (Figure 17-5) are similar to circular sedimentation basins. The primary objective of a thickener is to provide a concentrated sludge underflow related to the mass loading (lb/[ft^2/d], or kg/[m^2/d]) or to the unit area (ft^2/[lb/d], or m^2/[kg/d]).

The rake arms rotate like the sludge collector on a solids-contact clarifier and gently stir the sludge blanket to allow water to escape to the surface. Sludge is often thickened before a mechanical dewatering process, such as vacuum filtration or centrifugation, is used (Figure 17-6), so that these processes will work more effectively. The sludge drawn from a thickener ranges between 10 and 40 percent solid particles.

Gravity Belt Thickeners

Gravity belt thickeners (GBTs) use belt press technology to thicken sludges with solids concentrations less than 2 percent. A GBT consists of a gravity belt that moves rollers driven by a variable-speed drive.

Figure 17-7 illustrates a typical process flow diagram of a GBT. It shows the gravity drainage zone and discharge zone that encompass the system. In the gravity zone, most of the free water produced from the flocculation process (polymer conditioning) drains through a porous belt. Once the thickened slurry

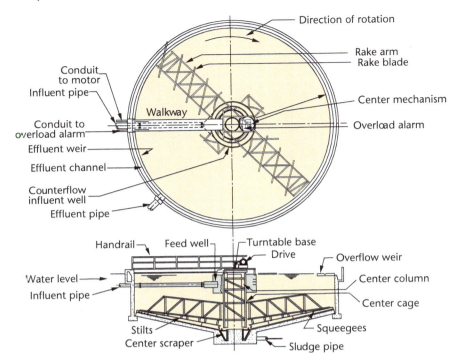

Figure 17-5 Gravity sludge thickener

Courtesy of Dorr-Oliver EIMCO.

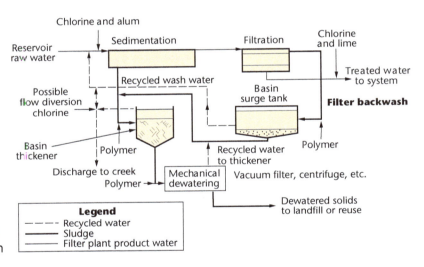

Figure 17-6 Mechanical sludge-dewatering system

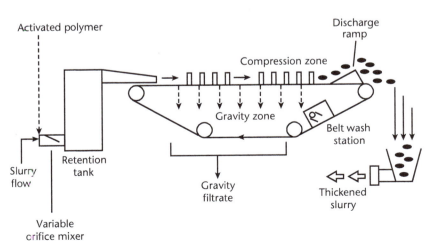

Figure 17-7 Gravity belt thickener process flow diagram

reaches the end of the gravity zone, it comes in contact with a discharge ramp that allows the thickened slurry to roll over itself, further removing water before it is discharged from the unit. (Depending on the manufacturer, this ramp can be adjustable while in operation to change its angle of elevation or it could be fixed at a certain angle.) Being able to change the angle while the unit is in operation increases the unit's efficiency. Several process variables affect the performance of a GBT. The four primary variables that govern operation are described as follows:

1. *Thickened solids concentration*, measured as percent total solids (abbreviated percent TS)
2. *Polymer dosage and cost*, measured as pounds of polymer (dry or liquid) per short ton of dry solids thickened (abbreviated pounds per ton)
3. *Solids loading*, measured as pounds of dry solids thickened per hour per meter belt width (abbreviated pounds per hour per meter)
4. *Capture*, measured as percent solids retained in the process

Vacuum Filters

A vacuum filter (Figure 17-8) consists of a cylindrical filter drum covered with a porous fabric woven from fine metal wire or from fiber such as cotton or nylon. Sludge fills the feed tank and is pulled onto the filter fabric by a vacuum within the filter drum. This vacuum pulls water away from the sludge and into the drum, leaving a thin, dry cake of sludge on the filter fabric (Figure 17-9). This cake is continuously removed from the fabric by a scraper.

The filter cake is hauled to a waste disposal site, and the water removed from the sludge is returned to the thickener. Vacuum filters normally are not efficient for processing alum sludge unless precoating is used. The process works better on lime-softening sludge, but it is difficult to achieve a filter cake dry enough for landfill disposal.

Belt Filters

A belt filter press (BFP) is a continuous-feed dewatering technology that uses polymer conditioning, gravity drainage, and rollers to apply increasing pressure to permeable belts to dewater water treatment plant sludges.

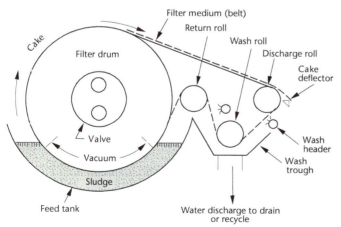

Figure 17-8 Sludge-dewatering vacuum filter (belt type)

Figure 17-9 Dry sludge on filter fabric

Figure 17-10 shows the gravity, low-pressure, and high-pressure zones included in a BFP.

In the first stage of the process, the sludge is drained by gravity to a nonfluid consistency. In the second stage, pressure is applied and gradually increased. In the last stage, the cake is sheared from the belt and further dewatered. The sludge must be conditioned so that it will stay on the belt. This is usually accomplished by the addition of a polymer.

Filter Presses

A filter press can also be used to dewater sludge for final land disposal. The sludge must first be conditioned, usually by a polymer. It is then forced into contact with the filter cloth. The older types of units were called *plate and frame filters*, in which the sludge was forced through the filter under pressure. With the newer diaphragm filters, sludge is initially filtered through the cloth for about 20 minutes, after which compressed air is applied to squeeze water out of the sludge. The cake is then dislodged by shaking or rotating the cloth.

Centrifuges

A centrifuge is a sedimentation bowl that rotates at high speeds to help separate sludge solids from the water by centrifugal force. As shown in Figure 17-11, sludge enters the rotating bowl through a stationary feed pipe. The rotating bowl causes the solids in the sludge to be thrown outward to the wall at a force equal to 3,000 to 10,000 times the force of gravity.

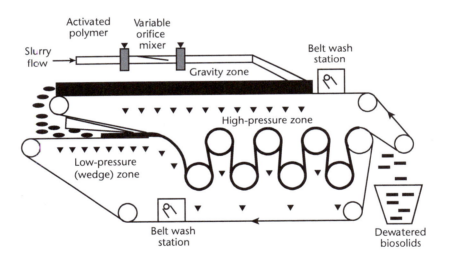

Figure 17-10 Belt filter press process flow diagram

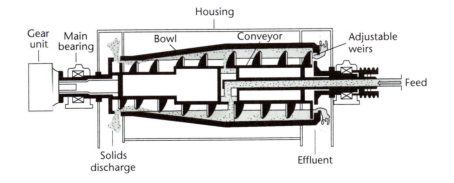

Figure 17-11 Sludge-dewatering centrifuge

Courtesy of Bird Machine Company, Inc.

The solids that settle against the wall of the bowl are scraped forward by a screw-type rotating conveyor and are discharged as shown in Figure 17-11. Clear water, called *centrate*, discharges through controlled outlets and is returned to the thickener. The sludge cake produced by a centrifuge is 50–65 percent solid particles.

Study Questions

1. In the precipitative softening plant, what percentage of solids sludge is produced?
   a. 1%
   b. 5%
   c. 10%
   d. 30%

2. Which sludge disposal method is most economical for lime–soda ash softening plants?
   a. Disposal into the sewage system
   b. Sand drying beds
   c. Lagoons
   d. Landfill the sludge

3. What process is used to concentrate sludge?
   a. Sand bed
   b. Solar lagoon
   c. Thickener
   d. Centrifuge

4. What process is used to dewater sludge?
   a. Wash water basin
   b. Sand bed
   c. Thickener
   d. Reclamation basin

5. In a rectangular basin, which type of collector uses suction pipes or a submersible pump to withdraw the sludge from the basin?
   a. Traveling-bridge collector
   b. Continuous sludge collector
   c. Chain-and-flight collector
   d. Floating-bridge siphon collector

6. In a drying bed, the sand is usually how deep?
   a. 1–2 in. (25–50 mm)
   b. 2–8 in. (50–200 mm)
   c. 4–12 in. (100–300 mm)
   d. 10–12 in. (250–300 mm)

7. A(n) _____ is a sedimentation bowl that rotates at high speeds to help separate sludge solids from the water by centrifugal force.
   a. activator
   b. inducer
   c. solenoid
   d. centrifuge

8. What is the cheapest way to transport the sludge to the lagoons?

9. What are the primary variables affecting the operation of gravity bed thickeners?

Chapter 18
Instrumentation and Control Systems

Flow, Pressure, and Level Measurement

Measurements of flow, pressure, and fluid levels are essential to the operation of a treatment system.

Flow Measurement

Treatment plants measure flow for the following primary reasons:

- The flow rate through the treatment processes needs to be controlled so that it matches distribution system use.
- It is important to determine the proper feed rate of chemicals added in the processes.
- The detention times through the treatment processes must be calculated. These times are particularly important to surface water plants that must meet $C \times T$ values required by the Surface Water Treatment Rule.
- Flow measurement allows operators to maintain a record of water supplied to the distribution system for periodic comparison with the total water metered to customers. This record provides a measure of "water accounted for," or conversely, the amount of water wasted, leaked, or otherwise not paid for.
- Flow measurement allows operators to determine the efficiency of pumps. Pumps that are not delivering their designed flow rate are probably not operating at maximum efficiency, and power is therefore being wasted. Pumps that do not produce at their designed rate should be checked to determine if they are worn or are not operating at their designed head.
- For well systems, it is very important to maintain records of the volume of water pumped and the hours of operation for each well. The periodic computation of well pumping rates can identify problems, such as worn pump impellers and blocked well screens.
- Reports that must be provided to the state by most water systems must include records of raw- and finished-water pumpage. Reporting may be required monthly for very small systems and daily for larger systems. The reports also should include a record of the rate of chemical application, which must be computed based on the treated-water pumpage figures.
- Wastewater generated by a treatment system must also be measured and recorded. This can be done directly by measuring the waste stream or indirectly by measuring the raw-water flow and subtracting the treated-water flow.

- Individual meters are often required for the proper operation of individual pieces of equipment. For instance, the makeup water to a fluoride saturator is always metered to assist in tracking the fluoride feed rate.
- Flow measurement is also important to allow for making tightly controlled flow changes that minimize hydraulic shocks.

Pressure Measurement

It is always necessary to measure pressure at several points in a treatment plant. The most important pressure is usually the plant discharge pressure, which governs the pressure being maintained on the distribution system. Pressure gauges are usually also provided at intermediate points in the treatment process, on special treatment units, and on the discharge of each pump so that each pump's operation can be monitored.

Pressure gauges usually operate with a bellows, diaphragm, or Bourdon tube that is linked to an indicator, as illustrated in Figure 18-1. A cutaway of a typical Bourdon tube pressure gauge is shown in Figure 18-2. Circular or strip-chart recorders, illustrated in Figure 18-3, are used for important pressure functions to provide a record of operations and a visual indication of operating trends. For example, when an operator sees on the distribution system pressure chart that the pressure has been slowly falling for the past hour, he or she can anticipate when additional pumps will have to be turned on to boost system pressure.

Distribution system pressure is also often transmitted to the treatment plant from one or more points on the distribution system. Common transmission points are at the base of elevated tanks and public buildings. The larger the system, the more important it is to have information on system pressure at remote locations.

Level Measurement

It is also usually necessary to measure the level of liquids at several points in the treatment process. The level-measuring devices that have been used for years are mechanical floats, bubbler tubes, and pressure gauges. There are also now a number of instruments that operate based on electrodes that sense the conductivity of the water or ultrasonic waves that are bounced off the water surface. Figure 18-4 illustrates several types of devices that measure liquid levels.

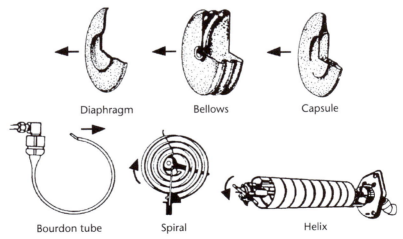

Diaphragm   Bellows   Capsule

Bourdon tube   Spiral   Helix

Figure 18-1 Typical pressure elements used to operate pressure gauges (arrows indicate movement with increased pressure)

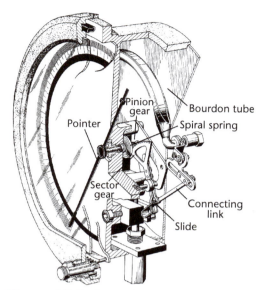

Figure 18-2 Cutaway view of a Bourdon tube pressure gauge

Illustration supplied by Dresser Instruments, Dresser, Inc.

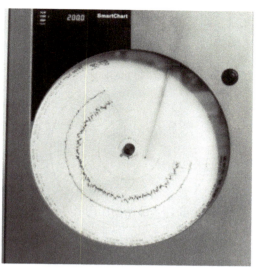

Figure 18-3 Circular chart recorder

Courtesy of Neptune Technology Group.

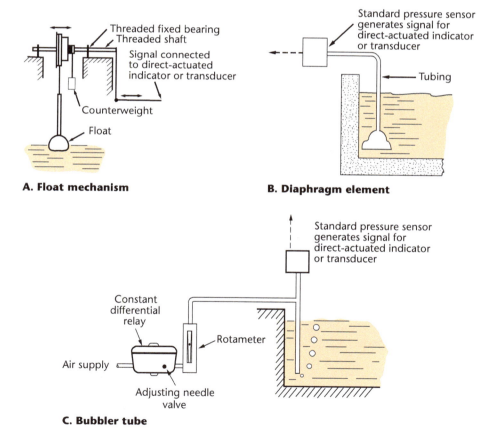

Figure 18-4 Types of liquid-level measuring devices

(continued)

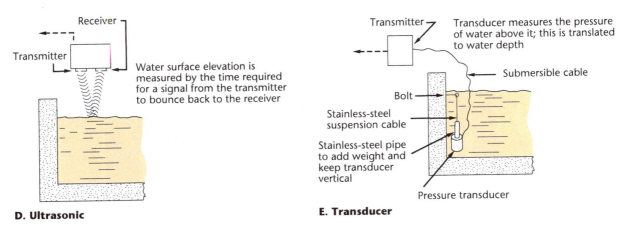

D. Ultrasonic

E. Transducer

Figure 18-4 Types of liquid-level measuring devices (continued)

Other Operational Control Instruments

Numerous instruments can be used to operate pieces of equipment automatically, monitor processes, alert the operator of malfunctions, and perform analyses automatically. This section discusses some of the principal instruments that are commonly used.

Chlorination

Reliable equipment is available that will monitor the chlorine residual and automatically regulate chlorine feed to compensate for changes in both flow rate and chlorine demand. Equipment is also available to provide an alarm if the residual falls outside preset limits, the chlorine gas pressure fails, or chlorine gas is detected in the air.

Filtration

Equipment available for controlling coagulation includes zeta potential meters, streaming current detectors, pH monitors, particle counters, and filterability test equipment. These devices either can alert the operator when the process is not operating within preset limits or can directly compensate for changes by regulating chemical feed rates or making other process adjustments.

Automatic turbidity monitoring is not only a good operational tool, but is required for meeting the effluent turbidity requirements of the Interim Enhanced Surface Water Treatment Rule. Monitors can be provided with recorders to ensure continuous compliance with requirements. They will also alert the operator of any turbidity breakthrough that occurs and provide trending capability to help predict the need to take a filter out of service.

Corrosion Control

Water systems required to install special corrosion control to meet the federal Lead and Copper Rule requirements will normally find that automation of the process is necessary to ensure continuous compliance. Continuous monitoring and recording of the pH of water entering the distribution system should be provided. In addition, software is available to compute the Langelier saturation index quickly and easily. Chapter 11 discusses corrosion control in detail.

Automation

Automation of a water treatment system can be minimal or extensive. In its simplest form, automation means that individual pieces of equipment operate and adjust to changes on their own. In more elaborate forms, many whole treatment plants and distribution systems are computer controlled to operate with little or no human supervision.

Example of Automated Equipment Operation

Figures 18-5 through 18-8 demonstrate the possible degrees of automation for chlorinator control. The same basic principles can be applied to some degree to most other treatment plant processes. A reliable automatic analyzer must be available to provide the necessary control. If the process being controlled is essential to the chemical or microbiological safety of the water, the automation should also include a monitor that will sound an alarm or shut down the system if dangerous under- or overfeed occurs.

Start–Stop Control

The simplest form of automation, start–stop control, is intended only to turn the chlorinator on and off at the same time as a pressure pump (see Figure 18-5). The chlorinator feed rate must be set manually. This type of arrangement would be adequate for a single-well system where the pump rate is fixed and the chlorine demand does not vary significantly.

Proportional Pacing

Proportional pacing represents the next degree of automation (Figure 18-6). This type of arrangement may be used where the pipeline flow is variable—for example, where several different pumps discharge varying amounts, but the chlorine demand is always relatively constant. The feed rate must be set manually, but the chlorinator will modulate to feed the proper concentration in relation to the total flow.

Residual Control

In cases where the chlorine demand varies but the flow is constant, an automatic residual analyzer can be set up to sample the chlorinated water some distance downstream from the chlorine application point (Figure 18-7). In these cases, the operator sets the desired chlorine residual, and the analyzer will automatically adjust the feed to provide the correct amount (even though the flow rate or chlorine

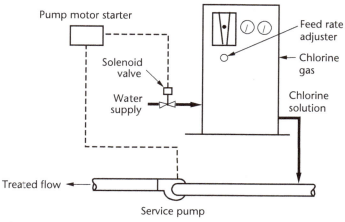

Figure 18-5 Start–stop control of chlorine feed

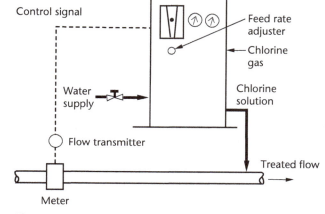

Figure 18-6 Proportional pacing of chlorine feed

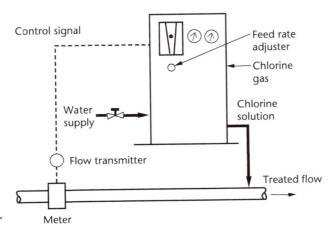

Figure 18-7 Residual control of chlorine feeder

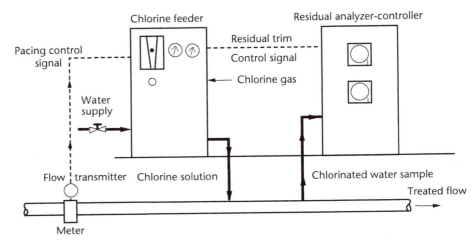

Figure 18-8 Combined flow and residual control of a chlorine feeder

demand will vary). The only problem with this system is that it tends to overreact when there are abrupt changes in flow rate, causing wide swings in the residual until it eventually stabilizes.

Combined Flow and Residual Control

When both the flow rate and the chlorine demand vary, the best type of system is one that uses combined flow and residual control (sometimes referred to as a *compound PID loop*, see Figure 18-8). In this case, the chlorinator is designed primarily to vary feed based on a signal from the flowmeter. If the flow rate is doubled, the amount of chlorine fed is immediately doubled. The analyzer then monitors the residual and varies the chlorinator feed rate, mostly in relation to the chlorine demand of the water.

Computerization

Over the past 20 years, computers have become nearly ubiquitous in water systems. It is also much easier to find plant personnel who are familiar with different types of computer equipment and will make full use of their capability.

One of the first uses of computers in water systems was "data logging" (i.e., keeping track of system information). The intention was to eliminate most or all of the pressure, level, and other types of charts in a plant, and instead to record all operating data in one place on an automatic printer. In addition to keeping records for reports and future review, the system could sound an alarm and provide a special printout whenever certain parameters exceeded the desired limits.

Further degrees of computerization have included computer control of specific equipment or processes and storage of plant records, such as piping diagrams, equipment operating and repair instructions, and maintenance records.

The final step has been complete computer control of treatment plants and distribution system pressure. Remote monitoring software now allows treatment plant personnel to monitor processes remotely, such as at their homes. This software can send an alarm via pager or cellular phone to the operator or monitoring company. This technology is especially suited for membrane plants, such as microfiltration membranes facilities. In some cases, personnel are still available to monitor the process. At other plants, the computer is allowed to operate the system without personnel during a night shift. Within limits, the computer is capable of starting and stopping pumps, adjusting chemical feed rates, and performing other functions. This operation, of course, requires a high degree of sophistication and reliability on the part of automatic analysis equipment, as well as a fail-safe capability. Systems operating without any on-site personnel are equipped to shut down if necessary and alert supervisors by phone when there are problems that the computer cannot correct.

State authorities are understandably cautious about allowing complicated treatment systems to function without an operator present. Only a human operator can determine certain elements of plant operation, such as those involving sight, sound, or smell. State authorities still have to be thoroughly convinced that there is no public health danger that could result from the failure of an automation system.

Particularly good sources of additional information on water system automation are AWWA Manual M2, *Instrumentation and Control*, and *WSO: Water Transmission and Distribution*.

 WATCH THE VIDEO
Instrumentation and Control (www.awwa.org/wsovideoclips)

Study Questions

1. Which type of continuous control will maintain control action when signal is lost?
   a. Floating proportional control
   b. Proportional control
   c. Proportional plus reset control
   d. Proportional plus reset plus derivative control

2. Which type of continuous control requires meters, transmitters, and transducers?
   a. Feedback control
   b. Feedforward control
   c. Floating control
   d. Proportional control

3. Which type of continuous control has a controller that constantly tries to change the output if there is deviation from the set point?
   a. Proportional plus reset plus derivative control
   b. Proportional control
   c. Floating proportional control
   d. Proportional plus reset control

4. Which flow measuring device is mechanical?
   a. Weir
   b. Flume
   c. Venturi
   d. Loss of head meter

5. Which of the following is *not* a component of a Bourdon tube pressure gauge?
   a. Pointer
   b. Pinion gear
   c. Bellows
   d. Slide

6. Only a human operator can determine certain elements of plant operation, such as those involving
   a. water pressure.
   b. water temperature.
   c. sight, sound, or smell.
   d. chlorine levels.

7. In controlling corrosion, which of the following should be continuously monitored and recorded as water enters the distribution system?
   a. pH of the water
   b. Water color
   c. Water temperature
   d. Water pressure

8. What is usually the most important pressure reading taken in a treatment plant?

9. What is the simplest form of automation, intended only to turn the chlorinator on and off at the same time as a pressure pump?

Chapter 19
Centrifugal Pumps

There are many types of pumps used to move water. Two primary classifications are velocity and positive displacement pumps. Positive displacement pumps are used mostly for precisely feeding liquid chemicals. Velocity pumps use a spinning impeller to accelerate water to high velocity within the pump casing. There are two main kinds of velocity pumps widely used in water systems: centrifugal and turbine. Although both of these are commonly encountered in water systems, centrifugal pumps find the most utility because of their low cost and versatility.

Operation of Centrifugal Pumps

The procedures for centrifugal pump operation vary somewhat from one brand of pump to another. The manufacturer's specific recommendations should be consulted before operating any unit. The procedures described in this chapter are typical and will serve as a guide if manufacturer's instructions are not available.

Pump Starting and Stopping

A major consideration in starting and stopping large pumps is the prevention of excessive surges and water hammer in the distribution system. Large pump-and-motor units have precisely controlled automatic operating sequences to ensure that the flow of water starts and stops smoothly. General procedures for starting pumps are as follows:

- Check pump lubrication.
- Prime the pump:
  - Where head exists on the suction side of the pump, open the valve on the suction line and allow any air to escape from the pump casing through the air cocks.
  - Where no head exists on the suction side and a foot valve is provided on the suction line, fill the case and suction pipe with water from any source, usually the discharge line.
  - Where no head exists on the suction side and no foot valve is provided, the pump must be primed by a vacuum pump or ejector operated with steam, air, or water.
- After priming, start the pump with the discharge valve closed.
- When the motor reaches full speed, open the discharge valve slowly to obtain the required flow. To avoid water hammer, do not open the valve suddenly.

centrifugal pump
A pump consisting of an impeller on a rotating shaft enclosed by a casing that has suction and discharge connections. The spinning impeller throws water outward at high velocity, and the casing shape converts this high velocity to a high pressure.

- Avoid throttling the discharge valve; this wastes energy.
- Before shutting down the pump, close the discharge valve slowly to prevent water hammer in the system.

Pump Starting

Centrifugal pumps do not generate any suction when dry, so the **impeller** must be submerged in water for the pump to start operating. If a pump is located above water level, a foot valve is often provided on the suction piping to hold the pump's prime (i.e., to keep some startup water in the pump). The foot valve, which is a type of check valve, prevents water from draining out of the pump when the pump is shut down.

Pump prime can also be maintained by placing a vacuum connection connected to both the pump suction and the high point of the pump, as illustrated in Figure 19-1. The priming valve automatically removes any air that accumulates and keeps the pump completely full of water at all times.

Controlling water hammer is important when a pump is being started. Large pumps are furnished with a valve on the discharge that is opened slowly after the pump gets up to speed. As a result, the surge of water does not produce a serious shock in the distribution system.

Pump Stopping

A check valve is usually installed in the discharge piping of small pumps to stop flow immediately after the pump stops. This immediate stoppage will prevent reverse flow through the pump. However, the sudden shutdown of a pump may cause water hammer. Relief valves or surge chambers may be installed to absorb the pressure shock.

On large pumps, smooth shutdown is ensured by closing the discharge valve slowly while the pump is still running and then shutting off the pump just as the valve finally closes. In this manner, the pumping unit is eased off the system. Some form of power-activated valve is necessary to obtain slow valve closure. Figure 19-2 shows power-operated discharge valves installed on large pumps. The valve operators are equipped with handwheels for manual operation in the event of a power failure.

Whenever a power failure occurs, the motor will stop while the discharge valve is still open, so there must be a way for the valve to close very rapidly before the pump reverses itself and begins to run backward. Battery power or

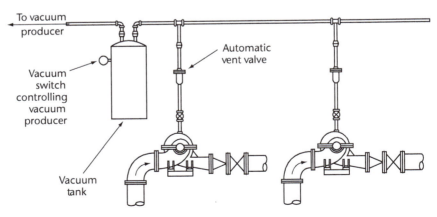

Figure 19-1 Vacuum-controlled central automatic priming

Reproduced with permission of McGraw-Hill Companies from *Pump Handbook* by Karassik et al., 2001. Published by McGraw-Hill.

impeller
The rotating set of vanes that forces water through a pump.

Figure 19-2 Pump discharge valves
Courtesy of Henry Pratt Company.

an emergency hydraulic system is usually provided to operate the valves in an emergency.

Whenever a pump must be shut down for more than a short period in freezing weather, the pump and exposed suction and discharge piping must be drained of water to prevent freezing. If the pump will be out of service for an extended time, the pump and motor bearings should be flushed and regreased, and packing should be removed from the stuffing box. The units should also be covered to prevent moisture damage to the motor windings and bearings.

 WATCH THE VIDEO
Pumps: Starting and Stopping (www.awwa.org/wsovideoclips)

Flow Control

Pumps are usually operated at constant speed. System pressure is controlled by having various sizes of pumps start or stop as necessary. Other ways to control the flow rate include throttling the discharge valve and using variable-speed motors or pump drives.

A major disadvantage of cycling pumps on and off as a means of controlling output is excessive motor wear. Medium-sized motors should not be cycled (i.e., started and stopped) more frequently than every 15 minutes, and larger motors should be cycled even less frequently. Frequent starting also increases power costs. Frequent cycling of pumps is an indication that the system probably does not have adequate distribution system storage.

Throttling the discharge valve in an attempt to approximate the required system flow should be done only when elevated storage is not available or when other, smaller pumping units are out of operation. In general, throttling should be avoided because it wastes energy. It is also necessary to make sure the valves used for throttling are appropriate for this purpose. Gate valves should not be used for throttling because the gate is loose in its guides and will vibrate when it is not fully open or shut. The best valves for throttling are plug, ball, self-actuating, or altitude valves. Butterfly valves can be used for throttling for short periods of time, but extended use may damage them.

If the system design requires continually varying pump discharge rates, variable-speed drives should be provided. Numerous variable-speed package drives are on the market, including continuously variable and stepped-speed motors, as well as constant-speed motors driving variable-speed electrical, hydraulic, and mechanical speed reducers coupled to the pump. Pumps can also be driven by variable-speed motors, which have either variable-voltage or variable-frequency controls.

Monitoring Operational Variables

A primary requirement at every pumping station is to measure the amount of water pumped and provide a record of water delivered to the system. It is also usually necessary to monitor pressure in the system and elevated tank levels as a way to control pump operation. Pumping station production records also provide the basis for the plant maintenance schedule. Past records are usually reviewed to determine the need for equipment replacement.

Suction and Discharge Heads

Pressure gauges should be connected to both the suction and discharge sides of a pump at the pressure taps supplied on the pump. The gauges should be mounted in a convenient location so the operator can frequently check pump performance. The pressure readings can also be electronically transmitted to a control room.

Bearing and Motor Temperature

The most common way to check bearing and motor temperatures in a small- to medium-sized plant is by feel. Experienced operators check pump operation by putting a hand on the motor and the pump bearing surfaces. They know how warm the surfaces should be. If a surface is substantially hotter than normal, the unit should be shut down and the cause of excessive heat investigated.

Special thermometers or temperature indicators are also available for monitoring the temperature at critical points in the pump and motor. It is particularly wise to have these monitors installed on equipment at unattended pump stations. The monitors will sound an alarm and automatically shut down the unit if the temperature gets too high.

Vibration

As with temperature, experienced operators get to know the normal feel and sound of each pump unit. They should investigate any change they notice. Vibration detectors are sometimes used on large pump and motor installations to sense equipment malfunctions, such as misalignment and bearing failure, that will cause excessive vibration. The detectors can also be used to shut down the unit if vibration increases beyond a preset level.

Speed

Monitoring the pump speed of variable-speed pumps is important because these pumps may experience cavitation (the creation of vapor bubbles) at low speeds. Centrifugal-speed switches can be installed on the pumping unit or contacts can be provided on a mechanical speed-indicating instrument to sound alarms or shut off the system if the speed goes too high or too low. Other systems use a tachometer generator that generates a voltage in proportion to speed. This voltage is used to drive a standard indicator near the pump or at a remote location. Underspeed and overspeed alarms can be activated by the speed-sensing device.

General Observations

An operator should also monitor surge-tank air levels, recording meters, and intake-pipe screens. Pumps with packing seals should be adjusted so that there is always a small drip of water leaking around the pump shaft. Idle pumps should be started and run weekly. All operations and maintenance should be recorded in log books.

Finally, operators must remain attentive to the general condition of the pump on a day-to-day basis. Unusual noises, vibrations, excessive seal leakage, hot bearings or packing, or overloaded electric motors are all readily apparent to the alert operator who is familiar with the normal sound, smell, sight, and feel of the pump station. Reporting and acting on such problems immediately can prevent major damage that might occur if the problem were allowed to remain until the next scheduled maintenance check.

Centrifugal Pump Maintenance

A regular inspection and maintenance program is important in maintaining the condition and reliability of centrifugal pumps. Bearings, seals, and other parts all require regular adjustment or replacement because of normal wear. General housekeeping is also important in prolonging equipment life.

Mechanical Details of Centrifugal Pumps

Size and construction may vary greatly from one volute-type centrifugal pump to another, depending on the operating head and discharge conditions for which the pumps are designed. However, the basic operating principle is the same. Water enters the impeller eye from the pump suction inlet. There it is picked up by curved vanes, which change the flow direction from axial to radial. Both pressure and velocity increase as the water is impelled outward and discharged into the pump casing. The major components of a typical volute-type centrifugal pump are described in the following paragraphs.

Casing

Water leaving the pump impeller travels at high velocity in both radial and circular directions. To minimize energy losses due to turbulence and friction, the casing is designed to convert the velocity energy to additional pressure energy as smoothly as possible. In most water utility pumps, the casing is cast in the form of a smooth volute, or spiral, around the impeller. Casings are usually made of cast iron, but ductile iron, bronze, and steel are usually available on special order.

Single-Suction Pumps

Single-suction pumps are designed with the water inlet opening at one end of the pump and the discharge opening placed at right angles on one side of the casing. Single-suction pumps, also called end-suction pumps, are used in smaller water systems that do not have a high volume requirement. These pumps are capable of delivering up to 200 psi (1,400 kPa) pressure if necessary, but for most applications they are usually sized to produce 100 psi (700 kPa) or less.

The impeller on some single-suction pump units is mounted on the shaft of the motor that drives the pump, with the motor bearings supporting the impeller (Figure 19-3A). This arrangement is called the close-coupled design. Single-suction pumps are also available with the impeller mounted on a separate shaft, which is connected to the motor with a coupling (Figure 19-3B). In this design, known as the frame-mounted design, the impeller shaft is supported by bearings placed in a separate housing, independent of the pump housing.

The casing for a single-suction pump is manufactured in two or three sections or pieces. All housings are made with a removable inlet-side plate or cover, held

Figure 19-3 Single-suction pumps

A. Close-coupled pump

B. Frame-mounted pump

in place by a row of bolts located near the outer edge of the volute. Removing the side plate provides access to the impeller. The pump does not have to be removed from its base for the side plate to be removed. However, all suction piping must be removed to provide sufficient access.

Some manufacturers cast the volute and the back of the pump as a single unit. Other manufacturers cast them as two separate pieces, which are connected by a row of bolts, similar to the inlet side plate. In units with separate backs, the impeller and drive unit can be removed from the pump without having to disturb any piping connections.

Double-Suction Pumps

Water enters the impeller of a double-suction pump from two sides and discharges outward from the middle of the pump. Although water enters the impeller from each side, it enters the housing at one location (usually on the opposite side of the discharge opening). Internal passages in the pump guide the water to the impeller suction and control the discharge water flow.

The double-suction pump is easily identified because of its casing shape (Figure 19-4). The motor is connected to the pump through a coupling, and the pump shaft is supported by ball or roller bearings mounted external to the pump casing.

The double-suction pump is usually referred to as a horizontal split-case pump. The term horizontal does not indicate the position of the pump. It refers to the fact that the housing is split into two halves (top and bottom) along the center line of the pump shaft, which is normally set in the horizontal position. However, some horizontal split-case pumps are designed to be mounted with the driveshaft in a vertical position, with the drive motor placed on top. Double-suction pumps can pump over 10,000 gpm (38,000 L/min), with heads up to 350 ft (100 m). They are widely used in large systems.

Removing the bolts that hold the two halves of the double-suction casing together makes it possible to remove the casing's top half. Most manufacturers

Figure 19-4 Double-suction pump casing shape

Courtesy of Ingersoll-Dresser Pump Company.

place two dowel pins in the bottom half of the casing to ensure proper alignment between the halves when they are reassembled. It is important that the machined surfaces not be damaged when the halves are separated.

Impeller

Most pump impellers for water utility use are made of bronze, although a number of manufacturers offer cast iron or stainless steel as alternative materials. The overall impeller diameter, width, inlet area, vane curvature, and operating speed affect impeller performance and are modified by the manufacturer to attain the required operating characteristics. Impellers for single-suction pumps may be of the open, semiopen, or closed design, as shown in Figure 19-5. Most single-suction pumps in the water industry use impellers of the closed design, although a few have semiopen impellers. Double-suction pumps use only closed-design impellers.

 WATCH THE VIDEO
Pumps: Centrifugal Pumps (www.awwa.org/wsovideoclips)

Wear Rings

In all centrifugal pumps, a flow restriction must exist between the impeller discharge and suction areas to prevent excessive circulation of water between the two. This restriction is made using wear rings. In some pumps, only one wear ring is used, mounted in the case. In others, two wear rings are used, one mounted in the case and the other on the impeller. The wear rings are identified in Figure 19-6.

The rotating impeller wear ring (or the impeller itself) and the stationary case wear ring (or the case itself) are machined so that the running clearance between the two effectively restricts leakage from the impeller discharge to the pump suction. The clearance is usually 0.010–0.020 in. (0.25–0.50 mm). Rings are normally machined from bronze or cast iron, but stainless-steel rings are available. The machined surfaces will eventually wear to the point that leakage occurs, decreasing pump efficiency. At this point, the rings need to be replaced or the wearing surfaces of the case and impeller need to be remachined.

Shaft

The impeller is rotated by a pump shaft, usually machined of steel or stainless steel. The impeller can be secured to the shaft on double-suction pumps using a key and a very tight fit (also called a shrink fit). Because of the tight fit, an arbor press or gear puller is required to remove an impeller from the shaft.

In end-suction pumps, the impeller is mounted on the end of the shaft and held in place by a key nut. The end of the shaft may be machined straight or with

Semiopen   **Closed**

Figure 19-5 Types of impellers
Courtesy of Goulds Pumps, ITT Industries.

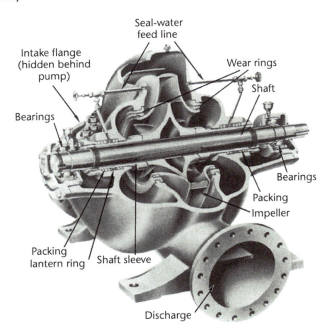

Figure 19-6 Double-suction pump

Courtesy of Ingersoll-Dresser Pump Company.

a slight taper. However, removing the impeller usually will not require a press. Several other methods are also used for mounting impellers.

Shaft Sleeves

Most manufacturers provide pump shafts with replaceable sleeves for the packing rings to bear against. If sleeves are not used, the continual rubbing of the packing can eventually wear out the shaft, which would require replacement. A shaft could be ruined almost immediately if the packing gland were too tight. Where shaft sleeves are used, operators can repair a damaged surface by replacing the sleeve, a procedure considerably less costly than replacing the entire shaft. The sleeves are usually made of bronze alloy, which is much more resistant than steel to the corrosive effects of water. Stainless-steel sleeves are usually available for use where the water contains abrasive elements.

Packing Rings

To prevent leakage at the point where the shaft protrudes through the case, either packing rings or mechanical seals are used to seal the space between the shaft and the case. Packing consists of one or more (usually no more than six) separate rings of graphite-impregnated cotton, flax, or synthetic materials placed on the shaft or shaft sleeves (Figure 19-7). Asbestos material, once common for packing, is no longer used on potable water systems. The section of the case in which the packing is mounted is called the stuffing box. The adjustable packing gland maintains the packing under slight pressure against the shaft, stopping air from leaking in or water from leaking out.

To reduce the friction of the packing rings against the pump shaft, the packing material is impregnated with graphite or polytetrafluoroethylene to provide a small measure of lubrication. It is important that packing be installed and adjusted properly.

Lantern Rings

When a pump operates under suction lift, the impeller inlet is actually operating in a vacuum. Air will enter the water stream along the shaft if the packing does not provide an effective seal. It may be impossible to tighten the packing sufficiently

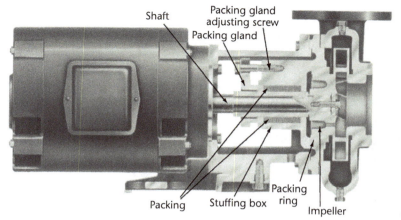

Figure 19-7 Pump packing locations

Courtesy of Aurora Pump.

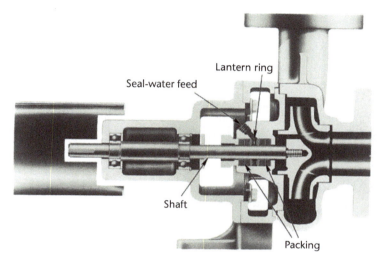

Figure 19-8 Lantern ring placed in the stuffing box

Courtesy of Aurora Pump.

to prevent air from entering without causing excessive heat and wear on the packing and shaft or shaft sleeve. To solve this problem, a lantern ring (Figure 19-8) is placed in the stuffing box. Pump discharge water is fed into the ring and flows out through a series of holes leading to the shaft side of the packing. From there, water flows both toward the pump suction and away from the packing gland. This water acts as a seal, preventing air from entering the water stream. It also provides lubrication for the packing.

Mechanical Seals

If the pump must operate under a high suction head (60 psig [400 kPa (gauge)] or more), the suction pressure itself will compress the packing rings, regardless of operator intervention. Packing will then require frequent replacement. Most manufacturers recommend using a mechanical seal under these conditions, and many manufacturers use mechanical seals for low-suction-head conditions as well. The mechanical seal (Figure 19-9) is provided by two machined and polished surfaces: one is attached to and rotates with the shaft; the other is attached to the case. Contact between the seal surfaces is maintained by spring pressure.

The mechanical seal is designed so that it can be hydraulically balanced. The result is that the wearing force between the machined surfaces does not vary regardless of the suction head. Most seals have an operating life of 5,000 to 20,000 hours. In addition, there is little or no leakage from a mechanical seal; a leaky mechanical seal indicates problems that should be investigated and repaired. A major advantage of mechanical seals is that there is no wear or chance of damage to shaft sleeves.

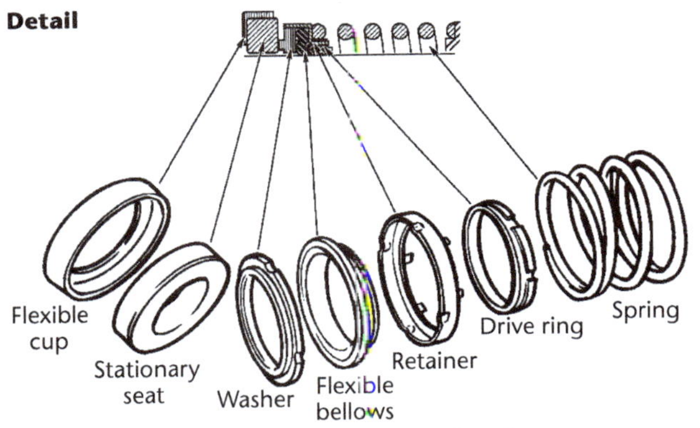

Detail

Flexible cup

Stationary seat

Washer

Flexible bellows

Retainer

Drive ring

Spring

Parts of a mechanical seal

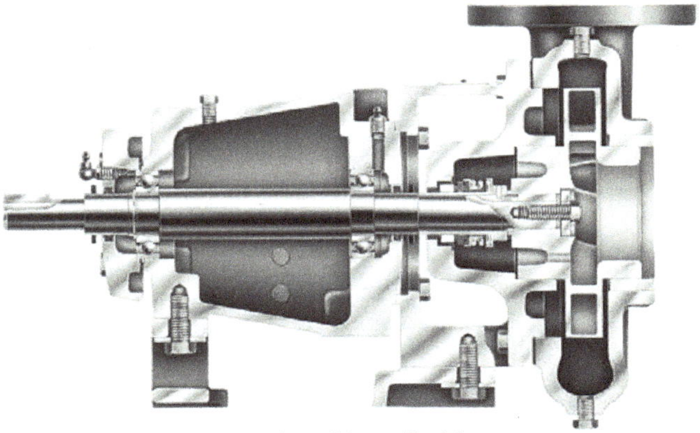

Mechanical seal installed in a pump

Figure 19-9 Mechanical seal parts and placement

Courtesy of Aurora Pump.

A major disadvantage of mechanical seals is that they are more difficult to replace than packing rings. Replacing the mechanical seal often requires removing the shaft and impeller from the case. Another disadvantage is that failure of a mechanical seal is usually sudden and accompanied by excessive leakage. Packing rings, by contrast, normally wear gradually, and the wear can usually be detected long before leakage becomes a problem. Mechanical seals are also more expensive than packing.

Bearings

Most modern pumps are equipped with ball-type radial and thrust bearings. These bearings are available with either grease or oil lubrication and provide good service in most water utility applications. They are reasonably easy to maintain when manufacturer's recommendations are followed, and new parts are readily available if replacement is required. Ball bearings will usually start to get noisy when they begin to fail, enabling operators to plan a shutdown for replacement.

Couplings

Frame-mounted pumps have separate shafts connected by a coupling. The primary function of couplings is to transmit the rotary motion of the motor to the pump shaft. Couplings are also designed to allow slight misalignment between the pump and motor and to absorb the startup shock when the pump motor is switched on. Although the coupling is designed to accept a little misalignment, the more accurately the two shafts are aligned, the longer the coupling life will be and the more efficiently the unit will operate (Figure 19-10).

Various coupling designs are supplied by pump manufacturers. Couplings may be installed dry or lubricated. Most couplings are of the lubricated style and

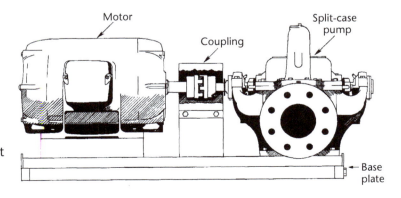

Figure 19-10 Alignment of motor and pump at coupling

require periodic maintenance, usually lubrication at 6-month or annual intervals. Dry couplings using rubber or elastomeric membranes do not require any maintenance, except for periodic visual inspection to make sure they are not cracking or wearing out. The rubber or elastomer used for the membrane must be carefully selected for the pump, because the corrosive chemicals used in water treatment plants could affect the life and operation of the coupling.

 WATCH THE VIDEO
Pumps: Seals, Rings, Bearings (www.awwa.org/wsovideoclips)

Inspection and Maintenance

A well-defined inspection and maintenance schedule is necessary to ensure long and reliable pump service and to preserve warranty rights on new units. It is important to maintain complete records of inspections and any service performed. An operator can best evaluate pump condition by comparing a pump's current performance to its performance when it was first installed. Therefore, complete testing immediately following installation is important.

The following should be checked periodically to ensure maximum operating efficiency and minimum maintenance expenditures:

- Priming
- Packing and seals
- Bearings
- Vibration
- Alignment
- Sensors and controls
- Head (pressure gauges)
- Cavitation

Priming

Pumps must be checked before startup to be sure they are primed. Capacity will be reduced and water-lubricated internal wear rings may be damaged if a pump is only partially primed. If a pump is primed from an overhead tank or other gravity supply, it may be started as soon as water shows at the top vent cocks. If it is primed by a vacuum pump, the action of the device itself will indicate when the casing is filled with water. Some pumps are equipped with a float switch that will not complete the circuit for starting the motor unless the pump is completely primed.

If the pump does not have a vacuum priming system, petcocks on top of the pump case should be opened routinely during operation to bleed off any air that might collect there. Continuous bleeds or air-release valves should be installed on pumps at unattended pump stations. All valves in the suction line must be fully open.

Packing

A correctly packed and adjusted stuffing box should be trouble-free. Packing should be inspected annually if a pump is run on a regular basis. The easiest way to do this is to remove the top of the casing or the packing gland. When the packing wears or is compressed to the point where it is impossible to tighten the gland further, a new set of rings should be installed. It is generally not considered good practice to make up for wear and compression by adding new rings on top of old ones. However, manufacturer's maintenance specifications may allow the addition of one more ring.

New packing must be installed with care to extend its life. The following guidelines should be observed:

- During disassembly, keep parts in the order in which they are removed, including the number of rings before and after the lantern ring.
- Clean the stuffing box.
- Check the shaft sleeve. Replace it if it is badly worn.
- Use packing and sleeve material that is the proper size and compatible with the expected service. For a severely abrasive or corrosive service, consult the pump manufacturer. Although precut packing is available, most operators purchase bulk packing and cut what they need for each job.
- Replace parts in the opposite order that they were removed.
- Replace all packing rings. The ends of the strips used to form rings may be cut diagonally or square. The ends should be carefully butted, with joints staggered at least 90 degrees. Four or more rings are usually placed with their ends at 90-degree intervals. Be sure to replace the lantern ring in its original sequence.
- Never overtighten the packing. Draw each ring down firmly. After all rings are installed, back off on the gland nuts about one full turn before the pump is started. Before its initial startup with new packing, the pump should rotate fairly easily by hand, unless it is very large. If it does not turn over, the packing should be loosened until it does, provided there are no other problems.
- At startup, the packing should be allowed to leak freely until conditions stabilize and it is evident that there are no hot spots in the area of the packing or bearings. Then gradually tighten the packing gland until the packing allows a slow drip.
- While the pump is running, adjust the gland by tightening the gland-bolt nuts. Tighten the nuts evenly, and tighten each nut no more than one-sixth of a turn every 20–30 minutes.
- Never tighten packing glands to the point where there is no leakage. Doing so will cause premature packing wear and scored shaft sleeves.
- After the initial installation and adjustment, check the packing regularly and adjust the gland whenever leakage increases. The leakage rate should be checked daily if possible.
- If cooling or sealing water is injected into the box, set its flow pressure to 15–20 psi (100–140 kPa) more than the pressure on the inboard end of the stuffing box. Excessive pressure will cause increased wear of the packing and sleeve.

Mechanical Seals

The operating temperature of a mechanical seal should never exceed 160°F (71°C). If there is a possibility that a seal will exceed this limit, it should be water cooled. The water can be supplied by the pump discharge or from an external source. The water must not contain dirt, grit, or other abrasive materials that could damage the seal. If, for instance, the pump is used for pumping raw water, seal water must be supplied from the filtered water system with adequate precautions taken to prevent a cross-connection.

It is important that the mechanical seal be designed for the stuffing-box pressure at which it will operate. A pump that develops a partial vacuum on the suction side must be fitted with a close-clearance bushing between the seal and the suction passage of the pump in order to maintain lubrication, cooling, and flushing fluid at the seal.

Bearings

Regular inspection and lubrication of bearings is essential to efficient pump and motor operation. Lubrication points should be checked and lubricated at the intervals prescribed by the manufacturer. The following checks are important:

- Oil level in bearing housings
- Free movement and proper operation of oil rings
- Proper oil flow for pressure-feed systems
- Proper type of grease for grease-lubricated bearings
- Proper amount of grease in the housing
- Bearing temperature

Oil-Lubricated Bearings   The bearing housing of oil-lubricated bearings should be kept filled with a good grade of filtered mineral oil. The oil should be changed after the first month of operation of a new pump. After the first oil change, oil changes should be performed every 6–12 months, depending on operating frequency and environmental conditions. Whenever oil is changed, and especially after the first oil change, the oil should be inspected for signs of bearing wear or excessive dirt. The following viscosities are recommended for use with various antifriction bearings:

- Ball and cylindrical roller bearings—70 Saybolt standard units (SSU) oil rated at operating temperature.
- Spherical roller bearings—100 SSU oil rated at operating temperature.
- Spherical roller and thrust bearings—150 SSU oil rated at operating temperature.

It is important that the bearing housing not be overfilled with oil. Most housings have the correct oil level indicated on a sight glass.

Grease-Lubricated Bearings   For grease-lubricated bearings, similar maintenance is required. During the initial run-in period, bearing temperatures must be closely monitored. The housing of bearings that are within an acceptable temperature range can be touched with the bare hand. After about 1 month of initial full-service operation, all bearings should be regreased.

Bearings should not be overgreased. Because of the internal friction caused by the churning grease, a bearing will run hotter if the grease pocket is packed too tightly. The grease used for lubricating antifriction pump bearings should be a sodium soap-base type that meets American Bearing Manufacturers Association group 1 or 2 classifications. Bearings should be regreased every 3–6 months according to the following procedures:

1. Open the grease drain plug at the bottom of the bearing housing.
2. Fill the bearing with new grease until grease flows from the drain plug.
3. Run the pump with the drain plug open until the grease is warm and no longer flows from the drain.
4. Replace the drain plug.

Grease does not need to be added to bearings between intervals unless the bearing seals are bad and grease has been lost. If bearings run hot, it is usually because the grease is packed in too tightly. In this case, the grease drain plug should be opened and some grease allowed to drain out. If the bearing has been disassembled, cleaned, and flushed, the housing should be refilled to one-third of its capacity.

Bearing Replacement   The life of a ball bearing will vary with the conditions of load and speed. Most pump bearings have a minimum operating life of 15,000 hours (about 1½ years), but they may last much longer. As a general rule, if a pump has been operating continuously for 1 or 2 years, the bearings should be replaced if the pump is taken out of service for any other repairs. It is more economical to replace bearings while other work is being performed than to wait until excessive heat or noise warns that the bearings are about to fail.

Operators should be extremely careful when removing or installing bearings. Appropriate bearing pullers and hydraulic presses should be used, especially if there is a chance that bearings will have to be reused. It is often good practice to leave serviceable bearings mounted until replacements have been obtained. Proper installation procedures, as directed by the manufacturer, should be followed to prevent damage to new bearings during installation.

Vibration Monitoring

On high-speed or large pump units, operators may use instruments to periodically monitor vibration in the vertical, lateral, and axial planes. These observations can give an early indication of possible future problems. When problems occur, an experienced operator will usually be able to detect undesirable vibration merely by listening to or touching the unit. In general, on units for which periodic vibration checks will be performed, a vibration test should be conducted immediately after the pump is installed to establish a baseline condition. This test should be followed by periodic measurements of vibration at intervals recommended by the manufacturer. For critical equipment that is operated continuously, monthly checks may be required. Less critical equipment may be checked annually.

Vibration sensors may be installed on pumping units or portable equipment may be used to make the measurements. Vibration readings should be taken on the shafts in or near the bearings. Any change from the baseline measurements of vibration magnitude or frequency indicates potential problems.

When operators observe vibration or overheating of a bearing or coupling, they should inspect the pump's shaft coupling and impeller immediately for possible imbalance. The imbalance could result from the presence of foreign material caught in the impeller, the accumulation of scale, mechanical breakage, or loss of metal due to corrosion or cavitation.

Alignment

Excessive vibration may also be caused by misalignment. Alignment should be checked in pumping units after they are first installed and brought up to operating temperature (often called a "hot alignment" check). This check is particularly important for frame-mounted pumps. A record of the initial readings should be made using a dial indicator gauge, as shown in Figure 19-11. Alignment should then be checked again periodically to ensure that the initial readings have not changed. Vibration due to misalignment is a common cause of bearing failures.

Sensors and Controls

Operating personnel should know how and why each part of an automatic control system functions. They should understand how each part affects the operation

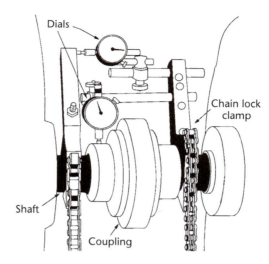

Figure 19-11 Dial indicator alignment gauge

of the system. The normal sequence and timing of each operation should be determined and any deviations identified for correction. Technicians familiar with each type of equipment should repair and adjust all of the more complex control devices. Pressure and float sensors may require adjustment seasonally or with changes in water demand.

Head

Pump discharge and suction heads should be checked and compared with baseline performance figures for when the pump was first installed. Some wear will necessarily reduce performance, but major reductions in capacity should be identified and corrected before the inefficient unit begins to waste large amounts of energy. Strip or circular chart recordings of pump performance can be helpful in identifying reduced pump capacity.

Operators can check an installed centrifugal pump for wear by closing the discharge valve and then reading the discharge pressure. This pressure can be compared with the original pump characteristics, after appropriate deductions are made for suction pressure. If the shutoff head is close to the original value, the pump is not greatly worn.

Cavitation

Under certain circumstances, a pump can pull water so hard that some of the water becomes small bubbles of vapor. This is called cavitation. The bubbles explode against the impeller, creating the sound of marbles in the pump. The consequences of cavitation may be only the bothersome noise, but over time the impeller will usually be damaged and the metal eroded.

To avoid cavitation, a pump should not be continuously operated at a rate much higher or lower than its designed rate. In addition, the suction requirements for the pump must be met. Turbine and submersible pumps must be provided with a deep enough sump to keep them submerged. Also, intake screens should be routinely cleaned if necessary to avoid suction restrictions.

Major Repairs

Major repair jobs that may be required for pumps at infrequent intervals include replacement of bearings, wearing rings, shaft sleeves, and impellers. These jobs require that the pump be removed from service for a period of time. Planning is essential to ensure that the work goes smoothly and quickly. If a pump is not too large, it can be returned to the manufacturer or the utility shop for repair. If the repair must be made in the field, it is especially important to have all parts on hand before the pump is removed from service.

Care should be taken to ensure that the correct replacement parts are used. If the identical part is not available, a substitute part may be used if it meets or exceeds the standards of the original part. The parts themselves or accurate specifications can be obtained from the pump manufacturer before the repairs are begun.

In addition to replacement parts, any necessary special tools should be on hand before the job is started. For example, bearing removal will usually require a bearing puller or a hydraulic press, possibly fitted with special collars.

Record Keeping

An adequate record of equipment specifications and maintenance performed will assist the operator in scheduling inspections and needed service work, evaluating pump equipment, and assigning personnel. An appropriate system could be based on data cards. Each card should list the make, model, capacity, type, date and location installed, and other information for both driver and driven unit. The remarks section should include the serial or part numbers of special components (such as bearings) that are likely to require eventual replacement.

A separate operating log should be kept that lists all pumping units along with a record of the operating hours. This record is an essential feature of any reasonable periodic service or maintenance schedule. In addition, a daily work record should be kept on each piece of equipment.

Many water systems now use a computer database to store these records, making the information easier to store and access.

Pump Safety

Specific pump safety precautions must be recognized and observed with respect to the following:

- Moving machinery
- Handling materials
- Using personal protection equipment
- Using hand tools
- Working with or near electrical devices
- Ensuring fire safety

Machinery should always be stopped before it is cleaned, oiled, or adjusted. The controlling switchgear should be locked out before any work begins, so that another person cannot start the machinery. A conspicuous tag should also be posted on or over the control panel, giving notice that the equipment is under repair and should not be restarted. The name of the person who locked the equipment out should also be listed.

Before a machine is restarted, check to be sure that all personnel are clear of danger and that working parts are free to move without damage. When disconnecting equipment, follow the manufacturer's instructions for disconnecting and securing drive and rotating equipment.

Guards over rotating parts should be secured in place to protect workers who are near equipment when it is in operation. Guards with hinged or movable sections should be provided where it is necessary to change belts, make adjustments, or add lubricants. If a guard or enclosure is within 4 in. (100 mm) of a moving part, the maximum opening in the screen should not exceed 1/2 in. (13 mm) across. (This will prevent an operator's fingers from getting too close to moving parts.) Guards placed more than 4 in. (100 mm) but less than 15 in. (380 mm) from a moving part can have openings no more than 2 in. (50 mm) across. The guard should be strong enough to provide complete safety, and guard structures should be constructed so that they cannot be pushed or bent against moving parts. Guards should be removed for maintenance only when the machinery is not in operation.

Study Questions

1. Which device applies an even pressure to the packing such that it compresses tight around the pump shaft?
   a. Lantern ring
   b. Mechanical seal
   c. Packing gland
   d. Seal cage

2. When using four packing rings, the rings should be staggered at
   a. 45 degrees.
   b. 90 degrees.
   c. 120 degrees.
   d. 180 degrees.

3. How is velocity head expressed mathematically?
   a. V/g
   b. V^2/g
   c. $V^2/2g$
   d. $V^2/32.2 \text{ ft} / \text{sec}^2$

4. Which type of centrifugal pump impeller is used for pumping medium-sized solids?
   a. Open
   b. Closed
   c. Semi-open
   d. Radial

5. The shaft's main function is to transmit _____ from the motor to the impeller.
   a. centrifugal force
   b. torque
   c. kinetic energy
   d. thrust

6. A major consideration in starting and stopping large pumps is the prevention of excessive surges and _____ in the distribution system.
   a. water hammer
   b. turbidity
   c. suction loss
   d. rust

7. The head of a double-suction pump can reach
   a. 350 ft (100 m).
   b. 100 ft (30 m).
   c. 1,000 ft (300 m).
   d. 500 ft (150 m).

8. What type of pump is designed with the water inlet opening at one end of the pump and the discharge opening placed at a right angle on one side of the casing?

9. What device is used when it is not possible to tighten the packing sufficiently to prevent air from entering the pump without causing excessive heat and wear on the packing and shaft or shaft sleeve?

10. The primary function of couplings is to transmit the rotary motion of the motor to what component?

Chapter 20
Treatment Plant Safety and Security Practices

Treatment Plant Safety Review

The discussion of plant safety presented in this chapter is not intended to be a comprehensive guide on the subject. It is the responsibility of utility personnel to follow safety regulations and to be properly trained to perform their duties safely. The information provided in this chapter should be adequate for most operator certification test examinations.

Safety and safe working conditions are the responsibility of each individual associated with a water utility. The degree of responsibility varies among upper management, supervisors, and operations personnel, but all play a role and all are responsible for their individual actions. Safety regulations guide those actions, but the creation of policies, procedures, and methods to meet safety requirements is the result of a collaborative effort within the water utility.

Utility managers are responsible for ensuring safe working conditions, maintaining the physical condition of their facilities, and supporting policies that encourage the safe performance of work duties. Supervisors at all levels directly control all work conditions and are responsible for the activities of the personnel they supervise. Each supervisor, foreman, crew chief, or lead staffer is responsible for ensuring that all work is done in compliance with safety practices, utility policies, and regulations.

Employees have a particular responsibility when it comes to safety. That responsibility is to correctly use the safety equipment provided and follow all safety policies and procedures. All employees must be aware of the safety requirements and help guard against unsafe acts and conditions.

Safety Regulations

The primary reason for developing and maintaining safe working conditions and practices is to eliminate injuries. Beyond the personal cost to an injured employee are the costs to the utility from lost time, medical expenses, and possible legal judgments. Other considerations include damage to equipment and property and resulting repair costs, and the potential need to hire and train new employees to perform the work duties of the injured employee.

Occupational Safety and Health Administration Act

Another reason for developing safety policies and procedures is the federal Occupational Safety and Health Administration (OSHA) Act (Public Law 91-596). Passed in 1971, the act established OSHA and compiled numerous safety and health standards that are applicable to every industry. Specific standards have been developed for most work activities and chemical substances that an employee may be exposed to in the course of his or her employment. These are minimum standards that must be followed; their requirements are itemized in the Code of Federal Regulations (CFR) (29 CFR, 1910 and 1926). The act provides for monetary penalties, which are escalated depending on the seriousness of the safety violation, and the potential for incarceration if OSHA safety standards are violated.

If a specific requirement does not exist, OSHA relies on the general duty clause, which requires each employer to provide employment, and a place of employment, that is free of recognized hazards that are likely to cause death or serious physical harm. The general duty clause also requires employees to comply with safety practices and procedures. In addition to OSHA, many states and cities have their own safety requirements. In most cases, these standards mirror the OSHA standards; however, in some cases, state or local requirements are more stringent and carry additional penalties for failing to comply.

Confined-Space Rules

A common hazard encountered in the water industry is confined spaces. Examples of confined spaces in the water and wastewater industry include access holes for valves, meters, and air vents; chemical storage tanks or hoppers; wet wells; digesters; sedimentation basins; filters; and reservoirs.

One of the most sobering statistics related to confined-space safety is one relating to the death rate for those entering confined spaces. Although the death rate relating to confined spaces continues to decrease, the percentage of those who die while attempting a confined-space rescue remains roughly the same. Each year about two-thirds of those who die in confined spaces are would-be rescuers. The cause of death is usually asphyxiation or the result of an atmospheric hazard that could not be seen.

Requirements

Under OSHA regulations, there are two basic types of confined spaces—permit-required and nonpermit-required. For a location to be classified as a confined space, the following three criteria must be met: the space must have limited means of entry and exit, the space must not be designed for continuous human occupancy, and the space must be of a size and configuration that allows humans to enter the space to perform work. If all three criteria are met, it is necessary to assess the space and determine if the space is also permit-required. A permit-required confined space meets one or more of the following criteria:

- The space contains, or has the potential to contain, a hazardous atmosphere.
- The material within the space has the potential to engulf the entrant.
- The internal configuration is such that it could trap the entrant (i.e., downward-sloping and converging floors).
- The space contains any other recognized serious health or safety hazard.

If the only hazard is atmospheric and the hazard can be controlled with ventilation, the space may be reclassified as nonpermit-required. Figure 20-1 is a decision tree that can be used to identify the need for permitting a confined space.

A written program is required when it is necessary to enter permit-required confined spaces. This program must contain a mechanism for identifying and controlling the hazard, have a written entry permit, provide for the identification and labeling of confined spaces, and provide employee training, among other requirements. A key component of this program is the requirement that before an employee enters a permit-required confined space, the internal atmosphere of the

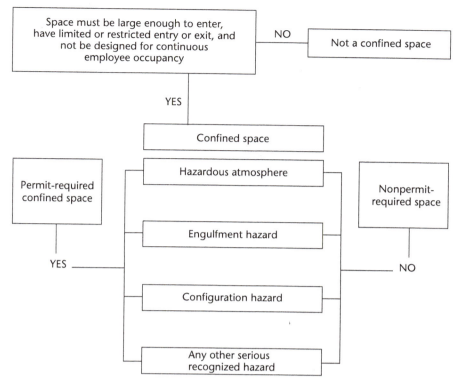

Figure 20-1 Confined-space decision tree

Source: Melinda Raimann, Cleveland, Ohio, Division of Water.

space be tested with a calibrated, direct-reading instrument capable of measuring oxygen content, presence of flammable gases and vapors, and potentially toxic air contaminants. In addition to these preentry tests, the atmosphere must be retested while work is being performed inside the space to ensure that acceptable conditions are being maintained. More details regarding the required program can be found in 29 CFR 1910.146.

An employer is required to provide its employees with the specialized equipment that is required for confined-space entry. This equipment includes the following:

- Testing and monitoring equipment
- Ventilation equipment (to eliminate or control atmospheric hazards)
- Communications equipment, as necessary
- Personal protective equipment (PPE)
- Lighting equipment, as necessary
- Barriers and shields to prevent unauthorized entry, as necessary
- Ladders for safe entry and exit
- Fall-prevention equipment required because of the difference in elevation
- Rescue and emergency equipment, including harnesses and hoists, self-contained breathing apparatus (SCBA), stretchers or backboards, and supplies

WATCH THE VIDEOS
Confined Spaces and Confined Spaces—Alternative Procedures
(www.awwa.org/wsovideoclips)

Personnel

There are three classes of employees, as they relate to confined spaces: the entry supervisor, the authorized entrant, and the authorized attendant.

The *entry supervisor* is responsible for knowing the conditions within the confined space, verifying that all equipment and procedures are in place prior to entry, verifying the availability of rescue services and the means of summoning them, terminating entries and canceling permits, and determining that acceptable conditions as specified in the permit continue for the duration of the entry.

The *authorized entrant* is responsible for knowing what hazards are to be faced, recognizing the signs and symptoms of exposure and understanding the potential consequences, knowing how to use any needed equipment, communicating with the attendant as necessary, alerting the attendant when a warning symptom or other hazardous condition exists, and quickly exiting when ordered or alerted.

Authorized attendants are responsible for being aware of the confined-space hazards and the behavioral effects of exposure; maintaining a count and the identity of the authorized entrants; preventing unauthorized entry; remaining outside the space until relieved; communicating with the entrants and monitoring activities inside and outside of the permit space; and ordering exit, summoning rescuers, and performing nonentry rescue as required.

General Plant Safety

Right to Know and MSDS

In order to follow safety practices and procedures that will protect an employee and fellow workers, an employee must be aware of the hazards to which he or she is exposed. A primary method of doing this is through a right-to-know program developed by the employer. The main premise of a right-to-know program is that employees have the right to be advised of the hazards associated with the chemicals and materials that they work with or that exist nearby in the course of their daily job activities.

Chemicals, as defined under the Hazard Communications standard (29 CFR 1910.1200), include those used for water treatment, lab procedures, cleaning, and maintenance applications. A **material safety data sheet (MSDS)** lists a chemical's identity, composition, and type or types; the hazard(s) associated with its use (is it flammable, an oxidizer, poisonous?); how it enters the body; the effects of exposure; health effect (short term, long term); the permissible exposure limit; how to handle spills or releases; and other related information. An MSDS for every chemical that is used in a water treatment plant should be kept in an organized fashion in a location that is readily accessible to all employees.

Employees are to be trained on the hazards of the chemicals they use or are exposed to and in the methods used to protect themselves from those hazards.

Risk Management and Emergency Response

The US Environmental Protection Agency (USEPA) issued the Risk Management Rule on July 20, 1996, in an effort to protect public health and safety. OSHA had previously released the Process Safety Rule in February 1992 (29 CFR 1910.119), which focused on employee health and safety but did not address the safety of nonemployees. According to the Risk Management Rule, employers must not store or use chemicals in quantities greater than the threshold quantity, which is defined by USEPA (e.g., more than 1,500 lb [680 kg] of chlorine), and must develop release scenarios, conduct hazard assessments, determine means and methods to limit public exposure should a spill or release occur, and provide the public with information about their plan.

---

material safety data sheet (MSDS)

A product description listing a chemical's identity, composition, and type or types; the hazard(s) associated with its use (is it flammable, an oxidizer, poisonous?); how it enters the body; the effects of exposure; health effect (short term, long term); the permissible exposure limit; how to handle spills or releases; and other related information.

Emergency response at a water treatment facility generally means responding to a chemical release. This response is governed under the Risk Management Rule and under OSHA (29 CFR 1910.1). A number of decisions must be made regarding emergency response, the first of which is to determine who will respond in the event of a chemical release. If the utility elects to have its employees respond to chemical releases, the utility is required to comply with the requirements of OSHA's Hazardous Waste Operations and Emergency Response (HAZWOPER) regulations.

Compliance requires the employer to provide specialized chemical response suits and respirators, medical exams to determine the fitness of the employees designated as responders, spill-containment materials, and annual training. Initial training is usually 24–40 hours in duration, depending on the responsibilities the response team is expected to assume; an annual 8-hour refresher course is also required. Even if a utility chooses to have others act as the designated emergency response team, the HAZWOPER regulations, as they relate to awareness-level training for those who may be exposed during a chemical release, are to be followed. This awareness-level training is an 8-hour course with an annual refresher that focuses on the hazards of the chemicals used and evacuation procedures, routes, and staging areas to be used.

A key component of any emergency response plan is for the responder to be aware of the hazards of the chemicals that may be released and to understand the precautions that must be taken to ensure the safety and well-being of all designated responders. This information is usually provided on the specific MSDS or as part of the facility's risk management or hazard communications program.

Standard practice for emergency response is to follow the incident response system first developed by fire and safety forces and designate someone in charge of the emergency site. Red (hot), yellow (decontamination), and green (clean) zones should be designated, and strict controls are used to ensure that no unauthorized or unprotected entrant enters the chemical-release area. A spill or release area should never be entered alone; one standby person, fully suited in the appropriate level of PPE, should be assigned to each entrant into the hot zone.

First Aid

Water treatment plants often provide first-aid stations for their employees. These stations are stocked with emergency supplies that are used when operators sustain minor injuries on the job. The stations (Figure 20-2) may include oxygen,

Figure 20-2 First-aid and safety station

Source: Conneaut, Ohio, Water Department.

bandages and antiseptics, fire blankets, and eyewash solutions. SCBA equipment may also be stored here if the location is near the chlorine area. Be sure to document use of all supplies immediately after use.

Chlorine Safety

Chlorine safety is a subset of chemical safety at many water treatment plants. Chlorine is a toxic chemical and is an irritant to the eyes, skin, mucous membranes, and respiratory system. Effects of exposure generally are evident first in the respiratory system and then in the eyes. The impact of exposure to chlorine is dependent on both concentration and time. The very young, the elderly, and people with respiratory problems are most susceptible to chlorine's effects. As the duration of exposure or the concentration increases, the affected individual may become apprehensive and restless, with coughing accompanied by throat irritation, sneezing, and excess salivation. At higher levels, vomiting associated with labored breathing can occur. In extreme cases, difficulty in breathing can progress to the point of death through suffocation. An exposed person with a preexisting medical or cardiovascular condition can have an exaggerated response.

Anyone exhibiting these symptoms should see a qualified health-care provider immediately, as his or her condition will deteriorate over the next few hours. The physiological effects from exposure to various levels of gaseous chlorine are presented in Table 20-1.

To protect employees who work with or around chlorine, the employer must ensure that chorine equipment is kept in safe working order, regular maintenance is performed on all chlorine equipment and the equipment used to handle it such as overhead cranes, and employees have the proper tools and PPE.

Table 20-1 Physiological effects of chlorine exposure

Exposure Level, ppm	Effects
0.02–0.2	Odor threshold (varies by individual)
<0.5	No known acute or chronic effect
0.5	ACGIH (American Conference of Governmental Industrial Hygienists Inc.) 8-hour time-weighted average
1.0	OSHA ceiling level
1–10	Irritation of the eyes and mucous membranes of the upper respiratory tract. Severity of symptoms depends on concentrations and length of exposure.
3	ERPG-2 (Emergency Response Planning Guidelines as values developed by the American Industrial Hygiene Association) is the maximum airborne concentration below which it is believed that nearly all individuals could be exposed for up to 1 hour without experiencing or developing irreversible or other serious health effects that could impair an individual's ability to take protective action.
10	Immediately dangerous to life and health (per the National Institute of Occupational Safety and Health)
20	ERPG-3 is the maximum airborne concentration below which it is believed that nearly all individuals could be exposed for up to 1 hour without experiencing or developing life-threatening health effects.

Source: The Chlorine Institute.

Chlorine cylinders of 150 lb (68 kg) are equipped with fusible plugs that will melt at temperatures between 157°F and 162°F (69–72°C) and release the contents of the cylinder (Figure 20-3). Consequently, cylinders must be stored in a temperature-controlled area. Labels identifying the contents of any cylinder or tank car should be legible and prominently displayed. Placards with Department of Transportation (DOT) codes identifying the contents and their hazards are to be posted on the exterior of chlorine rooms so that emergency response teams can identify and address the hazards they will be exposed to in the event of a release. All cylinders should be secured and protected from accidental contact by moving equipment or vehicles.

WATCH THE VIDEO

Chlorine Safety: Hazardous Properties (www.awwa.org/ wsovideoclips)

Personal Protective Equipment

OSHA requirements governing PPE are found in 29 CFR 1910 subpart I (see appendix E in the CFR), which also references various American National Standards Institute guidelines. As a general rule, PPE is to be considered as a last resort. The elimination of the hazard through engineering changes is the optimal solution, followed by administrative changes. If these methods do not eliminate or minimize the hazard to the appropriate level, PPE is to be used. The PPE must be appropriate for the hazard it is being provided to protect the employee against. This can be determined only through a hazard assessment and equipment evaluation. In general, PPE is provided by the employer, unless it is used voluntarily or by an employee outside the work environment. In addition to providing the equipment, the employer is responsible for ensuring that it is maintained in a sanitary and reliable condition.

Figure 20-3 Safely stored 150-lb (68-kg) chlorine cylinders. Caps are on and the chain is set.

Source: Conneaut, Ohio, Water Department

PPE can be provided for nearly every portion of the body. Occupational foot protection, or safety shoes, should be used when there is the potential for injury to the foot from falling, rolling, or piercing objects or when extreme thermal conditions exist. Different types of shoes or devices provide different types of foot protection. There are steel toe caps, both internal and external to the shoe; metatarsal guards; steel shanks; and various combinations of these devices. They are commonly worn in machine shops and during construction activities.

Another common piece of PPE is the hard hat, which is required when there is the potential for injury to the head from impact or falling objects. In a water treatment plant, hard hats are usually worn when an overhead crane is in use or during construction projects. They may be required if there are low vertical clearances and an employee has to crouch or duck to travel through the area, such as in a pipe gallery or subbasement location. There are also specific types of hard hats designed to provide protection from high-voltage shocks and burns.

Treatment plants typically have high-noise areas, usually on pump floors or in machine shops. If the employer is unable to lower the noise levels or limit employee exposure, hearing protection must be provided. Hearing protection comes in various forms, from earplugs to muffs, and offers varying degrees of protection. Noise-level sampling is required to determine what level of noise reduction is needed, which will be marked on the protective devices as a noise reduction reading. As a general rule of thumb, if an employee must raise his or her voice to be heard in the area in question, hearing protection is probably required. Exposures to noise levels above an 8-hour time-weighted average of 85 dBA (decibels measured on the A scale) require a hearing-conservation program, which includes monitoring. Specific requirements are found in 29 CFR 1910, subpart G.

Eye and face protection may also be required when handling chemicals or doing work that produces flying debris. Impact-resistant eye and face protection (glasses, goggles, face shields) are required if flying debris is likely to be present. Chemical splash–resistant and/or vapor-resistant equipment is required for chemical handling, as specified on the chemical MSDS. Body protection, such as aprons or coats, may also be required for handling chemicals. Again, refer to the MSDS for the appropriate type and composition of protective device to be selected. The employer is responsible for providing the equipment and ensuring that it is maintained. The employee is responsible for using it as required.

WATCH THE VIDEO
Personal Protective Equipment (www.awwa.org/wsovideoclips)

Self-Contained Breathing Apparatus

Self-contained breathing apparatus (SCBA) are a type of respirator that supplies safe, grade D (or better) breathing air to the wearer. These units are commonly used in the water and wastewater industries for emergency response or for employee protection when maintenance activities can reasonably be expected to result in the release of a hazardous material. They are also required in oxygen-deficient atmospheres or where the existing atmosphere is immediately dangerous to life and health. The other type of respirator that supplies breathing air is an airline respirator. These devices are

self-contained breathing apparatus (SCBA)

A type of respirator that supplies safe, grade D (or better) breathing air to the wearer.

not commonly used for emergency response because the length of the airline supplying the breathing air is only 300 ft (91 m). This type of respirator will not be discussed in this section.

The major components of an SCBA include an air tank, a harness for wearing the tank on the back, a pressure regulator, hoses, a face mask, and a low-air alarm; all components require routine inspection and maintenance. Refer to the manufacturers' literature for specific maintenance intervals. DOT also has testing, labeling, and maintenance requirements for air cylinders. OSHA inspection requirements mandate inspection during routine cleaning if respirators are worn routinely and at least monthly and after each use when used for emergency purposes.

OSHA requires employees to be medically fit to wear an SCBA. Wearing an SCBA puts additional strain on the human body; therefore, it is important to ensure that there are no underlying medical conditions that would limit an employee's ability to use the device. An employee must be medically evaluated before using an SCBA and on a periodic basis thereafter. Annual medical exams are no longer required, but annual review of an employee's fitness by a medical provider should be performed.

In addition to medical surveillance, OSHA requires fit testing of tight-fitting respirators. The two types of fit testing are qualitative and quantitative. Qualitative fit testing involves determining whether or not the employee can detect the odor of a test medium. Irritant smoke and banana oil are commonly used for this type of test. *Qualitative* fit testing is subject to the impression of the wearer and may not always yield an accurate result. *Quantitative* fit testing involves using a probed face mask identical in brand and size to the one assigned to the employee. The testing device measures the particles in the air outside the mask and the particles inside the mask and compares the levels to determine an acceptable fit. A number of manufacturers produce these testing devices. If an appropriate fit cannot be obtained with the assigned mask, another brand or size of respirator face mask may be needed to provide the required level of protection. Specific requirements regarding fit testing can be found in 29 CFR 1910.1050 (see appendix E in the CFR).

Lab Safety

General safety rules for laboratory workers or those who use chemical reagents in their duties can be found in the latest addition of *Standard Methods for the Examination of Water and Wastewater*. All water treatment plant employees should be familiar with the general rules. The previous discussion on PPE applies to laboratory workers in particular.

General Safety

Water plant structures can present opportunities to implement safety practices, and operators are urged to use a commonsense approach to identify these opportunities. Operators should work with designers of their treatment plant structures (Figures 20-4 and 20-5) to ensure that safe accommodations are part of the overall safety plan.

Figure 20-4 Guardrails at the 3-mgd Conneaut Water Treatment Plant. Ramp leads to chlorine storage area; 150-lb cylinders can be wheeled up the ramp.
Source: Conneaut, Ohio, Water Department.

Figure 20-5 Safety ladder arrangement at Santa Fe's lime silo. Note that tank climbing requires the use of a safety belt, which can be clamped to the railing as the climber ascends and descends.
Source: Sangre de Cristo Water Division.

Plant Security

Water Utility Security Initiatives

The Department of Homeland Security has taken steps in partnership with all public and private stakeholders to ensure the protection and resilience of water services. These water sector partners collaborate to be better prepared to prevent, detect, respond to, and recover from terrorist attacks and other intentional acts, in addition to natural disasters. This effort is known as the *all-hazards approach*.

Homeland Security Presidential Directive 7 (HSPD-7) identifies 18 critical infrastructure and key resources sectors and assigns an agency lead for each. The USEPA is the designated agency for the water sector.

All-Hazards Approach

Water utilities have taken steps to protect critical water supply facilities from theft and sabotage and have planned for response and recovery from events that interrupt water service. An all-hazards approach to security and emergency preparedness mirrors the time-tested multibarrier approach used in water treatment.

The Water Sector Coordinating Council consists of representatives from a cross-section of water agencies and professionals. The council interacts with federal agencies regarding homeland security issues. In this role, the council has developed several critical resources for water utilities.

- *Water Sector–Specific Plan: Annex to National Infrastructure Protection Plan*
- *All-Hazard Consequence Management Planning for the Water Sector*
- *Roadmap to Secure Control Systems in the Water Sector*
- *Recommendations and Proposed Strategic Plan: Water Sector Decontamination Priorities*

In addition, standards have been developed as part of the AWWA/ANSI process to assist water systems in their efforts to improve facility security and emergency preparedness.

- G430-09: Security Practices for Operations and Management
- G440-11: Emergency Preparedness Practices
- J100-10: Risk Analysis and Management for Critical Asset Protection (RAMCAP) Standard for Risk and Resilience Management of Water and Wastewater Systems

Water plant operators should be aware of the requirements for the storage and transportation of hazardous substances. Specifically, chlorine gas and large quantities of sodium hydroxide (caustic soda) have extensive requirements for emergency response and security. Many treatment plants have elected to abandon using chlorine gas due to these requirements and concerns. On-site generation of sodium hypochlorite has gained popularity as a way to avoid these hazards.

Treatment Plant Security Measures

Protecting a public utility's assets to ensure an uninterrupted, safe water supply is a fundamental concern of all levels of management. Plant operating staff should participate in the development of a master plan that will establish a level of protection for their areas of responsibility. A utility can implement a fully developed master plan to provide for overall security for the utility. These plans are site specific but contain common components. By breaking down these components into areas of responsibility, a utility can assure an integrated approach to plant security development. The major features of a security master plan are described next.

Physical Requirements

The physical requirements of the plan focus on actions that terrorists, saboteurs, or criminals might take to damage or contaminate the water supply or sabotage system infrastructure and the electric transmission system. Preventing acts of sabotage at water treatment plants, electric substations, maintenance facilities, and other utility facilities and in the distribution system requires physical barriers

Figure 20-6 Decorative security fence

to prevent intruders from entering a site and reaching vital areas or equipment (Figure 20-6).

Integrated Systems Approach

An integrated systems approach that combines people, equipment, and procedures is used to protect against possible threats and intrusions. The security plan has multiple layers of protection that provide several ways to monitor sites and provides deterrence measures for a multitude of possible threats. It also monitors employee activities and documents the presence of contractors, visitors, and other invitees to the facilities through the use of swipe cards and proximity card readers.

Detection Systems

Electronic detection systems, which are more reliable and predictable than systems that depend on people, provide alarms to alert security and plant personnel that the plant or facility is being approached or an intrusion attempt is under way. Many cost-effective devices are available to detect movements and forced entry, and even to record any events and times. Video surveillance is another useful detection system that can aid operators to prevent unlawful entry to the plant facility.

Delaying Tactics

Barriers, such as chain-link fences around perimeters, gates, and proximity card readers, are arranged in layers to create physical and psychological deterrents and to delay intrusion long enough for security to detect the intrusion and respond. Barriers channel entry and exit through specified control points, facilitating efficient identification and control of authorized people and equipment.

Assessment Systems

Closed-circuit television cameras monitor facility entrances and special secured areas and allow security to assess an alarm received while limiting exposure until assistance arrives. These systems can be linked with other security systems to control entry and record images for use by authorities.

Communications and Response Systems

Security assessment and monitoring devices are linked to communications systems to transmit information to a local monitoring site, a central monitoring site, and the security force. As the security officer or facility personnel respond, they remain in communication with a central monitoring site and other site personnel at all times by several means of alternate communication, such as 800-Mhz radios, cell phones, walkie-talkies, pagers, and other communication devices available at the site.

Training and Qualifications

Security and facility personnel effectiveness must be maintained through training. Training exercises against mock adversaries should be conducted periodically to add realism to the training. In addition, training on system maintenance and reporting and policy enforcement must be conducted annually for effective system management.

Access Control

Employees are issued identification badges that must be carried at all times. Access control systems at the entrance gate and within buildings will confirm a person's identity and authorization before entrance is granted. Visitors are also issued badges upon entry to the facility.

Reliability of Personnel

Security awareness will become part of each employee's initiation to the work site. Thorough background investigations should be conducted in accordance with department policies as the primary method for learning about any past actions that may indicate problems with individual employees.

Testing, Maintenance, and Auditing

Periodic testing of all security equipment is required to make sure it is operating correctly. If problems are found, corrective measures are required to ensure the security function provided by the equipment is immediately restored. Physical security programs should be audited periodically to be certain requirements are being met.

Contingency Planning

Plant operators and facility managers must plan and provide for all reasonable contingencies, such as strikes, natural disasters, and on-site emergencies that require quick support and coordination from outside agencies and security personnel. Backup personnel should be identified to respond to these unusual situations. These plans should be reviewed periodically to ensure that contact information is current.

Study Questions

1. Chlorine gas
   a. will burn in the presence of oxygen and moisture.
   b. will explode in the presence of oxygen and an ignition source.
   c. will support combustion.
   d. will conduct electricity.

2. Chlorine storage rooms should
   a. have sealed walls and doors that open inward.
   b. be fitted with chlorine-resistant power exhaust fans ducted out at ceiling level.
   c. have lights and fans on the inside and wired to the same switch.
   d. have a window in the door so an operator can look into the room to detect any abnormal conditions.

3. Before an authorized entrant enters a permitted confined space,
   a. he or she must don SCUBA gear.
   b. the internal atmosphere of the space must be tested to measure oxygen content, presence of flammable gases and vapors, and potentially toxic air contaminants.
   c. verify the availability of rescue services.
   d. ensure there is adequate lighting in the space to conduct the necessary work.

4. If a substantial chlorine leak incident occurs, which agency should be called for actual hands-on assistance?
   a. The Occupational Safety and Health Administration (OSHA)
   b. The Chemical Transportation Emergency Center
   c. The Transportation Emergency Institute
   d. The Chlorine Institute

5. Sites are required to do a site assessment under the process safety management (PSM) regulations (OSHA) if the facility in a single process has more than how many pounds of chlorine?
   a. 1,000 lb
   b. 1,500 lb
   c. 2,000 lb
   d. 4,000 lb

6. Which organization issued the Risk Management Rule on July 20, 1996, in an effort to protect public health and safety?
   a. AWWA
   b. NIOSH
   c. OSHA
   d. USEPA

7. Chlorine cylinders must be stored
   a. lying on a flat surface.
   b. in a temperature-controlled area.
   c. with at least 10 ft (1.5 m) of space on all sides.
   d. in an area above 160°F (71°C) to ensure chemicals do not settle.

8. Which individual in a confined-space operation is responsible for knowing the conditions within the confined space, verifying that all equipment and procedures are in place prior to entry, verifying the availability of rescue services and the means of summoning them, terminating entries and canceling permits, and determining that acceptable conditions as specified in the permit continue for the duration of the entry?

9. The Department of Homeland Security has partnered with all public and private stakeholders to be better prepared to prevent, detect, respond to, and recover from terrorist attacks and other intentional acts, in addition to natural disasters. What is the term for this effort?

10. A material safety data sheet should exist for what type of chemicals?

Chapter 21
Administration, Records, and Reporting Procedures

Records of plant operations and maintenance activities provide a basis for predictive decision making and proof of compliance. Records also help operators to efficiently use chemicals (minimize waste) and equipment.

Process Records

In addition to the records that are required by the local regulatory agency, a water treatment plant should maintain records of activities as they relate to each process or discipline. The following sections describe the types of records operators should keep when performing their duties.

Coagulation and Flocculation

Maintain records of past performances for coagulants, including dosages, turbidities (sedimentation basin and filtration), particle counts, filterability indices, and zeta potentials as well as the temperatures, pH values, and color values at which the coagulants performed. For softening plants, record which softening chemicals are used. Also record detention times and overflow rates in basins used for softening as well as residual quantities produced. Keep equipment records such as hours run for each mixer and flocculator, basin-cleaning activities, maintenance performed on motors and gears, and electrical use records.

Sedimentation

Calculate and record surface overflow rate, weir overflow rate, turbidity performance of each basin, and quantity of sludge removed from each basin (with a calculation for percent solids).

Maintain records of any operations and maintenance activities for each unit, including blowdown, cleaning, raking, and so on. Safety records for confined space entry may be necessary.

The Partnership for Safe Water suggests that the turbidity performance (and particle count, if available) in each sedimentation basin be recorded periodically (daily, every 4 hours, or hourly). Record the settled water turbidity as a combined flow, but realize that it cannot characterize the performance of an individual basin.

Filtration

For each filter, record the rate of flow, in million gallons per day or gallons per minute; head loss; length of run and unit filter run volume (UFRV); turbidity (in accordance with the Interim Enhanced Surface Water Treatment Rule [IESWTR]); particle counts; and dosage of any filter aid used (Table 21-1). At each backwash,

Table 21-1 A sample filter record

Time	Rate of Flow, mgd or gpm	Head Loss, ft	Hours in Service	Total Volume Filtered (cumulative)	UFRV, gal/ft²	Turbidity, ntu and particle counts	Backwash Start	Length of Backwash, minutes and amount used (gal)	Surface Wash, amounts and minutes	Filter Aid Applied? Dosage?	Filter to Waste? ntu at Startup
0000											
0200											
0400											
0600											
0800											
1000											
1200											
1400											
1600											
1800											
2000											
2200											

record the amount of backwash water and surface wash water used, the length of the backwash, any observations during backwash, and time of filter-to-waste if used. Record the percentage of water treated that the backwash water represents (make note of amounts greater than 4 percent). All information can be included on one form.

On a quarterly basis, record bed expansion measurements, condition and depth of media, and an evaluation of the bed surface. On a yearly basis, record the results of solids retention analysis before and after backwash and the results of sieve analysis of the media. Record the addition of any media to the bed, including amounts and types. Keep a record of underdrain inspections and any work that was done on any appurtenances.

Record results of any testing required for filter-exceptions reporting under the IESWTR (this is regulated). Many states have required forms for plant operators to submit for filter performance. Consult the state regulatory agency for specific requirements.

Chlorination or Disinfection

Maintain records of the type of disinfectant used, including ordering information (phone number, address, and shipment amounts and container types), as well as information on costs and current dosage rates. Also results of bacteriological testing should be recorded for each process that uses a disinfectant, the water temperature and pH, and any unusual conditions that may occur. Keep records of safety training.

Precipitative Softening

Record the amount of softening chemicals used and the results of jar testing and lab analyses that justify these amounts. Record the results of analysis for alkalinity, pH, hardness, and related parameters for each shift, preferably every two hours for raw water and for all steps in the process. Keep a record of the chemical feeder setting and record the amount of water treated. Put everything on one form and maintain it daily.

Record the amounts of sludge produced. If a solids contact unit is used, record the sludge blanket depth and results of settling tests. Keep records of the amounts and rates of solids returned to the process, if applicable.

Ion Exchange

Record the hardness, alkalinity, magnesium, calcium, and pH of the source and treated waters. Record the total amount of water treated and the amount of water treated by each softening unit, including bypass water. Also, keep a record of the amount of salt used for regeneration and the length of each cycle between regenerations. Record the amount of backwash water used and the amount of waste sent to sewer or other disposal places.

Aeration

Keep process records of raw water and finished water quality to determine if processing is accomplishing its goals. Also, record the daily quantity of water treated, the details of safety and maintenance procedures, and changes in other treatment processes that may be affected by aeration. Document the process operating conditions including the air-to-water ratio to calculate performance and cost information.

Adsorption

Records kept regarding activated carbon differ depending on the type of carbon used.

- Powdered activated carbon (PAC): Record the types of tastes and odors being experienced, the threshold odor number (TON), and the dates of taste-and-odor occurrences. Keep a record of the coagulant and PAC dosage predicted from jar tests as well as the actual dosage used.

- Granular activated carbon (GAC): Record the number of filter hours; UFRV; any losses of GAC from the filter, measured monthly; dates of installation; and periodic TON from each filter. Also record when backwashing was performed and the amount of backwash water used as well as the raw-to-finished organic content removal.

Iron and Manganese Removal

If using ion exchange, the previously referenced records can be used. In addition, record raw and finished iron and manganese levels and results of distribution systems analyses performed. If permanganate is used, record amounts and dosages, including those predicted by jar testing.

If a sequestrant is used to control iron and manganese, record the amount and type used and the data from distribution system flushing efforts.

Fluoridation

Keep a daily record of the amount of fluoride in the finished water and the raw water, if necessary (a weekly raw water analysis may be sufficient if the regulator allows). Also, keep a daily record of the chemical dosages used and record the chemical feed rate each hour of each shift. Compute and record the daily fluoride chemical feed and compare it to the lab analysis as a double-check. Record safety classes and lectures that were provided to operators.

Corrosion Control

Maintain records of corrosion chemical dosages and any data available as to their performance, such as coupon testing results, Corrator® readings, temperatures, flushing activities, and lead and copper compliance results (i.e., 90th percentile levels and maximum levels). If phosphates are used, keep strict records of dosages and amounts of total and dissolved phosphate in the system. Flush mains regularly, especially dead-end mains; sample for phosphates at 5-, 10-, and 15-minute intervals into the flush; and record results. Record the main flushing velocities, in feet per second. Track any customer complaints that may relate to use of corrosion chemicals.

Water Main Flushing

Main flushing records are best kept in a searchable database. Record main location and size; size of flushing pipe or hydrant; flow rate at flush; gallons used for flushing; velocity, in feet per second; chlorine residual; amount of phosphate used, if any; and results of biological testing, such as heterotrophic plate count quantity or quantity of coliform. Query the database by main size and show gallons used and time needed, along with any chemical and biological data, to generate monthly reports. In this way, a profile will emerge that shows the amount of water necessary to flush each size main, the time it takes, and the results obtained.

Reporting

Reporting requirements are generally a function of the compliance reports that local and federal regulators require. In general, compliance testing results must be reported no later than 10 days after the end of the month in which they were accumulated. It is good practice for water treatment plant staff to provide reports to upper management. Even if the administration does not require these reports, they should be generated and kept on file for future reference.

Like a diary, a written report is an excellent reminder of events that took place and the methods used to handle these events. Internal memos of safety issues, personnel and disciplinary actions, purchases, and maintenance events can actually help protect staff from allegations of nonfeasance. Reports should be written as soon as possible after an event takes place so that details can be easily recalled. Do not write reports or internal memos in anger; take time to cool down before recalling an event.

Most compliance reports must conform to a specific format. This enables the regulator to assemble large amounts of data from many utilities in an orderly fashion. Internal or outsourced reports, other than compliance reports to regulators, should include the following:

- Times and dates of the audit and any description of incidents or accomplishments
- Financial or inventory considerations
- Personnel involved
- Conclusions drawn, with supporting data and references
- List of report recipients, with acknowledgment of receipt, if appropriate
- An executive summary if the report is lengthy and involved

The report should be filed in a logical order (e.g., file reports under general categories such as research, safety, finance, and so on) so that the report can be easily retrieved in the future. As time permits, reread all reports to gauge progress in reaching goals and to learn from past mistakes.

Reporting Treatment Incidents

There are specific requirements for the time and method of reporting a regulatory violation. These are described in the drinking water regulations for the applicable regulatory agency. However, there are times when it is not clear if there has been a violation or when there is a water quality incident that may affect consumers. The water plant personnel should have a predetermined plan of action to deal with these situations.

The notification and action plan for incident reporting should include plant management, utility management, other utility officials, and regulatory agencies. The utility should have discussions with the regulatory agency representatives prior to any incident to establish an understanding of how they will handle this type of communication. The agency may have technical resources to assist the plant personnel with a response. Also, the agency may be helpful when dealing with mass communication media regarding the incident.

Plant Performance Reports

Periodic (monthly, quarterly, and annual) plant performance reports should be prepared, and these should be examined to detect trends and identify opportunities for improvement. The operating reports previously described are the basis for these performance reports. Performance reports summarize the operating results and usually include various charts and calculated performance measures. These summary reports do not need to be lengthy and, indeed, can be added to the operating reports if desired. The purpose of the performance reports is to examine the plant results over longer time periods to see if there are trends that need to be investigated. All plant operating personnel should review these reports to provide feedback and to participate in efforts to meet plant operating goals.

Public Relations

It is important for a water utility to maintain public confidence. Satisfied customers will pay their bills, are less likely to complain if they are temporarily inconvenienced during system maintenance and construction, and are more likely to be supportive of rate increases and bond issues. Operations personnel are often engaged with customers when they are investigating a complaint. Not only is this an opportunity to solve a problem and gain customer confidence, but also important information can be obtained on the system's condition.

Dealing With the Media

On occasion, distribution system employees may be approached by reporters from the local media asking about the work they are doing. Although one must be courteous, the general rule in talking to reporters is, *Don't unless you absolutely have to!* There are many opportunities for being misquoted, and it is quite embarrassing for the utility to see the wrong information in print.

The best policy is to give a very brief explanation and offer no more information than is requested. Beyond that, the workers should say they are not qualified to go into more detail. Large municipalities and water utilities maintain a public relations department whose job is to deal with the media, so reporters should be referred there for additional details. If such a department does not exist, the reporter should be referred to the utility manager, city manager, or mayor.

Customer Complaint Response

Responding to customer complaints is an important element in assessing system performance. Information from customers about water quality, reliability of service, and security can be helpful when these topics are of concern. A professional interaction with customers is often the most effective method to convey a positive view of the utility and gain customer satisfaction. Also, customer complaint reporting procedures may be required by some state or local agencies.

The objectives of a customer complaint response program are to (1) address the customer's concerns, (2) gather information that can be used to identify system problems, and (3) catalog data that can be used to assess long-term system performance goals. System operators need specific training to achieve these objectives.

Study Questions

1. What is the rationale for establishing the three categories of public water systems?
   a. Billing purposes
   b. Federal funding
   c. State and local funding
   d. Exposure differences to contaminants

2. The American Water Works Association recommends that water treatment chemicals be purchased in quantities that maintain a minimum supply at all times of
   a. 14 days.
   b. 15 days.
   c. 20 days.
   d. 30 days.

3. The Partnership for Safe Water suggests that the _____ (and particle count, if available) in each sedimentation basin be recorded periodically (daily, every 4 hours, or hourly).
   a. turbidity performance
   b. floc count
   c. sludge ratio
   d. aeration efficacy

4. In general, compliance testing results must be reported no later than _____ after the end of the month in which they were accumulated.
   a. 24 hours
   b. 1 week
   c. 10 days
   d. 1 month

5. Most state regulatory agencies require water plant reporting according to what schedule?

6. After repairing a small chlorine gas leak, what would be your next action?

7. Filter performance can be determined by reviewing what reports?

8. How often must treatment plants record the results of solids retention analysis before and after backwash and the results of sieve analysis of the media?

9. In communicating with the media, what is the best policy?

10. List the three objectives of responding to customer complaints.

Chapter 22
Additional Study Questions

As you advance through the *WSO Water Treatment* series, it is important that you maintain your competence at lower certification levels. Consider revisiting material covered in the Grade 1 book that is not covered in this Grade 2 book. A review of Chapters 4, 5, and 6 is particularly encouraged. You can check your knowledge of this material by answering the following questions.

Water Sources and Treatment Options

1. What is the water table?

2. What are the advantages and disadvantages of a confined aquifer for a water supply?

3. Is rain pure water?

4. What is the multiple-barrier approach used for most surface water treatment systems?

5. What processes are used to remove particles from surface water?

6. Which term refers to the level of the water surface in an aquifer?
   a. Runoff
   b. Water table
   c. Channel flow
   d. Watershed

7. What is a key disadvantage of using a confined aquifer for a water supply?
   a. It may be easily depleted.
   b. It is under pressure, making pumping costs excessive.
   c. It is not protected from contamination.
   d. It cannot be used in conjunction with other aquifers to supply a well.

8. Should rain water be considered pure?
    a. No, it may contain pollutants from the air.
    b. Yes, it is free of pollutants until reaching the ground.
    c. It depends on the proximity to a major city.
    d. It depends on weather conditions.

9. What term refers to a series of treatment steps that are redundant in their ability to remove contaminants?
    a. Air stripping
    b. Flocculation
    c. Ion exchange
    d. Multiple-barrier approach

10. Which of the following is *not* a process used to remove particles from surface water?
    a. Coagulation
    b. Percolation
    c. Sedimentation
    d. Filtration

Groundwater Quality and Wells

1. Immediately below the water table is the
    a. piezometric surface.
    b. capillary zone.
    c. saturated zone.
    d. recharge zone.

2. Which type of wells are commonly used near the shore of a lake or near a river?
    a. Monitoring wells
    b. Bedrock wells
    c. Gravel wall wells
    d. Radial wells

3. Many streams would dry up after a rain if it were not for
    a. human construction influences.
    b. surface runoff.
    c. capillary action.
    d. groundwater flow.

4. An artesian aquifer could occur in a(n)
    a. confined aquifer.
    b. unconfined aquifer.
    c. water table aquifer.
    d. shale formation.

5. Bored wells, constructed with an auger driven into the earth, are limited in depth to approximately
   a. 20 ft.
   b. 30 ft.
   c. 50 ft.
   d. 60 ft.

6. Because harmful bacteria do not penetrate very far into the soil, wells are generally free of harmful organisms if
   a. exposed to direct sunlight.
   b. more than 50 ft (15 m) deep.
   c. connected to multiple aquifers.
   d. insulated with limestone.

7. At the surface, all wells should have a sanitary seal, which prevents contamination from entering the
   a. well casing.
   b. surrounding ground.
   c. well slab.
   d. submersible pump.

8. What term refers to the level of the water surface in the well when no water is being taken from the aquifer?
   a. Pumping level
   b. Drawdown
   c. Static water level
   d. Water table

9. Which type of well is commonly used near the shore of a lake or near a river to obtain a large amount of relatively good-quality water from adjacent sand or gravel beds?
   a. Radial well
   b. Natural well
   c. Bedrock well
   d. Rotary well

10. Which of the following is *not* a method by which aquifer performance can be evaluated?
   a. Drawdown
   b. Sequestration
   c. Specific-capacity
   d. Recovery

Surface Water Source Treatment

1. Which type of algae would most likely be found in nutrient-rich waters?
   a. Blue-green algae
   b. Green algae
   c. Diatoms
   d. Brown algae

2. The quantity of dissolved oxygen in water is a function of
   a. pH, alkalinity, temperature, and total dissolved solids.
   b. temperature and alkalinity.
   c. pH and temperature.
   d. temperature, pressure, and salinity.

3. At which temperature does the maximum density of water occur?
   a. −0.4°C
   b. 0.0°C
   c. 0.4°C
   d. 4.0°C

4. Destratification of a lake will often reduce
   a. turbidity.
   b. nutrients.
   c. fish.
   d. algae.

5. Zebra mussels multiply and spread so fast because
   a. the female produces over 250,000 eggs in a single season.
   b. the larvae are free swimming for 2 or 3 months before attaching themselves.
   c. they commonly adhere to boats so they move around and disperse eggs.
   d. they can live out of water for as long as 3 months.

6. Which of the following is *not* a principal factor affecting how rapidly surface water runs off the land?
   a. Rainfall intensity
   b. Soil moisture
   c. Vegetation cover
   d. The water's pH

7. Water is drawn into a water supply system from a surface impoundment or stream through
   a. a diverted aqueduct.
   b. a natural lake.
   c. a backwash discharge.
   d. an intake structure.

8. A circumstance in which it is more economical or practical to provide treatment at the source, versus in a water treatment facility, is referred to as
   a. destratification.
   b. recharging.
   c. in situ treatment.
   d. biological infusion.

9. Which of the following is *not* a problem commonly associated with algae?
   a. Taste and odor
   b. Filter clogging
   c. Supply shortage
   d. Toxicity

10. Division of water into three layers, each of a different temperature, is referred to as
   a. salinization.
   b. siltation.
   c. infiltration.
   d. stratification.

Study Question Answers

Chapter 1 Answers

1. **d.** moles of solute per liter of solution.
2. **c.** chemical formula
3. **b.** molecular weight.
4. **a.** Normality
5. **b.** solute.

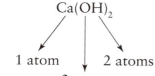

6.   2 atoms
7. 16.1 lb of $CaCO_3$
8. 5.21 lb of chemical
9. 31.11 mL
10. 6

Chapter 2 Answers

1. **c.** 2.52 hr

   First determine the volume in gallons for the clarifier:

   Volume, gal = $(0.785)(\text{diameter})^2(\text{depth})(7.48 \text{ gal/ft}^3)$

   = $(0.785)(112.2 \text{ ft})(112.2 \text{ ft})(10.33 \text{ ft})(7.48 \text{ gal/ ft}^3) = 763{,}584.83$ gal

   Then, convert mgd to gallons per hour, as detention time is asked for in hours:

   $(7.26 \text{ mgd})(1{,}000{,}000/1 \text{ M})(1 \text{ d}/24 \text{ hr}) = 302{,}500$ gph

   Equation: Detention time, hrs = $\dfrac{\text{Volume, gal}}{\text{Flow rate, gal/hour (gph)}}$

   Detention time, hrs = $\dfrac{763{,}584.83 \text{ gal}}{302{,}500 \text{ gph}}$ = **2.52 hr**

2. **c.** 69.1 mL/min

   Know: 1 day = 1,440 min.

   $(26.3 \text{ gal/d})(3{,}785 \text{ mL/gal})(1 \text{ d}/1{,}440 \text{ min}) = $ **69.1 mL/min**

3. **a.** 33.7 hr

First, calculate the number of gallons per foot in the tank:

Equation: gal/ft = (0.785)(diameter, ft)^2(1 ft)(7.48 gal/ft)

$= (0.785)(119.8 \text{ ft})(119.8 \text{ ft})(1 \text{ ft})(7.48 \text{ gal/ft}) = 84,272 \text{ gal/ft}$

Next, determine how many feet of water can be used from the tank:

Number of usable feet of water = 27.6 ft – 16.0 ft = 11.6 ft

Next, multiply the number of usable feet of water by the gal/ft to get the total number of usable gallons:

Total usable gallons = (11.6 ft)(84,272 gal/ft) = 977,555.2 gal

Lastly, determine the hours of supply left:

$$\text{Supply, hr} = \frac{(977,555.2 \text{ gal})}{(483 \text{ gpm})(60 \text{ min/hr})} = \textbf{33.7 hr}$$

4. **b.** 37 lb/d, F

First convert gpm to mgd:

mgd = [(1,750 gpm)(1,440 min/d)] ÷ (1,000,000) = 2.52 mgd

Next, determine the fluoride (F) required:

F, required = F desired – F raw water

= 1.20 mg/L – 0.15 mg/L = 1.05 mg/L F

$$\text{Equation: Fluoride feed rate} = \frac{(\text{Dosage, mg/L})(\text{mgd})(8.34 \text{ lb/gal})}{(\text{Percent F ion})(\text{Chemical purity})}$$

$$= \frac{(1.05 \text{ mg/L F})(2.52 \text{ mgd})(8.34 \text{ lb/gal})}{(60.6\%/100\%)(98.1\%/100\%)}$$

= 37.12 lb/day, round to **37 lb/day, F**

5. **b.** 4.8 gpm/ft^2

First, find the surface area of the filter:

Filter surface area = (24 ft)(28 ft) = 672 ft^2

$$\text{Equation: Filtration rate} = \frac{\text{Flow rate, gpm}}{\text{Filter surface area, ft}^2}$$

$$= \frac{3,250 \text{ gpm}}{672 \text{ ft}^2} = 4.836 \text{ gpm/ft}^2\text{, round to } \textbf{4.8 gpm ft}^2$$

6. **b.** head.

7. **a.** feed rate.

8. area of surface × third dimension

9. $(0.785)(D^2)$

10. 4,187 ft^3

Chapter 3 Answers

1. **b.** 0.2 mg/L
2. **c.** 1.3 mg/L
3. **c.** At point(s) where water enters the distribution system, including all filter effluents if a surface water treatment plant
4. **a.** coliform samples.
5. **b.** 0.060 mg/L
6. **a.** benchmarking
7. Zero
8. IESWTR
9 GWR
10. TCR

Chapter 4 Answers

1. **c.** 6.5 to 7.2
2. **a.** van der Waals forces
3. **a.** Allowing excessive flocculation time
4. **c.** More dense floc
5. **b.** turbidity
6. **a.** milligrams per liter.
7. Any four of the following: type and concentration of contaminants, temperature, alkalinity, pH, turbidity, color, total organic carbon
8. As the water temperature decreases, the viscosity of the water increases, which slows the rate of floc settling.
9. 30 minutes

Chapter 5 Answers

1. **c.** Use a multibarrier approach—coagulation, flocculation, sedimentation, and filtration.
2. **c.** Poor inlet baffling
3. **a.** less than 0.5-log
4. **d.** algae.
5. **b.** a sludge layer develops.
6. 10 gpm/ft
7. Any four of the following: sludge pipe, scum trough, flights, collector drive, baffle, effluent adjustable weirs
8. SWTR
9. Test for turbidity
10. Lime sludge

Chapter 6 Answers

1. **d.** Floc breakthrough
2. **c.** a sand boil.
3. **d.** Sedimentation
4. **c.** Schmutzdecke
5. **b.** rectangular and constructed of concrete.
6. **a.** before
7. 3–8 gpm/ft^2
8. Pipe lateral collectors
9. IESWTR
10. Diatomaceous earth

Chapter 7 Answers

1. **d.** has little pH effect.
2. **c.** sometimes produce disinfection by-products known to be carcinogenic.
3. **b.** concentration of chlorine, contact time, pH, and temperature.
4. **b.** Langelier saturation index
5. **c.** Vacuum
6. **c.** Diffusers
7. **b.** temperature of the chlorine liquid.
8. The Chlorine Institute
9. A venturi device that pulls chlorine gas into a passing stream of dilution water, forming a strong solution of chlorine and water
10. A heating device used to convert liquid chlorine to chlorine gas
11. Any four of the following: chlorine leaks, stiff container valves, hypochlorinator problems, tastes and odors, sudden change in residual, trihalomethane formation
12. 2.5 times as dense

Chapter 8 Answers

1. **d.** Shorter filter runs
2. **c.** potassium permanganate solution during backwashing.
3. **d.** Glauconite sand coated with manganese dioxide
4. **c.** 10–12 gpm/ft^2
5. **d.** adding a layer of anthracite above the greensand.
6. **c.** low level of dissolved iron and manganese and no dissolved oxygen.
7. D/DBP
8. Manganese greensand filter
9. Detention or contact chamber
10. Intermittent

Chapter 9 Answers

1. **c.** every day.
2. **b.** 4 mg /L
3. **c.** Every day
4. **d.** After water has received complete treatment
5. **d.** 6 in. (150 mm)
6. Both before and after fluoride is added
7. The manufacturer's recommendations
8. Low fluoride readings
9. Any three of the following: daily analyses of raw-water fluoride concentration (unless it is known to be stable), daily analysis of finished water fluoride concentration, daily records of the amount of chemical fed (in pounds or kilograms), records by the operating shift of the chemical feed rate setting, daily computation of the theoretical concentration

Chapter 10 Answers

1. **b.** Grab sample
2. **c.** 8
3. **a.** <2.0.
4. **a.** $Na_2S_2O_3$
5. **d.** blockage.
6. **b.** Various points in the system
7. **c.** free carbon dioxide
8. **a.** Potassium permanganate
9. Composite sample
10. Jar testing

Chapter 11 Answers

1. **b.** Langelier saturation index
2. **b.** $CaCO_3$
3. **b.** CO_2
4. **d.** Polyphosphates
5. **c.** corrosion.
6. **b.** almost any metal that is exposed to water.
7. Magnesium
8. Localized and uniform
9. Any two of the following: lime, sodium carbonate, sodium hydroxide

Chapter 12 Answers

1. **c.** lime.
2. **d.** calcium carbonate.
3. **c.** 1 month's

4. **b.** Pellet reactor
5. Magnesium hydroxide
6. To serve as additional carbonate to form calcium carbonate
7. It gets hot and forms hydrated lime in a slurry.
8. To correct the pH and stabilize the water prior to filtration
9. Lead and Copper Rule

Chapter 13 Answers

1. **d.** Rock salt or pellet-type salt
2. **d.** exhausted.
3. **c.** 50–75%
4. **b.** iron
5. **a.** 6.
6. **d.** its small surface area does not allow it to dissolve fast enough.
7. Positively charged ions, such as calcium and magnesium, that migrate toward the cathode
8. Operational cycle
9. Total organic carbon

Chapter 14 Answers

1. **b.** organic compounds responsible for tastes and odors.
2. **c.** cannot be prevented by
3. **b.** They lose their capacity to adsorb.
4. **c.** It should be covered with a fine mist of spray from a hose or a chemical foam extinguisher.
5. At exactly the feed point
6. Fiberglass

Chapter 15 Answers

1. **c.** Cascade aerator
2. **d.** Draft aerator
3. **c.** 25:1
4. **a.** Surface
5. **d.** Mechanical aerator
6. **c.** splash aprons
7. Any five of the following: central feed pipe, splash guard, catch bowl, deflector plate, collector pan, overflow baffle, spray bottle, outlet, inlet, aluminum flange
8. Diffused oxygen level

Chapter 16 Answers

1. **a.** Reverse osmosis membrane
2. **b.** 95; 97
3. **a.** High-pressure pump
4. Recovery
5. Feedwater

Chapter 17 Answers

1. **b.** 5%
2. **c.** Lagoons
3. **c.** Thickener
4. **b.** Sand bed
5. **d.** Floating-bridge siphon collector
6. **c.** 4–12 in. (100–300 mm)
7. **d.** centrifuge
8. Via pipeline
9. Polymer dosage and cost, thickened solids concentration, solids loading, and capture

Chapter 18 Answers

1. **a.** Floating proportional control
2. **a.** Feedback control
3. **d.** Proportional plus reset control
4. **d.** Loss of head meter
5. **c.** Bellows
6. **c.** sight, sound, or smell.
7. **a.** pH of the water
8. Plant discharge pressure
9. Start–stop control

Chapter 19 Answers

1. **c.** Packing gland
2. **b.** 90 degrees.
3. **c.** $V^2/2\,g$
4. **c.** Semiopen
5. **b.** torque
6. **a.** water hammer
7. **a.** 350 ft (100 m).
8. Single-suction pump
9. Lantern ring
10. Pump shaft

Chapter 20 Answers

1. **c.** will support combustion.
2. **d.** have a window in the door so an operator can look into the room to detect any abnormal conditions.
3. **b.** ensure the internal atmosphere of the space has been tested to measure oxygen content, presence of flammable gases and vapors, and potentially toxic air contaminants.
4. **b.** The Chemical Transportation Emergency Center
5. **b.** 1,500 lb
6. **d.** USEPA
7. **b.** in a temperature-controlled area.
8. Entry supervisor
9. The all-hazards approach
10. Every chemical that is used in a water treatment plant

Chapter 21 Answers

1. **d.** Exposure differences to contaminants
2. **d.** 30 days.
3. **a.** turbidity performance
4. **c.** 10 days
5. Monthly reports are most often required and immediate notice of Tier 1 violations.
6. Review the situation and determine the cause. Establish procedures to prevent reoccurrence. Notify utility management so that it can be determined if health authorities should be notified.
7. Filter run length, turbidity profiles, and head-loss trends
8. Every year
9. Give a very brief explanation and offer no more information.
10. Address the customer's concerns, gather information that can be used to identify system problems, and catalog data that can be used to assess long-term system performance goals.

Chapter 22 Answers

Water Sources and Treatment Options

1. The level of the water surface in an aquifer.
2. Advantages—protected from contamination and probably under pressure so less pumping costs. Disadvantages—may be easily depleted.
3. No, it may contain pollutants from the air.
4. A series of treatment steps that are redundant in their ability to remove contaminants
5. Coagulation, sedimentation, and filtration
6. **b.** Water table

7. **a.** It may be easily depleted.
8. **a.** No, it may contain pollutants from the air.
9. **d.** Multiple-barrier approach
10. **b.** Percolation

Groundwater Quality and Wells

1. **c.** saturated zone.
2. **d.** Radial wells
3. **d.** groundwater flow.
4. **a.** confined aquifer.
5. **d.** 60 ft.
6. **b.** more than 50 ft (15 m) deep.
7. **a.** well casing.
8. **c.** Static water level
9. **a.** Radial well
10. **b.** Sequestration

Surface Water Source Treatment

1. **a.** Blue-green algae
2. **d.** temperature, pressure, and salinity.
3. **d.** 4.0°C
4. **d.** algae.
5. **c.** they commonly adhere to boats so they move around and disperse eggs.
6. **d.** The water's pH
7. **d.** an intake structure.
8. **c.** in situ treatment.
9. **c.** Supply shortage
10. **d.** stratification.

References

Inversand Company. n.d. *Manganese Greensand*. Available at http://www.inversand.com/our-product/technical-data/manganese-greensand.

Glossary

absolute pressure   The total pressure in a system, including both the pressure of water and the pressure of the atmosphere (about 14.7 psi, at sea level). Compare with *gauge pressure*.

activated alumina (AA)   The chemical compound aluminum oxide, which is used to remove fluoride and arsenic from water by adsorption.

adsorbent   Any material, such as activated carbon, used to adsorb substances from water.

adsorption   The water treatment process used primarily to remove organic contaminants from water. Adsorption involves the adhesion of the contaminants to an adsorbent, such as activated carbon.

aeration   The process of bringing water and air into close contact to remove or modify constituents in the water.

air binding   A condition that occurs in filters when air comes out of solution as a result of pressure decreases and temperature increases. The air clogs the voids between the media grains, which causes increased head loss through the filter and shorter filter runs.

air scouring   The practice of admitting air through the underdrain system to ensure complete cleaning of media during filter backwash. Normally an alternative to using a surface wash system.

alkalinity   A measurement of water's capacity to neutralize an acid. Compare *pH*.

alum   The most common chemical used for coagulation. Also called *aluminum sulfate*.

anion   A negative ion.

anion exchange   Ion exchange involving ions that have negative charges, such as chloride.

average daily flow (ADF)   A measurement of the amount of water treated by a plant each day. It is the average of the actual daily flows that occur within a period of time, such as a week, a month, or a year. Mathematically, it is the sum of all daily flows divided by the total number of daily flows used.

average flow rate   The average of the instantaneous flow rates over a given period of time, such as a day.

$C \times T$ value   The product of the residual disinfectant concentration C, in milligrams per liter, and the corresponding disinfectant contact time T, in minutes. Minimum $C \times T$ values are specified by the Surface Water Treatment Rule as a means of ensuring adequate kill or inactivation of pathogenic microorganisms in water.

cation   A positive ion.

cation exchange   Ion exchange involving ions that have positive charges, such as calcium and sodium.

centrifugal pump   A pump consisting of an impeller on a rotating shaft enclosed by a casing that has suction and discharge connections. The spinning impeller throws water outward at high velocity, and the casing shape converts this high velocity to a high pressure.

chemical equation   A shorthand way, using chemical formulas, of writing the reaction that takes place when chemicals are brought together. The left side of the equation indicates the chemicals brought together (the reactants); the arrow indicates in which direction the reaction occurs; and the right side of the equation indicates the results (the products) of the chemical reaction.

chemical formula   Using the chemical symbols for each element, a shorthand way of writing what elements are present in a molecule and how many atoms of each element are present in each of the molecules. Also called a *chemical formula*.

chlorinator   Any device that is used to add chlorine to water.

chlorine cylinder   A container that holds 150 lb (68 kg) of chlorine and has a total filled weight of 250–285 lb (110–130 kg).

coagulation   The water treatment process that causes very small suspended particles to attract one another and form larger particles. This is accomplished by the addition of a chemical, called a coagulant, that neutralizes the electrostatic charges on the particles that cause them to repel each other.

coefficient   An indication of the relative number of molecules of the compound that are involved in the chemical reaction.

composite sample   A series of individual or grab samples taken at different times from the same sampling point and mixed together.

concentration   In chemistry, a measurement of how much solute is contained in a given amount of solution. Concentrations are commonly measured in milligrams per liter (mg/L).

contactor   A vertical, steel cylindrical pressure vessel used to hold the activated carbon bed.

conversion   The process of changing from one unit of measure to another (e.g., from gallons to liters).

corrosion   The gradual deterioration or destruction of a substance or material by chemical action. The action proceeds inward from the surface.

decant   To draw off the liquid from a basin or tank without stirring up the sediment in the bottom.

detention time   The average length of time a drop of water or a suspended particle remains in a tank or chamber. Mathematically, it is the volume of water in the tank divided by the flow rate through the tank. The units of flow rate used in the calculation are dependent on whether the detention time is to be calculated in minutes, hours, or days.

dewatering (of sludge)   A process to remove a portion of water from sludge.

diatomaceous earth filter   A pressure filter using a medium made from diatoms. The water is forced through the diatomaceous earth by pumping.

diffuser   (1) A section of a perforated pipe or porous plates used to inject a gas, such as carbon dioxide or air, under pressure into water. (2) A type of pump.

disinfection by-product (DBP)   A new chemical compound formed by the reaction of disinfectants with organic compounds in water. At high concentrations, many DBPs are considered a danger to human health.

dissolved air flotation (DAF)   A clarification process in which gas bubbles are generated in a basin so that they will attach to solid particles causing them to rise to the surface. The sludge that accumulates on the surface is then periodically removed by flooding or mechanical scraping.

effluent   Water flowing from a basin.

empty bed contact time (EBCT)   The volume of the tank holding an *activated carbon* bed, divided by the flow rate of water. The EBCT is expressed in minutes and corresponds to the detention time in a GAC filter bed.

equivalent weight   The weight of a compound that contains one equivalent of a proton (for acids) or one equivalent of a hydroxide (for bases). The equivalent weight can be calculated by dividing the molecular weight of a compound by the number of H^+ or OH^- present in the compound.

feedwater   Raw water that has undergone pretreatment (acidification or scale inhibitor addition) prior to entering the membrane arrays.

filter backwash rate   A measurement of the volume of water flowing upward (backwards) through a unit of filter surface area. Mathematically, it is the backwash flow rate divided by the total filter area.

filter loading rate   A measurement of the volume of water applied to each unit of filter surface area. Mathematically, it is the flow rate into the filter divided by the total filter area.

filter media   Granular material through which material is collected and stored when water passes through it.

filter tank   The concrete or steel basin that contains the filter media, gravel support bed, underdrain, and wash-water troughs.

floc   Collections of smaller particles (such as silt, organic matter, and microorganisms) that have come together (agglomerated) into larger, more settleable particles as a result of the coagulation–flocculation process.

flocculation   The water treatment process, following coagulation, that uses gentle stirring to bring suspended particles together so that they will form larger, more settleable clumps called floc.

gauge pressure   The water pressure as measured by a gauge. Gauge pressure is not the total pressure. Total water pressure (absolute pressure) also includes the atmospheric pressure (about 14.7 psi at sea level) exerted on the water. However, because atmospheric pressure is exerted everywhere (against the outside of the main as well as the inside, for example), it is generally not written into water system calculations. Gauge pressure in pounds per square inch is expressed as "psig."

grab sample   A single water sample collected at one time from a single point.

granular activated carbon (GAC)   Activated carbon in a granular form, which is used in a bed, much like a conventional filter, to adsorb organic substances from water.

granular media filter   Materials used to filter raw water, including granular filter media sand, crushed quartz, garnet sand, manganese greensand, and filter coal, typical ranging in sizes from 0.25 mm to 1.20 mm, though some sizes can be larger or smaller than this range. Water treatment uses other granular filter media, but this list cites those most often used.

head (pressure)   (1) A measure of the energy possessed by water at a given location in the water system, expressed in feet (or meters). (2) A measure of the pressure (force) exerted by water, expressed in feet (or meters).

hydrated lime   Another name for calcium hydroxide, $Ca(OH)_2$, which is used in water softening and stabilization. Also called *slaked lime.*

hypochlorination   Chlorination using solutions of calcium hypochlorite or sodium hypochlorite.

impeller   The rotating set of vanes that forces water through a pump.

influent   Water flowing into a basin.

injector   The portion of a chlorination system that feeds the chlorine solution into a pipe under pressure.

instantaneous flow rate   A flow rate of water measured at one particular instant, such as by a metering device, involving the cross-sectional area of the channel or pipe and the velocity of the water at that instant.

ion exchange   A process used to remove hardness from water that depends on special materials known as resins. The resins trade nonhardness-causing ions (usually sodium) for the hardness-causing ions, calcium and magnesium. The process removes practically all the hardness from water.

iron   An abundant element found naturally in the earth. As a result, dissolved iron is found in most water supplies. When the concentration of iron exceeds 0.3 mg/L, it causes red stains on plumbing fixtures and other items in contact with the water. Dissolved iron can also be present in water as a result of corrosion of cast-iron or steel pipes. This is usually the cause of red-water problems.

jar test   A laboratory procedure for evaluating coagulation, flocculation, and sedimentation processes. Used to estimate the proper coagulant dosage.

lagooning   The disposal of sludge by transporting it to a lagoon.

manganese   An abundant element found naturally in the earth. Dissolved manganese is found in many water supplies. At concentrations above 0.05 mg/L, it causes black stains on plumbing fixtures, laundry, and other items in contact with the water.

material safety data sheet (MSDS)   A product description listing a chemical's identity, composition, and type or types; the hazard(s) associated with its use (is it flammable, an oxidizer, poisonous?); how it enters the body; the effects of exposure; health effect (short term, long term); the permissible exposure limit; how to handle spills or releases; and other related information.

maximum contaminant level (MCL)   The maximum permissible level of a contaminant in water as specified in the regulations of the Safe Drinking Water Act.

maximum contaminant level goal (MCLG)   Nonenforceable health-based goals published along with the promulgation of an MCL. Originally called *recommended maximum contaminant levels (RMCLs).*

milk of lime   The lime slurry formed when water is mixed with calcium hydroxide.

mole   The quantity of a compound or element that has a weight in grams equal to the substance's molecular or atomic weight. Used in this text generally as an abbreviation for *gram-mole.*

molecular weight   The sum of the atomic weights of all the atoms in the compound. Also called *formula weight.*

mudball   An accumulation of media grains and suspended material that creates clogging problems in filters.

normality   The number of equivalent weights of solute per liter of solution.

packed tower   A cylindrical tank containing packing material, with water distributed at the top and airflow introduced from the bottom by a blower. Commonly referred to as an air stripper.

pellet reactor   A conical tank, filled about halfway with calcium carbonate granules, in which softening takes place quite rapidly as water passes up through the unit.

percent by weight   The proportion, calculated as a percentage, of each element in a compound.

pH   A measurement of how acidic or basic a substance is. The pH scale runs from 0 (most acidic) to 14 (most basic). The center of the range (7) indicates the substance is neutral, neither acidic nor basic.

pipe lateral collector   A filter underdrain system using a main pipe (header) with several smaller perforated pipes (laterals) branching from it on both sides.

polymer   High-molecular weight, synthetic organic compound that is used to aid in the coagulation and flocculation process.

pounds per square inch (psi)   A measure of pressure.

pounds per square inch gauge (psig)   Pressure measured by a gauge and expressed in terms of pounds per square inch.

powdered activated carbon (PAC)   Activated carbon in a fine powder form. It is added to water in a slurry form primarily for removing those organic compounds causing tastes and odors.

pressure   The force pushing on a unit area. Normally pressure can be measured in pascals (Pa), pounds per square inch (psi), or feet of head.

quality assurance (QA)   A plan for laboratory operation that specifies the measures used to produce data of known precision and bias.

quality control (QC)   A laboratory program of continually checking techniques and calibrating instruments to ensure consistency in analytical results.

quicklime   Another name for calcium oxide, CaO, which is used in water softening and stabilization. Also called *unslaked lime*.

radial flow   Flow that moves across a basin from the center to the outside edges or vice versa.

rate-of-flow controller   A control valve used to maintain a fairly constant flow through the filter.

recarbonation   The reintroduction of carbon dioxide into the water, either during or after lime–soda ash softening, to lower the pH of the water.

recovery   The ratio of the volume of product water, or permeate produced, to raw water treated.

rectilinear flow   Uniform flow in a horizontal direction.

regeneration   The process of reversing the ion exchange softening reaction of ion exchange materials. Hardness ions are removed from the used materials and replaced with nontroublesome ions, thus rendering the materials fit for reuse in the softening process.

resin   In water treatment, the synthetic, bead-like material used in the ion exchange process.

reverse osmosis (RO)   A pressure-driven process in which almost-pure water is passed through a semipermeable membrane. Water is forced through the membrane and most ions (salts) are left behind. The process is principally used for desalination of sea water.

running annual average (RAA) The average of four quarterly samples at each monitoring location to ensure compliance with the Stage 2 DBPR.

rust Oxidized iron.

sand boil The violent washing action in a filter caused by uneven distribution of backwash water.

saturator A piece of equipment that feeds a sodium fluoride solution into water for fluoridation. A layer of sodium fluoride is placed in a plastic tank and water is allowed to trickle through the layer, forming a solution of constant concentration that is fed to the water system.

scaling Metal deposits left in pipelines and plumbing fixtures.

sedimentation The water treatment process that involves reducing the velocity of water in basins so that the suspended material can settle out by gravity.

sedimentation basin A basin or tank in which water is retained to allow settleable matter, such as floc, to settle by gravity. Also called a *settling basin*, *settling tank*, or *sedimentation tank*.

self-contained breathing apparatus (SCBA) A type of respirator that supplies safe, grade D (or better) breathing air to the wearer.

sludge The accumulated solids separated from water during treatment.

solids-contact basin A basin in which the coagulation, flocculation, and sedimentation processes are combined. The water flows upward through the basin. It is used primarily in the lime softening of water. It can also be called an upflow clarifier or sludge-blanket clarifier.

solution A liquid containing a dissolved substance. The liquid alone is called the solvent, the dissolved substance is called the solute. Together they are called a solution.

SPADNS method A procedure used to determine the concentration of fluoride ion in water; a color change takes place following addition of a chemical reagent. SPADNS is the chemical reagent used in the test.

specific ultraviolet absorbance (SUVA) A test that determines humic content by measuring the absorbance of UV light at 254 nm and divides that value by the dissolved organic carbon concentration.

stabilization The water treatment process intended to reduce the corrosive or scale-forming tendencies of water.

streaming current detector An instrument that passes a continuous sample of coagulated water. The measurement is similar in theory to zeta potential determination and provides a reading that can be used to optimize chemical application.

surface overflow rate A measurement of the amount of water leaving a sedimentation tank per unit of tank surface area. Mathematically, it is the flow rate from the tank divided by the tank surface area.

ton container A reusable, welded tank that holds 2,000 lb (910 kg) of chlorine. Containers weigh about 3,700 lb (1,700 kg) when full and are generally 30 in. (0.76 m) in diameter and 80 in. (2.03 m) long.

total organic carbon (TOC) The amount of carbon bound in organic compounds in a water sample as determined by a standard laboratory test.

tubercle A knob of rust formed on the interior of cast-iron pipes as a result of corrosion.

turbidimeter A device that measures turbidity, the amount of suspended particulate matter in the water.

turbidity A physical characteristic of water making the water appear cloudy. The condition is caused by the presence of suspended matter.

US Environmental Protection Agency (USEPA) A US government agency responsible for implementing federal laws designed to protect the environment. Congress has delegated implementation of the Safe Drinking Water Act to the USEPA.

wash-water trough A trough placed above the filter media to collect the backwash water and carry it to the drainage system.

weir overflow rate A measurement of the flow rate of water over each foot of weir in a sedimentation tank or circular clarifier. Mathematically, it is the flow rate over the weir divided by the total length of the weir.

Wheeler filter bottom A patented filter underdrain system using small porcelain spheres of various sizes in conical depressions.

zeta potential A measurement (in millivolts) of the particle charge strength surrounding colloidal solids. The more negative the number, the stronger the particle charge and the repelling force between particles.

Index

NOTE: *f* indicates a figure; *t* indicates a table.